T0189461

Communication Protocols

Drago Hercog

Communication Protocols

Principles, Methods and Specifications

 Springer

Drago Hercog
University of Ljubljana
Ljubljana, Slovenia

ISBN 978-3-030-50407-6   ISBN 978-3-030-50405-2   (eBook)
https://doi.org/10.1007/978-3-030-50405-2

This Springer imprint is published by the registered company Springer Nature Switzerland AG
The registered company address is: Gewerbestrasse 11, 6330 Cham, Switzerland

Preface

The author of this book has been teaching the closely interrelated subjects of communication protocols and communication networks for many years both at the University of Ljubljana, Slovenia, and at several high engineering schools in France. Communication protocols have also been the main topic of the author's research activities for many years. This book arises from these teaching and research experiences. Although the areas of communication networks and communication protocols strongly depend on each other, I tried to avoid the treatment of communication networks as much as possible in this book. A reader shall therefore have to rely on the extensive literature on the topics of communication networks which is available elsewhere.

It took a long, a too long time to write this book, especially because I retired while working on it. However, due to the fact that the emphasis of this work is on the principles on which communication protocols are based, rather than on the description of specific protocols, the author hopes that the book will be useful for both the students of communications and the engineers who are already working in industry, but still want to deepen and broaden their knowledge of this interesting area. It might even be read by some individuals who are interested in communication protocols, although their basic area of expertise is not communication technologies.

Nowadays, the great majority of communication devices are computers of different shapes, sizes and complexities; they communicate by exchanging messages. The exchange of messages must of course conform to certain rules, which must be agreed upon in advance; otherwise communicating devices could not understand each other. The rules for the exchange of communication messages are called communication protocols. Hence, communication protocols are a kind of artificial languages which allow different communication devices to communicate and to understand each other. The field of communication protocols can therefore be viewed as a topic that is related to communications, computer science, logic, even linguistics, and, of course, electrical engineering, as all communicating devices are electrical devices; one must also be aware that many problems which are being solved by communication protocols are in fact electrotechnical problems.

All the time, ever new communication protocols are being developed and put into use. A book describing only specific protocols would therefore very quickly become obsolete. Specific protocols, however, are based on communication methods that are much more general and can be used by different protocols in slightly different ways which are adapted to the conditions of use and the environment in which specific protocols are to operate; in this sense, communication protocols can be viewed as the implementations of communication methods. Of course, existing communication methods are also being modified and new communication methods are being developed; this process is, however, much slower than the development of specific protocols. The majority of this book is therefore devoted to the treatment of communication methods on which specific communication protocols are based. (We should remark here that many communication methods are often also referred to as communication protocols.) Readers who will know and understand communication methods will certainly have no difficulties to understand specific protocols based on them, both those existing and used already and those that are yet to be developed and put into use; it will also be much easier for them to study and understand new communication methods that are yet to be developed. Towards the end of this book, however, some specific protocols are briefly discussed, which should give readers some idea about how communication methods are used in practice. Nowadays, the communication networks whose operation is based on the protocol IP are becoming more and more common and are expected to prevail also in the near future; in the last part of this book, we therefore treat mainly protocols that are used in the Internet and in IP networks in general. Not many specific protocols are described and their descriptions are quite short; these descriptions must primarily be viewed as examples intended to additionally illustrate the main topics of this book, hence the communication methods.

It was already told that communication protocols are artificial languages which allow communication devices to communicate, hence to exchange information between them. A protocol is implemented in the form of a hardware or a software, which inside a communication device provides for the communication with other communication devices. Such devices may be designed and manufactured by different producers and can communicate only if communication protocols are unambiguously specified so that these specifications cannot be understood in different ways, and the devices are designed and produced strictly in accordance with these specifications. Specifications of protocols are therefore often defined in formal specification languages which allow specifications to be unambiguous (often standardisation organisations use such languages, too); besides this, formal specifications can serve as the starting point for the correctness verification or even for the implementation of protocols. Nowadays, it is therefore advisable that a communications engineer knows and can understand at least one formal language. There exists one additional reason why the knowledge of formalisms for the specification of protocols is important; a student or an engineer who is capable to formally specify a protocol in one of the formal languages will certainly have a much better idea of how a protocol really functions and operates. Up to now, many specification formalisms have already been defined and used; the choice of the formal language to be presented and used in this book was therefore not quite easy. The formalisms which are based on the finite state machine model are considered nowadays by many authors to have a too high implementation bias (it would be difficult to oppose this opinion) and also obsolete (which is not necessarily true), while those formalisms that are based on a process algebra are often viewed as modern. However, the former formalisms are more simple and easier to be understood and used by a practicing engineer, which is the reason that they found their use in the everyday industrial practice, while the latter formalisms are a favourite topic of research activities at universities and institutes. It is true that it is nowadays possible to formally prove the correctness of protocol specifications which are based on a process algebra; unfortunately, this is in practice only feasible for the cases which are simple enough; complex specifications are therefore still composed with formalisms which are based on the finite state machine model. These were the reasons why the formal language SDL, which is based on the extended finite state machine model, was chosen to be presented and used in this book; furthermore, many computer-aided design tools for protocol specification, verification (with simulation) and even implementation, which are based on SDL, have been developed and are used in the industrial environment. In this work, the language SDL is not only presented in a special chapter but also used to formally specify numerous communication methods and protocols which are discussed here. In spite of this, some other specification models and formalisms are also briefly overviewed in the first part of the book.

The contents of the book are partitioned into four parts, and each of these parts is further divided into several chapters. Part I contains three chapters discussing the specification of systems and protocols, including the description of the specification language SDL; SDL is presented with the sufficient detail which should allow a reader to understand the specifications that are presented in the continuation of this book. Part II consists of four chapters, which discuss a general definition and description of protocols, their role in a communication system, and their properties, as well as the concept of protocol stacks. A very important task of Parts I and II is to acquaint a reader with many basic concepts and the technical terminology, which are then extensively used in the continuation of the book. The seven chapters of Part III form the core, as well as the most important and the longest part of the book, as they describe and specify many communication methods, on which numerous specific protocols are based. At last, some specific communication protocols are briefly presented in twelve chapters of Part IV. A list of references is added which were consulted by the author and which could also be used by readers, and also a short summary of SDL language constructs. At the end of the book, there is the index of technical terms with page numbers for reference.

Any work unavoidably reflects its author; this can be seen in the style of writing, in the way the contents are presented, and, most importantly, in the choice of topics and the emphases which the topics and subtopics are given in the book. As was already pointed out, the author of this book spent quite a lot of time researching communication protocols; an important emphasis of these research activities was put on the field of sliding window protocols; these research activities also included the simulations of protocol performance. Prior to this, he was interested in the methodologies for the design of complex systems. One must therefore not be surprised to find out that the chapter devoted to sliding window protocols is quite long and includes some simulation results for these protocols. All the performance simulators which were used to generate these results were designed and implemented by the author. Also, the author could not resist the temptation to include a discussion on system and protocol design methodologies in the first part of the book. Some of the most important research achievements of the author are included both into the contents of the book and into the list of references; however, only those references are listed which are strongly related to the contents of this book and treat the topics that cannot be found elsewhere.

As was already mentioned, many communication methods and communication protocols in this book are formally specified in the specification language SDL. For this purpose, the software *Cinderella SDL* [41] was employed which includes both the editor of SDL specifications and the logical simulator of SDL specifications. While the editor verifies the syntactic correctness of a specification on the fly, the simulator simulates the behaviour of the specified system and draws the simulation results in the form of MSC diagrams, which allows a user to test the functional correctness of a specification. Of course, a simulation is possible only if the specification is completely formal. Furthermore, the whole system, including both protocol entities and users, must be specified; often, the model of a communication channel also had to be specified to model some peculiarities of the channel, such as delays and losses. In the description of a system to be simulated, there may be no informal specifications, no declarations may be missing. The syntactic and the functional correctness of all specifications that are presented in the book were verified with the editor and the simulator, respectively; for the presentation in the book, however, some specifications were somehow simplified. In most cases, only the specifications of protocol entities are shown, as only they are necessary to specify the protocol, while the specifications of users and channels are needed only to be able to simulate the system. Many specifications shown in the book are also generalised in the sense that the definitions of some data types and constants are either hidden in the specification of the system (which is not shown) or replaced with commentaries for the purpose of the presentation in the book. Where the results of logic simulations are presented in the form of MSC diagrams, the original MSC diagrams, as output by the SDL simulator, were also simplified by removing from them those processes that are not mandatory to understand the communication process and preserving only the exchange of messages between protocol entities.

The protocols which are used in practice are standardised in most cases. There are many organisations which develop and publish communication standards; some of the most important among them are *ISO, ITU, IEEE, ETSI* and *IETF*. In many cases, the identifiers of standards are not listed, unless they also serve as the names of protocols themselves. Of course, a reader can easily find the respective standards, especially because the web addresses of the organisations for standardisation are given in the list of references.

Towards the end of the book there is a list of important references which were consulted when writing this book; the readers of this book are also invited to see some of these references to find some additional information which cannot be found here, or to discover the topics of communication protocols from some other perspective as is offered by the author of this work. The list of references that is given here is of course not exhaustive, as there is a huge amount of literature on communications and communication protocols available. Technical books are the most numerous in the list, there are less technical papers and even less standards. This should, however, pose no problem for readers, as many search engines on the web allow nowadays anybody to easily find the technical literature of interest. Although the French language is by far not as widespread as the English language nowadays, one book in French [5] is also listed (probably because the author likes this book, and he likes the French language even more).

Forty-one references can be found in the list. They are arranged into several groups. References [1]–[5] list some of the numerous books which treat communication systems and within this frame also communication protocols. Some of these books are very popular, e.g. [1] and [2], and also [5]; the latter one can unfortunately not be read by everybody because of the language barrier. The above books have already been published in many editions; a reader will of course look for the latest one. References [6]–[10] are devoted to the special treatment of the theory of communication protocols. In references [11]–[18], the models and languages for the specification of protocols are discussed, while the language SDL is specially treated in references [19]–[23]. The performance properties of protocols are discussed in references [24]–[26], while reference [27] explains how the author developed the performance simulators to generate the results which are included in this book. The books [28] and [29] treat only the protocols which are used in IP networks. The book [28] in particular is considered by many communications engineers a kind of a »bible« for the world of Internet protocols. The papers [30]–[32] contain the results of the author's research on the theory of sliding window protocols which are also included in this book. References [33]–[37] refer to the web pages of some of the most important organisations for standardisation; these pages are excellent starting points to search for various standards. In reference [38], the specification language SDL is standardised, and in [39], the protocol PPP is standardised. The encyclopaedia »Wikipedia« [40] can also be used when working in the field of communications, but with great care to distinguish between good articles and those that are less good (which is of course true with all web resources). Finally, [41] refers to the web page of the developer of the software Cinderella SDL which was used by the author for editing and simulating the SDL specifications which are included in this book.

This work is not a scientific paper. References to already published work are therefore not included in the text. The majority of the material which is present in this book can also be found elsewhere, and in particular in various books that treat communication protocols. The research results of the author, which have not yet been much published but are yet included in this book, include the definition of the relative efficiency of protocols [26], some topics in the theory of sliding window protocols that are published in [30], [31] and [32], and also the methodology for the development of the performance simulators for protocols specified in SDL [27] (this methodology is not a topic of this book, but was used to develop the simulators to generate the performance simulation results which are inserted in the book). All the SDL models which are published in this book were also developed by the author.

The Index of acronyms appears before the main text of the book and lists the acronyms which are mentioned in this text. Although too many acronyms are often employed in technical texts, which can substantially deteriorate their readability, the use of acronyms was avoided as much as possible in this book. Of course, there are names of technologies and protocols which were originally coined as acronyms and can therefore not be avoided.

The short summary of SDL language constructs (titled SDL Keywords and Symbols) can serve as a quick reference to a reader when studying the SDL specifications which are included into this book.

The Index of technical terms contains the alphabetical list of technical terms which are used or mentioned in the text. With each term, one or more numbers are listed indicating the pages where this term is defined or importantly mentioned. With the terms which are specially related to SDL constructs, the word SDL is added within parentheses.

The author enjoyed the time spent when writing this book. He hopes that the readers who will read it will also enjoy it and profit from it, and that the book will help them to more deeply understand the very interesting field of communication protocols.

The contents of this book are the result of my pedagogic and research work in the area of communication protocols through many years. The final result is, however, also due to the endeavours of the publisher, the editors, the reviewers and the graphical designers. Thanks to all of you! The errors (I am sure there are still too many, in spite of my numerous and careful proofreadings) stay only my fault and responsibility.

I owe thanks also, and especially, to my family for their patience and understanding. The work is dedicated to my three sweet grandchildren—Svit, Matic and Tiara.

Ljubljana, Slovenia   Drago Hercog

Contents

Abbreviations

ACK	acknowledgment
ARP	Address resolution protocol
ARQ	automatic repeat request
ASCII	American standard code for information interchange
ASN.1	Abstract syntax notation one
BEC	backward error correction
BECN	backward explicit congestion notification
ber	bit error rate
BER	Basic encoding rules
CCS	Calculus of communicating systems
CDMA	code division multiple access
CIR	committed information rate
conf	confirm
CR	carriage return
CSMA	carrier sense multiple access
CSMA/CA	carrier sense multiple access with collision avoidance
CSMA/CD	carrier sense multiple access with collision detection
CSP	Communicating sequential processes
CTS	clear to send
DCF	distributed coordination function
DE	discard eligibility
DHCP	Dynamic host configuration protocol
DNS	domain name system
DTLS	Datagram transport layer security
EFSM	extended finite state machine
ENQ	enquiry
ETSI	European telecommunications standards institute
FDDI	Fibre distributed data interface
FDMA	frequency division multiple access
FDT	formal description technique

FEC	forward error correction
FECN	forward explicit congestion notification
FIFO	first-in-first-out
FSM	finite state machine
FTP	File transfer protocol
GSM	Global system for mobile communication
GUI	graphical user interface
HARQ	hybrid automatic repeat request
HDLC	High-level data link control
HTML	Hypertext markup language
HTTP	Hypertext transfer protocol
HTTPS	Hypertext transfer protocol secure
ICMP	Internet control message protocol
ICMPv6	Internet control message protocol version 6
IEEE	Institute of electrical and electronics engineers
IETF	Internet engineering task force
IFS	interframe space
IMAP	Internet Message access protocol
ind	indication
IP	Internet protocol
IPCP	IP control protocol
iPDU	information protocol data unit
IPv4	Internet protocol version 4
IPv6	Internet protocol version 6
IPV6CP	IPv6 control protocol
ISO	International organisation for standardisation
ITU	International telecommunication union
LAN	local area network
LAPB	Link access procedure—balanced
LAPD	Link access procedure on D channel
LAPDm	LAPD mobile
LAPF	Link access procedure for frame mode bearer services
LCP	Link control protocol
LF	line feed
LLC	logical link control
LOTOS	Language of temporal ordering specifications
MAC	medium access control
MAN	metropolitan area network
MIME	Multipurpose internet mail extensions
MIR	maximum information rate
MSC	Message sequence charts
MSS	maximum segment size
NAT	network address translator
NCP	Network control protocol

ND	Neighbor discovery protocol
OFDMA	orthogonal frequency division multiple access
OSI	Open systems interconnection
PAN	personal area network
PCF	point coordination function
PCI	protocol control information
PDU	protocol data unit
per	packet error rate
POP3	Post office protocol version 3
PPP	Point to point protocol
PPPoA	PPP over ATM
PPPoE	PPP over ethernet
QoS	quality of service
RAR	release after receive
RARP	Reverse address resolution protocol
RAT	release after transmit
RED	random early drop
req	request
resp	response
RFC	request for comment
RTCP	RTP control protocol
RTP	Real-time transport protocol
RTS	request to send
SAP	service access point
SCTP	Stream control transmission protocol
SDL	Specification and description language
SDP	Session description protocol
SDU	service data unit
SIP	Session initiation protocol
SLIP	Serial line internet protocol
SMTP	Simple mail transfer protocol
SSH	Secure shell
SSL	Secure sockets layer
TCP	Transmission control protocol
TDMA	time division multiple access
TFTP	Trivial file transfer protocol
TLS	Transport layer security
TLV	type-length-value
TTCN	Testing and test control notation
ttl	time to live
UDP	User datagram protocol
UML	Unified modelling language
URI	universal resource identifier
URL	universal resource locator

UTF-8	Unicode transformation format—8-bit
VoIP	Voice over IP
WACK	wait acknowledgment
WiFi	wireless fidelity
WLAN	wireless local area network
WWW	World Wide Web
XML	Extensible markup language

Part I
Specification of Communication Systems and Protocols

Communication (or telecommunication) systems are very complex. The design, implementation or maintenance of such systems is not an easy task, so it must be undertaken with care and following the methodologies which have been developed with the aim to ease these processes and, what is especially important, to reduce the probability of design and implementation errors which could have unpleasant, or even fatal consequences for a project. In the next few chapters a short presentation of these methodologies will therefore be given, as they involve the specification of systems and protocols, too.

The system design is of course not the main topic of this book. Our discussion about it should therefore be seen only as an introduction to the treatment of protocol design. Communication protocols are very important components of communication systems. Nowadays, communication protocols are themselves so complex that everything which will be told about systems is true in case of protocols, too.

A system or a protocol design is always begun with the specification of the services that are to be provided by the system/protocol, the environment in which it is to operate, and, of course, the system/protocol itself. The protocol specification is especially important because protocols are normally specified in standards. Communication engineers must therefore be able to read and understand protocol specifications, and some of them must even write them. A reader of this book will therefore be acquainted with some models and languages used to specify systems and especially protocols.

While systems in general are treated in Chap. 1, Chap. 2 is dedicated specifically to communication protocols; in this chapter, some languages for protocol specification will also be presented. In Chap. 3, the specification language SDL will be described in more detail; in the rest of the book, this language will be used to formally specify some protocols.

Chapter 1
System Specification and Design

Abstract The concept of a system is presented in this chapter. Then the basics of system design methodologies are explained. The notions of abstraction and implementation are presented, as well as the concept of system/module specification and implementation as the basic steps of a system design process. Systems are described from the viewpoint of their structure and functionality. The most important models used to model the structure, the information and the functionality of a system and its components are overviewed. The block diagrams are presented as the models of structure. The data types and event types are shortly described as the information models. Among functional models, the time-sequence diagrams, finite state machines, extended finite state machines, Petri nets, process algebras and temporal logic are overviewed. The model of extended finite state machine is presented here with a special care, as it is the theoretical basis for the functional modelling in the formal specification language SDL to be used later in this book. The chapter contains six figures.

1.1 System

When thinking about the world or about just a part of it, especially when thinking technically, one usually uses the term *system*. A system, in its simplest form, may be viewed as the one shown in Fig. 1.1. The term inputs in this view means a value, a sequence of values or a behaviour which are received by the system from its environment and are not, at least not directly, influenced by the system, while the word outputs in the figure means a value, a sequence of values or a behaviour that are determined by the system which affects its environment in this way.

The terms input and output *behaviour* mean that input and output values of the system change in time; the system behaviour can therefore also be observed in time. Such a system is called a *dynamic system*. A system with no inputs is referred to as an *autonomous system*. A *reactive system* responds to an input behaviour with its output behaviour. An input or output of an *analogue system* can assume any value within a bounded range and may change at any time, while an input or output of a *discrete*

D. Hercog, *Communication Protocols*, https://doi.org/10.1007/978-3-030-50405-2_1

Fig. 1.1 System

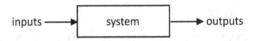

system can only assume values from a finite set of discrete values and may change only in discrete points in time. *Events* are changes that occur in discrete points in time on inputs, outputs or within the system itself; a system whose behaviour can be described in terms of input, output and internal events is referred to as *discrete-event system*. A reactive discrete-event system that must respond to input events within a bounded time period is called a *real-time system*.

The response of a *stateless system* (also referred to as *memoryless system*) in a point of time depends only on input values in that particular point of time; if, however, the response of a system in a point of time depends not only on the input in that particular point of time but also on the previous input behaviour, such a system is called a *stateful system* (or *system with memory*). A stateful discrete-event system can be found in different states; a *state* of a system can be seen as the abstraction (or as the concentrated image) of the previous system behaviour. A system is aware of its state; the concepts of state and memory are therefore closely related. The state of a system can change in time.

In this book exclusively discrete-event systems will be treated.

The *functionality* of a reactive system is determined with the dependency of the output behaviour on the input behaviour. The functionality of a system must be such as a system designer or a system user wants it to be; it must be well defined and is given as a part of the system *specification*. The system specification is considered correct if it defines the functionality which complies with the users' wishes and needs.

Up to this point only the functionality of a system has been discussed. However, the *external structure* of a system can also be seen in Fig. 1.1. The external structure determines where and how a system can be excited, and where and how it can respond to the excitation.

The complete specification of a system determines both the external structure and the functionality of the system.

One frequently speaks about *modules* or about *processes* instead of about systems. Modules and processes are system components, where the term module usually indicates a structural component, while the word process always indicates a functional component.

A system (or a module or a process) must be implemented (made) so that its behaviour conforms to its specification. The *implementation* of a system is therefore said to be correct if it behaves in conformance with its specification.

For a given system, there exist many combinations of input and output behaviours, a system can also be found in a great number of states. A system, its specification and its implementation can therefore be very complex. Hence, a system must be specified and implemented with great care, in order to possibly avoid design errors as much as possible. Unfortunately, design errors can happen nevertheless due to the limited ability of the human mind to cope with complexity. The correctness of the specification and implementation of a system must therefore always be verified, which allows one to detect errors and correct them. In order to carry out different design activities, such as specification, implementation and verification, some pre-scribed and proven design procedures are usually followed which are commonly referred to as the *design methodology*. One of the basic methods to reduce the design complexity is the system *structuring*. This means that a system is partitioned into components (modules, processes) with well-defined external structures and func-tionalities. The interaction of components must be as simple as possible, but also well defined. The structure that is most usually used is the *hierarchical structure* which is somehow similar to a tree. Another very important methodologic principle is the *abstraction*; a module can be abstractly presented with its external structure and functionality while (temporarily) ignoring its implementation; in such cases one usually thinks about a system as a *black box*. Hence, one can view the specification of a system as the abstraction of its implementation. Of course, when the system is implemented, its implementation must conform to its specification.

In short, a (sub)system can be described with its structure and its functionality; the structure description includes both external structure (inputs and outputs) and internal structure. The *internal structure* of the system describes how it is actually implemented with components, while the functionality describes its behaviour as observed on its external structure (inputs and outputs). A (sub)system is specified with its external structure and functionality, while it is implemented with its internal structure and the functionality of its components. If both specification and imple-mentation of a system/module/process are defined, they must conform one to another.

Although many different aspects of a system can be considered (such as its aesthetic appearance, colour, price...), only its structure and functionality will be considered here.

In this book, only the discrete-event real-time communication systems will be treated. The role of such systems is to process and transfer information.

1.2   System Design

The specification of a module defines the external structure (inputs and outputs) of the module and its functionality—i.e. the behaviours on its inputs and outputs, or, better to say, the dependence of the output behaviour on the input behaviour. The module specification should be as much independent of its possible implementations as possible; this property allows module designers for more freedom in module implementation, the module can be implemented in many different ways. The dependence of the specification on possible implementations is referred to as *implementation bias*. The implementation bias of a module specification should therefore be as low as possible which actually means that the specification should be as abstract as possible. A module specification is the abstraction of this module which can be used when implementing another module with this one as a component, even if the implementation of this module is not (yet) known. So, the module specification (abstraction) and its implementation represent the most important elements of a hierarchical system design.

The process of a complex system design is a sequence of design steps.

A design step begins with the specification of the module to be designed, hence with the module external structure and functionality specifications. To implement a module means to specify the set of its components (submodules), specifications of external structures and functionalities of submodules, as well as their interconnections. The functionality of the module can now be derived from the specifications of its submodules and their interconnection; if this derived functionality is consistent with the functionality that was specified for the module, the design step correctness is verified.

There are three possibilities for *correctness verification*.

The testing of a prototype seems to be the easiest method for correctness verification. Unfortunately, this is mostly not true. Before testing it, the prototype should be implemented as a hardware or a software module, and only then could it be tested to compare its behaviour with the behaviour specification. There are several drawbacks of this verification method. Firstly, the verifications of both specification and implementation are combined in a single step; this method does not work if the specification itself is not correct! Secondly, the cost of a prototype implementation can be very high and is certainly not justified if the specification was already erroneous. Furthermore, it is well known that testing can prove the presence of errors in implementation, but cannot prove the absence of them; unfortunately, real systems are nowadays so complex that it is virtually not possible to test all the states and transitions between them in a reasonable time (although some methodologies for the design for testability are available today).

The method of implementation correctness verification that is in most cases easier, faster, cheaper and more frequently used is simulation. The simulation model of a module specification and the simulation model of its implementation (submodules and their interconnection) are developed. Simulations of both specification and implementation can be run separately, and then the results of these simulations can be compared. A simulation model is simulated within the environment that is provided by the simulation program. The simulation of a module is somehow similar to the testing of a prototype; unfortunately, the design correctness cannot be proved with the simulation either. However, a simulation model development is easier and cheaper than a prototype development and production. A simulation model should mimic the module that is to be simulated as closely as possible—the better is the simulation model, the more accurate will be the simulation results. If a simulation model is qualified with the timing properties of the module and the simulation is run in a timed environment, efficiency simulations can also be carried out.

A system specification and implementation correctness can also be analytically verified. Such analysis is carried out by systematically considering different combinations or state transitions, or searching for possible traces through states of the system. In this way one can discover *deadlocks* (the states that cannot be left by the system) or *livelocks* (cyclic transitions through states), as well as *unreachable states* (the states which can never be reached from the initial state). If the specification is given mathematically, the correctness can be proved, too. Unfortunately, mathematical proofs are quite difficult for an average engineer; even for an expert engineer, they are the more difficult, the more complex is the system.

In recent years, a lot of research was dedicated to invent efficient methods for analytic verification and proving correctness of system specifications and implementations; a bulk of software was also developed to support such activities. However, the simulation is still the most frequently used, at least in case of complex systems.

A design process that is referred to as *top-down design* is carried out as follows. A system (or the module that is the highest in the hierarchy) is first specified, then it is implemented with submodules (with their abstractions specified), and the correctness of the implementation is verified; such a design step has already been described in the previous text. Thereafter this design step is performed for all submodules, then for all subsubmodules, and so on, until all sub...submodules that are needed are already implemented and available. The result of such a design process is a hierarchically organised system. The order in which design steps are carried out can also be modified. The design process can also be begun at the bottom of the hierarchical structure (*bottom-up design*), or some modules in the middle of the hierarchical structure are designed first.

1.3   Models for System Description and Specification

Discrete technical systems have been rapidly developing in last decades, with their complexity also increasing more and more. The need for informal and formal models to describe and specify such systems has therefore also been growing. Nowadays, many such models are available. Both informal description techniques and formal languages for specification and description of systems are based on these models. While informal specifications describe systems in natural languages with the aid of sketches and figures, formal languages allow for system descriptions that are nonambiguous and machine readable and may therefore be used to provide input to computer-aided design tools. A formal language may be based on one or more models with the addition of formally defined syntax and semantics. The dividing line between models and languages is therefore usually not clear. In spite of this fact, models and languages will be treated separately in this text, as several languages can be based on a single model, and a single language can be based on several models. In addition to those models and languages which are intended for protocol specifications, some others will also be mentioned that help system and protocol designers in some way.

Both the structure and functionality of systems can be modelled and specified with models and formal languages.

1.3.1 Structure Modelling

As was already told, a system or a module can be described with its external and internal structure. The external structure determines the points of interaction with the environment, i.e. inputs and outputs of the system/module. On the other hand, the internal structure shows how a system/module is composed of its components, referred to as submodules, and how submodules communicate between them and with the external world. A communication between modules or between modules and the environment can be either unidirectional or bidirectional. The external and internal structure must, of course, be compatible, in the sense of both contents and direction of transfer; the same must be true for the communication between modules. In Fig.1.2 an example of the external and the internal structure of a system is shown; while the system and the modules are shown as rectangles, the communication paths are drawn as lines with arrows (indicating the direction of information transfer). The external structure is drawn with bold lines, and the internal structure with thin lines. Both unidirectional and bidirectional communication paths are shown; however, where two paths are in contact (e.g. an external and an internal one) they must agree as the transfer direction is concerned.

Virtually all complex systems are hierarchically structured. A *hierarchical structure*, also referred to as a *tree structure*, is the structure in which the relation parent–child is defined between elements, so that any child has exactly one parent, and there exists exactly one element which is not a child because it has no parent—this element is called the root. In a hierarchical structure of modules, the system is implemented with modules, and any module can be implemented with submodules. Such a structuring may be nested to any depth. The system is well defined if any module that has no children (internal structure) has its functionality defined; these elements are referred to as the leaves of the hierarchical structure.

In Fig. 1.3, an example hierarchically structured system is shown; the root of this structure is the system that contains modules m1, m2 and m3, the module m1 contains submodules m11, m12 and m13, etc. Fig. 1.4 shows the same system where the hierarchical structure is shown so that any child is drawn directly below its parent.

The hierarchical modular structure is used in design of virtually all complex technical systems.

Fig. 1.2 Example of external and internal structure of system

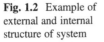 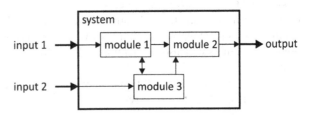

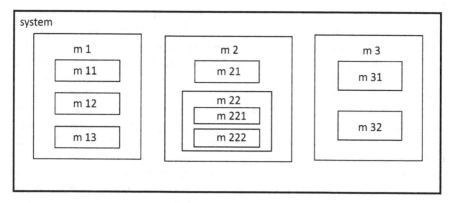

Fig. 1.3 Example of hierarchical system

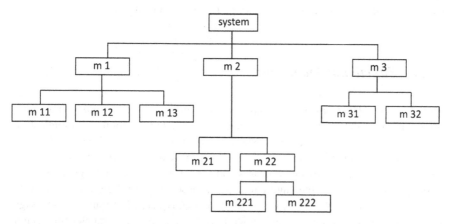

Fig. 1.4 Hierarchical view of the system in Fig. 1.3

1.3.2   Information Modelling

Data Types—A *data type* is a model of static information. They have been known
and used for a long time, as they form an essential element of any programming
language. A data type is a set of values which can be used to describe information.
The information is static in the sense that a value can be kept by a system for some
time, as it is stored in memory; one may ask »how many« or »which value« to
enquire about static information. Associated with a data type are operators that can
be applied on the values of that type. Besides basic data types (such as sets of integer
or real values) there are structured types (such as sets of sequences or composed
values that consist of several components) and abstract data types (it is up to a user to
define the functional model and the implementation of an abstract data type,
including its operators). An example of a basic data type is the set of integer values
with value examples such as -25 or 8402, the binary operators (that produce a result

from two operands) summation, subtraction, multiplication and division, and the unary operator (which produces a result from a single operand) of sign change.

Event Types—An *event type* is a model of dynamic information. Event types are used in modelling discrete-event systems. An event type is a set of events. An event must be »caught« in the moment when it occurs; therefore it can be thought about as a dynamic information. When considering an event, one can ask not only "which event" but also "when it occurs". Events are the elements of a model which allow for the synchronisation of different processes. An example event type is a set of message (such as an acknowledge) receptions; a certain message type can be received several times, so there are several instances of an event type.

An event may be parameterised with values that are associated to it. This yields the possibility of a combined event-value modelling of information. Related to the example in the previous paragraph is the reception of an acknowledgment of a specific message (or, in other words, the reception of an acknowledgment with an associated value that specifies the message to be acknowledged).

1.3.3 Functional Modelling

Time-Sequence Diagram, Timing Diagram—A *time-sequence diagram* is a graphic presentation of the exchange of messages between two or more communication entities. Communication entities are shown as vertical or horizontal lines with entity names indicated at the beginning of lines. The arrows that run from one entity line to another one represent message transfers and are furnished with message names and possibly with important message parameters; these lines can be drawn oblique to entity lines if one wants to emphasise the time that is needed to transfer a message from one entity to another, or perpendicular to entity lines if the transfer time is not considered important (either it is neglectable with respect to the general timing in the diagram or only the sequence of messages is to be indicated). In a diagram, the time usually runs from top to bottom or from left to right; however, this is not always explicitly shown. The events in time-sequence diagrams are in most cases not subjected to a strict timing scale, as the sequence of events is primarily of interest. A diagram in which message transfers are drawn according to a strict time scale is called a *timing diagram*. In Fig. 1.5 an example time-sequence diagram can be seen that shows message transfers between communication entities UA, A, B and UB; messages between entities A and B are shown as if they need some time to be

Fig. 1.5 Time-sequence diagram example

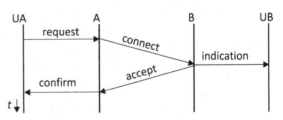

transferred, while the messages between UA and A, and UB and B, respectively, seem to be transferred in no time (physically this of course cannot be true, only the times to transfer these messages are neglectable if compared to the times to transfer messages between A and B).

Time-sequence diagrams offer a simple and clear way to graphically demonstrate message transfers. They are therefore frequently used in books and papers about protocols, and even as additions to informal protocol specifications to graphically explain some characteristic scenarios. One must however be aware that only a single *communication scenario* (message transfer sequence) can be shown in a time-sequence diagram; as there are many possible scenarios during the lifetime of a real system, a protocol specification that is based on time-sequence diagrams can never be complete.

Finite State Machine—A *finite state machine* (*FSM*) is a model frequently used for discrete-event system and module specifications. It is based on the fact, already mentioned in this text, that a discrete system can be found in different states; the number of states is finite, whence the name of the model.

A finite state machine may pass from one state to another one only when it has received an input event; only at this time it also can generate an output event. Exactly one of its states is declared as the initial state; it is in this state that the machine wakes up when it is turned on (either with the on/off button or after it has been powered up). If a finite state machine contains a final state (which is not necessary), that is the state from where no transitions are possible (if the machine enters the final state it stops working). Hence, a finite state machine is defined as a set of input and output events, a set of states and state transitions, an initial state and possibly a final state; to any transition is associated an input event that triggers that particular transition, and possibly zero or more output events that are triggered by the transition.

A finite state machine can be shown as a graph, where vertices represent states and directed edges represent state transitions. The initial and final states are also indicated, as well as the input and output events that are associated with transitions. Such a graph may also be referred to as a state transition diagram. Instead of a state transition diagram a state transition table can also be given to define a finite state machine.

In Fig. 1.6 an example of a finite state machine graph is shown. Circles denoted as S_i represent states, with the state S_0 being the initial state. Designations I_i denote the input events (that ignite transitions) and O_i represent the output events which are triggered by transitions. If the machine shown if Fig. 1.6 receives the input event I_0 while in the state S_0 it generates the output event O_0 and passes to the state S_2; if, however, it receives the input event I_2 while in the state S_0, it generates the output event O_2 and passes to the state S_1. If this machine receives the input event I_4 or the input event I_5 while in the state S_3, it passes to the state S_1 in both cases, but generates O_4 in the former case and O_5 in the latter case. If, however, the machine receives I_1 in the state S_0, there will be no reaction and no state change, as the reception of I_1 is not defined for the state S_0.

The finite state machine is the oldest formal model of discrete systems and has been in use for decades. The fact that a finite state machine can also be described as a graph allows for the analytic treatment of a finite state machine functionality, based

Fig. 1.6 Finite state
machine example

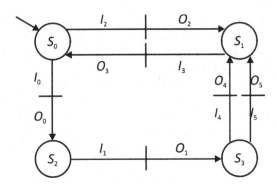

on the graph theory which is a mathematical discipline. Unfortunately, the finite state machine model has a big disadvantage, referred to as state explosion problem. The number of states rapidly increases with the complexity of the system that is modelled; the higher is the number of states, the less clear and the less manageable is the model.

A finite state machine is aware of its state (otherwise it would not be able to correctly react to input events). Hence, it can model a stateful system (a system with memory).

Several kinds of finite state machines have already been defined in past times. In this book we will not distinguish between them; for us, finite state machines will be interesting especially as the model on which the functional specification in the SDL language is based.

Extended Finite State Machine—An *extended finite state machine* (*EFSM*) is an extension of the finite state machine model; besides the states, an extended finite state machine uses an additional form of memory—variables in which it can store values. These values may be used in different actions and decisions that are carried out during state transitions (the state into which the machine will pass after the transition is finished may also depend on these values); they can also be modified during state transitions. Like FSM, an EFSM also can generate output events during state transitions. With an extended finite state machine, various values can be associated with input and output events and can thus be exchanged between the machine and its environment.

The real state of an extended finite state machine is determined with both its symbolic state (that is explicitly specified for the machine) and the values of all variables. The symbolic state can also be called the control state, as the machine behaviour is in general determined with the sequence of symbolic states and input events received in these states. Most usually, however, the symbolic or control states will be referred to simply as states. The formal separation of control states and various combinations of variable values is the principle that essentially simplifies the model complexity, especially because the variables and the values which are stored in them can be arbitrarily abstract (e.g. integer values, etc.); the number of control states and transitions between them can therefore be relatively small, even if the number of combinations of control states and variable values might be huge. This very property of EFSM solves the state explosion problem of the basic FSM. An extended finite state machine could also be thought of as a basic finite state machine, the states of which are parameterised with variable values.

Petri Nets—A *Petri net* is a formal model, based on mathematical sets and relations, suitable for modelling distributed discrete-event systems and proving their correctness. A Petri net can be described mathematically with sets and relations, or it can be presented graphically. Similarly as a finite state machine, a Petri net is based on the states of a system (which are called places in this context) and state transitions; during a transition, various activities can be carried out. Any finite state machine can be presented as a Petri net; any Petri net, however, cannot be shown as a finite state machine, as the concept of Petri nets is an extension of the finite state machine concept. The Petri net of a communication system can be more complex than the equivalent finite state machine.

Petri nets found their usage mainly in modelling and analytic correctness proving of computing systems, especially those which provide for parallel processing. Modelling communication systems, however, is more frequently based on finite state machine concept.

Abundant literature, both in books and papers, is dedicated to Petri nets.

Process Algebra—The *process algebra* is a family of formalisms which can be used for abstract modelling of concurrent processes, such as communication systems. The syntax and semantics of these mechanisms are well defined and allow for the combining processes that are running sequentially or parallelly (concurrently), the communication between processes (with message exchanges), the synchronisation of them, the abstraction (distinction between processes that can be viewed from their environment and those that cannot be seen from outside), the recursion (processes which can run in loops), and the selection (choice of two or more alternatives). The equivalence of processes is also defined. Processes may be used in process equations that can be specified and solved. All of this allows for the concise and nonambiguous specification of communication and other concurrent systems with no implementation bias, as well as the formal verification and proofs of specification correctness, or at least the absence of errors.

In general, the specifications that are based on a process algebra are more abstract (with less implementation bias) than those which are based on a finite state machine model. On the other hand, they are more difficult to be understood and conceived by an average engineer, primarily because of their abstractness and mathematical orientation.

Several process algebras have been defined in recent years. One of the oldest and best known is CCS (*Calculus of Communicating Systems*), defined by R. Milner. The process algebra CSP (*Communicating Sequential Processes*), introduced by C. A. R. Hoare, is also well known.

Temporal Logic—The *Temporal logic* is a discipline that was conceived as an extension of predicate logic. It uses operators with timing conditions, which allows the arrangement of logic values and events in time.

Chapter 2
Protocol Specification and Design

Abstract A communication protocol is first presented as a set of rules that determine the behaviour of a communication system in terms of the exchange of messages between communication entities. Then a protocol is also described as an artificial language which specifies this behaviour. The concepts of the abstract and the transfer syntax of protocol messages are explained and the importance of the rules for the behaviour of protocol entities is emphasised. The difference between informal and formal specification is explained, and the importance of the formal description techniques to assure nonambiguous specifications is emphasised. The basic techniques for informal and formal specification are presented. The difference between the functionality specifications that are based on the finite state machine model, and those which are based on process algebras, is pointed out. Then some of the most important formal languages are overviewed: ACT ONE and ASN.1 for message specification, SDL, Estelle and LOTOS for protocol specification, and also UML, MSC and TTCN for communication protocols design. The chapter contains five figures.

2.1 Communication Protocol

Devices that communicate in a communication system (a communication network) are referred to as *communication entities*. Communication entities communicate in accordance with the rules which are usually standardised; a set of rules which determine how two or more entities should communicate is called a *communication protocol*, a *telecommunication protocol* or simply a *protocol* (there also exist protocols other than communication protocols, which, however, will not be of interest in this book). The term protocol is normally used when discrete communication systems are treated, although there is fundamentally no reason why it should not be used in case of analogue systems, too; in this text, however, only protocols used in discrete communication systems will be considered. In such systems *protocol entities* (as communication entities are called in this context) communicate by exchanging *protocol messages*; a protocol specifies the kind of messages that may

D. Hercog, *Communication Protocols*, https://doi.org/10.1007/978-3-030-50405-2_2

be used, their format, and the rules which determine when a protocol entity may or must send a protocol message to another protocol entity. A message can generally be viewed as a discrete value; in most cases, this value is not a simple value, but rather a structured one which means it consists of several components, usually referred to as *fields*; the message structure may be hierarchical, as a field may consist of subfields... The simplest message that has no internal structure is a *bit*; often an *octet* (a *byte*) is also considered a basic message, although it is actually a sequence of eight bits. (Although the terms octet and byte are synonyms, the former is usually used in communications, while the latter is more often employed in computer science). A message may bring zero or more values with it. Message transmission and reception represent an output event of the sending entity and an input event for the receiving entity, respectively. A protocol defines the meaning of messages and values brought by them.

Protocol entities that wish to communicate one with another must of course use the same protocol, otherwise they will not be able to understand each other. In most cases protocols are standardised which allows entities that might be produced by different producers and managed by different operators to interoperate. Systems that employ standardised and publicly known protocols are usually referred to as *open systems*.

The rules that determine protocol message exchanges are most appropriately specified by specifying the operation of protocol entities as the components of a communication system, along with the specification of protocol messages which may be exchanged. The specification of the operation of protocol entities is most appropriate because a protocol is implemented by implementing protocol entities. Although time-sequence diagrams may be more obvious and also easier to read and understand for most engineers, they are not appropriate for protocol specification. A single time-sequence diagram can present only a single communication scenario; however, there may be a large number of possible communication scenarios in the lifetime of a protocol. Time-sequence diagrams are therefore mostly used to clarify textual descriptions and present some example scenarios. In this book, time-sequence diagrams will be used for this purpose.

2.2 Protocol as a Language

As a set of rules that determine the communication between machines, a protocol can be viewed as a language; of course, this is an artificial language. Like any language (natural or artificial), a protocol is defined by its *syntax* and *semantics*. The syntax of a protocol must be defined at three levels. The *abstract syntax* of protocol messages defines the set of messages that may be used, along with their parameters; the *transfer syntax* of protocol messages, which is also called *concrete syntax*, defines the format of messages in the sense of their internal structure, hence how messages are composed and transferred as sequences of fields, bits or octets; finally, the *supermessage syntax* specifies which sequences of protocol messages may be transferred between protocol entities during the course of a communication process. The supermessage syntax is specified with the rules which determine when and how protocol entities may or must send specific messages. As the logic correctness of a protocol is concerned, the message sequences are primarily important; the timing of message transfers, however, influences the protocol efficiency. The semantics of a protocol determine the meaning of protocol messages and sequences of them.

2.3   Informal and Formal Protocol Specification

The specification of a communication (or some other technical) system may be written in a natural language, such as English, German or French. Alternatively, an artificial language with well-defined and unambiguous syntax and semantics can be used for this purpose; many special artificial languages have already been defined for the specification of different kinds of technical systems.

A specification that is written in a natural language is referred to as *informal specification*. Natural languages have more or less well-defined syntaxes, especially for the word and phrase formation; the semantics of texts written in a natural language is, however, much more loosely and less precisely defined. A specific text written in a natural language may have more different meanings, as the meaning of a word or a phrase may depend on the context; on the other hand, several different words (referred to as synonyms) may have a similar or even equal meaning. Even a whole phrase may have a meaning that depends on its context! These properties of a natural language allow authors to express themselves in a free or even artistic manner; such texts are, however, sometimes difficult to be understood correctly (in accordance with what the author wanted to tell). If an informal specification is used, it is therefore quite difficult to guarantee it to be unambiguous. Different system designers may happen to understand such a specification in different ways; consequently, different implementations of the system may not be compatible and are therefore unable to communicate. One of the consequences of ambiguous specifications are also problems that are encountered in logical correctness verification of specifications and implementations. In order to avoid such problems, the authors of standards usually use a strange and complicated language that is hard to read and understand.[1]

A *formal specification* is a specification that is written in an artificial language with precisely and unambiguously defined syntax and semantics. The technique of specifying systems in *formal languages* is usually referred to as *formal description technique* and indicated with the acronym *FDT*. Of course, one cannot write poetry in a formal language; however, that is not its purpose! As the syntax and semantics of formal languages are unambiguously defined, formal specifications of protocols are

---

[1]Professor   Tanenbaum   from   Amsterdam   wittily   named   such   a   language   a »beaurocratspeaklanguage«

also unambiguous; this very fact allows for rigorous correctness verifications and efficiency assessments of specifications and implementations, regardless of the way they are carried out (analytically or with simulations). Correctness verifications that are based on analytical proofs must of course always be based on formal specifications, written in a mathematical way. In brief, formal specifications that are precise, readable and unambiguous present an excellent starting point for analysis, simulation model development and implementation of communication protocols. A specification which is formally defined is also machine readable which means that it may be used as the input to computer-aided design tools that allow for the automatic specification analysis (e.g. search for deadlocks and livelocks), simulation model generation or even generation of protocol implementation (either in the form of hardware or software modules).

In this chapter, some methods that can be used for informal protocol specification will first be described; then a few formal languages for communication protocol specification will be shortly overviewed. In the next chapter, the formal language SDL (Specification and description language) will be described in more detail. This language which was developed and standardised by the International Telecommunication Union (ITU) will be used in this book to formally specify some communication methods and mechanisms that are used by protocols.

2.4 Informal Protocol Specification Techniques

In this section, we will not be very specific. Specification techniques to be described here being informal, different authors use them with different styles and appearances. However, all of them generally follow the same guidelines which will be described here.

2.4.1 Communication System Structure Specification

Traditionally, so-called *block diagrams* have been used to informally describe structural components of communication (and other) systems; they graphically show protocol entities as well as connections between them and with their users. All the elements of a structural description (both blocks and their interconnections) may be assigned names that can be referenced in the accompanying textual description. Figure 2.1 shows an example block diagram specifying a communication system consisting of two users, two protocol entities, and interconnections between them.

Fig. 2.1 Block diagram
example

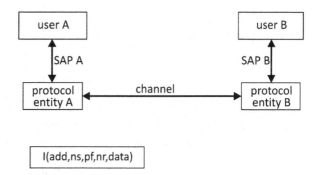

I(add,ns,pf,nr,data)

Fig. 2.2 Example of abstract syntax of protocol message

2.4.2 Specification of Protocol Messages

In Sect. 2.2 we explained the terms abstract syntax and transfer syntax of protocol messages.

The abstract syntax of protocol messages is informally defined as a list of *message types*; a protocol entity may transmit or receive arbitrarily many messages of the same type which are referred to as *instances* of that particular message type. The abstract syntax of a message type must specify the number and the types of message parameters (values), which can be brought by a message of this type. Figure 2.2 shows an example of the abstract syntax of a message type named I which brings five values (parameters) with it. The types of these values must also be specified; in the given example, the parameters ns and nr may assume integer values between 0 and 7, the parameter pf can assume logical values 0 and 1, the parameter add also may assume only two values, and the parameter data is a bit sequence of arbitrary length. (We should mention here that this example describes the message type I of the protocol LAPB).

The transfer syntax of a protocol message describes the internal structure of the message; hence it describes how that particular message is built as a sequence of fields, subfields, octets and bits (one must keep in mind that a message is always transferred through a network as a sequence of bits). Any element of this structure must have a specific and recognisable place within the message, so the receiver can recognise the fields within the received message. Any element in this structure (a field, a subfield…) also has its meaning which must be understood by the receiver; a sequence of bits by itself has no sense if the receiver does not know how to interpret it in the sense that it can recognise the value that is brought by this field and relate it to the value that was declared in the abstract syntax of the message.

The format of a protocol message is usually specified with the graphic presentation of its structure, including the placements and lengths of fields, as well as the implementation of their contents. At the lower layers of a protocol stack the structure is often shown linearly, as a sequence of fields together with their lengths and contents; this presentation may also be hierarchically organised so that fields are shown as sequences of subfields, and so on. Figure 2.3 shows an example transfer

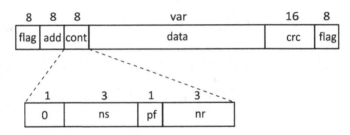

Fig. 2.3 Example of transfer syntax of protocol message in linear format

syntax which is partly hierarchically organised, although it is essentially linear; the transfer syntax shown in Fig. 2.3 is related to the abstract syntax of the message type I of the protocol LAPB shown in Fig. 2.2. The numbers that are shown above the fields and subfields in this figure indicate their lengths in bits. The detailed description of formats and necessary interpretations of all the fields must also be given. One can see that the definitions of the abstract and the transfer syntax of a message do depend one on the other: the fields ns and nr are three-bit fields which allows them to contain decimal integers between 0 and 7, as already specified in the abstract syntax; the field pf is a one-bit field, so it can contain a logical value; as the field add can contain one of two possible values, as specified by the abstract syntax, eight bits are of course not needed—the reason for this is the fact that the field add in the predecessor of the protocol LAPB could contain more values, while in LAPB it is not really significant. Comparing Fig. 2.2 and Fig. 2.3, one can easily see that each field in the transfer syntax does not necessarily correspond to a parameter defined in the abstract syntax, some fields having purposes other than to bring values of parameters. In the example shown in Fig. 2.3 the fields flag mark the beginning and the end of a message, the one-bit subfield of the field cont that contains the value 0 indicates the message type (the receiver can begin to interpret a message only after it has become aware it has to do with the message type I, as the messages of other types have different formats), while the field named crc allows the receiver to detect the errors which may have affected the message during the transfer.

In the higher layers of protocol stacks the message contents are usually organised by octets; hence it is more appropriate to present the message format in a 2-dimensional (rectangular) form, so that the vertical dimension (from top to bottom) shows the octets or even longer units, while the horizontal dimension (from left to right) shows the bits or octets in a line. Such a presentation is especially appropriate if the linear form would yield a picture which would be too long. The transfer syntax shown in Fig. 2.3 can also be shown in the 2-dimensional form which can be seen in Fig. 2.4. The numbers on the top of Fig. 2.4 indicate the bits within an octet, while the numbers on the right edge of the figure indicate the lengths of the fields or field groups in octets. Such a presentation is of course possible if the total message length equals an integer number of octets.

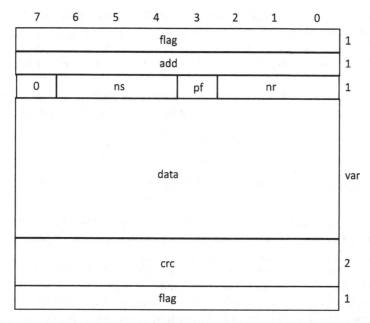

Fig. 2.4 Example of transfer syntax of protocol message in 2-dimensional format

The transfer syntax specifications that have been shown here can be used only if the message formats are relatively simple. In case of more complex message formats the formal specification may be preferred.

2.4.3   Functionality Specification

Informal specifications of protocol functionalities are often based on a finite state machine model or time-sequence diagrams, but are described in the textual form in a natural language. We have already explained in Sect. 2.1 why the functionality specification of protocol entities is more appropriate than the functionality specification of a whole communication system; this is one of the reasons why specifications that are based on a finite state machine model are preferable to those based on time-sequence diagrams for protocol specification.

2.5   Formal Languages for Protocol Specification

In past decades, a lot of different formal languages have been defined, used and implemented (in the form of computer-aided design software, such as specification editors, simulators and compilers into protocol implementations); some of them are used in industry and/or research institutions, while the others have been forgotten. Here only some of them will be shortly described. These languages may serve different purposes within the frame of a protocol design process, so this section is partitioned into three subsections, according to the basic purpose of the languages to be discussed.

The descriptions of selected languages will be very short in this chapter. In the next chapter, however, one of them, the language SDL, will be described in more detail.

2.5.1 Languages for Specification of Protocol Messages

ACT ONE—ACT ONE is the language which allows for the algebraic specification of abstract data types. Data types can be combined, they can also be parameterised; all of this allows for the specification of generic types.

ASN.1—ASN.1 (*Abstract Syntax Notation One*) is a standard language (it was standardised by the International Organisation for Standardisation (ISO)); it is used for the specification of the abstract syntax of protocol messages. This language is at least as powerful as the majority of programming languages. A protocol message is specified as a data type. Basic data types, such as BOOLEAN, NULL, INTEGER, ENUMERATED, REAL, and different arrays, can be used, as well as composed data types (SEQUENCE, SEQUENCE OF, SET, SET OF, CHOICE) and subtypes. Associated with the language ASN.1 are the rules for the definition of the transfer syntax of protocol messages, specified in ASN.1; some of these rules are standardised. The *Basic Encoding Rules* (*BER*) are the oldest and the most frequently used rules for the transformation of abstract syntax into transfer syntax; these rules are standardised, too. The language ASN.1 is mostly used for the specification of protocol messages in higher layers of a protocol stack (the application and presentation layers, according to the OSI reference model), although they may in principle be used in lower layers, too.

2.5.2 Protocol Specification Languages

SDL—In this section the language *SDL* (*Specification and Description Language*) will be very shortly presented; only its role and significance within the family of protocol specification languages will be explained. In the next chapter it will be discussed in more detail.

The International Telecommunication Union (ITU) has been developing the language SDL since 1980 and standardised it in the Recommendation ITU-T Z.100, which was therefore published in several versions; in general, the differences between versions are not very essential. As the development of the language is concerned, the most important versions are SDL-92 (published in 1992) and

SDL-2000 (published in 2000). The main purpose of the SDL language is to make possible the precise and unambiguous specification and description of telecommunication systems; however, the language allows for the specification and description of discrete reactive systems in general.

In SDL a system is structurally described as a hierarchical structure of blocks, subblocks, etc. The functionality of a block is specified with processes that communicate between them by exchanging signals. The functional specification of a process is based on the extended finite state machine model. The communication between processes is asynchronous as processes receive signals through input queues which are implicitly associated with them. Processes can store and process data, while signals can transport data between processes. In SDL, data are described with built-in data types, with abstract data types or in the ASN.1 language.

From the very beginning the SDL language has had two equivalent syntaxes: the text syntax (SDL/PR—SDL Textual Phrase Representation) and the graphical syntax (SDL/GR—SDL Graphic Representation). Due to its evidence and intuitiveness the graphical syntax is more comprehensible for a novice (it is also very similar to the so-called flow charts that are well known to most engineering students and engineers and often used in software engineering), so only the graphical syntax will be explained and used in this book. It is adequate for simple specifications (only such will be presented in this text); however, the engineers who use SDL to program more complex communication systems usually employ the textual syntax. There are also some SDL language elements that can be expressed only in textual form.

A specification presented in SDL describes the structure which possibly (although not necessarily) indicates the actual structure of the system to be designed, and the functionality which is based on the extended finite state machine model, so it is very close to the reasoning of an average engineer; SDL signals, too, model protocol messages very closely. Even the model of asynchronous interprocess communication closely resembles the usual communication between protocol entities. On one hand, all of these properties make SDL specification quite similar to possible implementations (they have a high implementation bias); on the other hand, SDL language is very user friendly and not difficult to be learned due to these properties. This user friendliness is even more emphasised if the graphical syntax is used which is so similar to flow charts that are a component of the basic training of engineers.

ITU recommendations recommend the SDL language to be combined with the languages ASN.1, MSC and UML.

Many computer-aided design tools are based on the SDL language, sometimes combined with the languages MSC, UML, ASN.1 and TTCN; many of them were developed with commercial intentions in mind. Mostly they are professional tools which may be sold at quite high prices. The majority of these tools include specification editors, simulators and compilers to implementations, sometimes also the tools for correctness verification. The best overview of these tools can be found at the SDL forum web site.

The SDL language found its use in the organisations for standardisation, such as ITU-T or ETSI, for the specification of protocols; in industry it is used as the specification and programming language for the design of communication systems; it also serves as a successful educational tool. It is still a research topic.

The SDL language is the topic of many books. Of course, the basic source of information about it is the ITU-T recommendation Z.100.

Estelle—The language *Estelle* (*Extended State Transition Language*) was developed and standardised by the International Organisation for Standardisation (ISO); however, this standard was withdrawn in 1997, so it ceased to be a valid standard. This language, too, can be used to specify and model not only communication systems but also all real-time systems and the systems with parallel operation of components.

The language Estelle was developed as an extension of the programming language Pascal. It allows for the structured system specification which is based on the hierarchical structure of modules. This structure may be dynamic, as modules may generate new modules or abolish existing ones. The static model of information is based on the data types which are already known in Pascal. The functional specification of a module is based on the extended finite state machine model. The external structure of a module is defined with interaction points through which modules can transmit or receive information. Interaction points of different modules may be bound dynamically. Any interaction point has an implicitly associated queue, so the intermodule communication is asynchronous. A module may also exchange information with its parent and children via shared variables.

The language Estelle did not find much use in industry and research institutions; hence, not much computer-aided design tools were based on it. Probably, these are the main reasons why the official Estelle standard was cancelled.

LOTOS—The language *LOTOS* (*Language Of Temporal Ordering Specifications*) was also developed and standardised by the International Organisation for Standardisation (ISO) with the purpose to allow for the specification and correctness verification of communication protocols conformable to the OSI reference model. However, the language can be used for the specification of communication and distributed systems in general. It may be used even for the specification of digital electronic circuits. The language was standardised in 1989.

The formally defined syntax and semantics of the language LOTOS are based on the process algebra, particularly on the algebra CCS (Calculus of Communicating Systems) which was already mentioned in Sect. 1.3.3. Data types, including abstract data types, are modelled according to the language ACT ONE. The functional specification of a system is based on the so-called behavioural expressions; in these expressions processes may be combined sequentially or parallelly, their activations can be enabled by temporal and logical conditions. An essential element of behavioural expressions is the synchronisation between processes which is carried out by events and data exchanges on ports that are common to processes (this communication mechanism is referred to as the »rendezvous« mechanism). Asynchronous communication may be modelled only with explicitly specified waiting queues. Processes may be structured which allows one to hierarchically decompose systems. The basic element of a behaviour specification is the event.

LOTOS provides for powerful specification mechanisms and allows for specifications with low implementation bias. The correctness of specifications can in principle be formally proved. While the low implementation bias is a very favourable property, this advantage must be payed for by the mathematical way of specification which is somehow user unfriendly for most engineers.

The language LOTOS is being constantly developed. One of the additions to the language includes the possibility of graphical specifications (G-LOTOS); E-LOTOS is an extended version of the language.

LOTOS is a popular research topic at numerous universities which have developed many LOTOS-based design tools. In industry it is less popular.

LOTOS is the topic of many books; the most relevant source for it is of course the standard.

2.5.3 Some Other Languages for Design Process Support

UML—UML (*Unified Modelling Language*) is a standardised graphically oriented language for system modelling. It allows for the specification, description and documentation of systems. It was standardised by ISO. Although it was primarily conceived as the aid for software design, communication systems, especially at higher levels of abstraction, can also be modelled with it. Both the structure and the functionality of a system can be described. This language is quite complex.

MSC—MSC (*Message Sequence Charts*) is the language, defined in the Recommendation ITU-T Z.120, used to formally describe the message exchange between communication entities. Hence, communication scenarios may be described in MSC; in this sense, MSC is similar to time-sequence diagrams we described in Sect. 1.3.3. In reality, the MSC language is very powerful and offers much more mechanisms than barely a formal description of time-sequence diagrams. It allows for the structuring of communication entities and even time-sequence diagrams, their parameterisation, enabling/disabling functionalities with logic and temporal conditions, combining of diagrams and value-based information modelling. A communication system can be described at different levels of abstraction. MSC is primarily a graphical language. Although a system operation can also be described in a textual form, the textual syntax is primarily used to store descriptions and transfer them between different design tools, while the graphical syntax is primarily used for the system specification and design. The language MSC is usually used in combination with the languages SDL and TTCN. Its most important uses are requirements specification and testing scenarios description. Some simulators of communication systems present simulation results in the form of MSC diagrams.

Figure 2.5 shows a simple MSC diagram for the scenario which was already informally shown in Fig. 1.5 as a time-sequence diagram.

TTCN—The language *TTCN* (*Testing and Test Control Notation*) is used for the specification of test sequences intended for the testing of a protocol implementation conformance with standards. The term *conformance testing* is used for this kind of

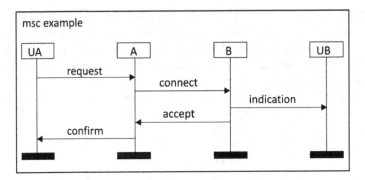

Fig. 2.5 Example MSC diagram, equivalent to Fig. 1.5

testing. The TTCN language was developed and standardised by the European Telecommunications Standards Institute (ETSI). The latest version of the language is TTCN-3. In older versions the acronym TTCN stood for Tree and Tabular Combined Notation.

Chapter 3
SDL Language

Abstract The standard specification language SDL is described with the sufficient detail which should allow a reader to understand the numerous specifications that are presented in the continuation of the book and also to compose simple specifications of their own. Although both textual and graphic syntaxes are defined in SDL, only the graphic syntax of SDL operators is used in this chapter and in the rest of the book. Many SDL mechanisms which are not to be used in the book are omitted from this chapter. At first, the basic language elements for the specification of structure, information and functionality are presented, namely the block, channel, signal, data type and process. The concept of a type and an instance is explained. It is shown how a block or a process instance is specified. The techniques to specify the functionality of a process are especially emphasised. At the end of the chapter, the simulation of SDL specifications is shortly discussed. There are 22 figures in this chapter, mainly showing graphic symbols of SDL language elements.

The language *SDL (Specification and Description Language)* was already briefly presented in Sect. 2.5.2; here it will be explained with much more detail. The goal of this chapter is to present the language to the extent that readers will get some idea about the language, about its purpose and its basic capabilities; above all, they will be able to understand numerous communication mechanism specifications that will be given in this book and even to compose some simple specifications by themselves. The complete language SDL provides for much more mechanisms than will be explained here; a reader willing to know more will find appropriate material in many books devoted to this topic, as well as in the Recommendation ITU-T Z.100.

No examples of specifications and descriptions in SDL will be given in this chapter. This is not necessary, as numerous SDL descriptions will be shown in next chapters to specify many communication methods and protocols. This chapter may therefore serve also as a language reference to the readers when they study specifications in the following chapters. The appendix named "SDL Keywords and Symbols" at the end of the book may also be used as a quick reference to language mechanisms and constructs.

D. Hercog, *Communication Protocols*, https://doi.org/10.1007/978-3-030-50405-2_3

3.1   Basic Presentation of the Language

The language SDL was developed and standardised by the International Telecommunication Union (ITU) as a specification language to be used in the design of telecommunication systems and protocols. Although this is actually the principal area of its use, the language can be used for the specification and description of any discrete reactive system that operates in real time.

In SDL both the structure and functionality of a communication system can be specified at different levels of abstraction (with different amounts of detail). The Recommendation ITU-T Z.100 does not explain the meaning of the terms specification and description; however, some authors consider the word description to indicate the structure and the term specification to mean the functionality, while the others pretend a system to be specified at higher levels of abstraction and described at lower levels of abstraction. In any case, it is possible to specify a system (which means that the implementation is not determined) or to describe an already (at least partially) implemented system. In this book, only the specification of protocols will be of interest; this means that the communication system structure will consist of users, protocol entities and channels interconnecting them, while the functionality will only be specified for protocol entities. Hence, SDL will be primarily used in this book to specify the operation of protocol entities.

The functionality specification, if given in textual form (SDL/PR), is similar to the usual procedural programming most students and engineers are already used to; however, the graphic form of specification (SDL/GR) resembles the well-known flow charts which are often used to sketch an algorithm before the program coding is even tackled. However, although the classical programming and the dynamic system specification share many similarities and common properties, there are yet some important differences between the two: the classical programming results in a sequential program which sequentially runs different operations and does not essentially depend on time, while a dynamic real-time system operation does depend on time. The language SDL therefore offers the construct »when something happens, then do something« which allows for the synchronisation of the operation of different system components that otherwise may operate asynchronously and independently. In addition, SDL offers the possibility of time measuring. The operation of sequential systems and the operation of those that run in real time essentially differ in their (in)dependence on time.

Although SDL is a formal language for system specification, it allows one to combine formal specifications with informal ones (the latter have the form of simple character sequences which are actually commentaries). Informal descriptions may have two purposes. Designers may want to formally specify only the basic outline of the system functionality and describe the details that are not mandatory for the basic operation understanding as informal commentaries. Alternatively, they may formally specify only the basic operation of a system with informal commentaries describing the rest at first; later they may refine the specification by replacing more and more commentaries with formal specifications and thus adding more and more detail to the system specification. One must however be aware that only the specifications that are entirely formal are machine readable. A specification that is to be automatically translated into a simulation model or even into an implementation using an appropriate computer-aided software must therefore be completely formal.

3.2 Language Elements

As was already told, SDL has two equivalent syntaxes, the graphical and the textual one. In this book, only the graphical syntax will be used. In this syntax, SDL operators are shown as standardised graphic symbols; some of these symbols will be explained in this chapter. The meaning of these symbols is, however, generic; textual specifications must be written into these symbols to give them more specific meaning. There are, however, some SDL language elements (declarations and definitions) which may be given only textually, as sequences of characters. Such textual constructs are written in the so-called *text frames*. Any language construct consists of the construct denotation and construct parameters; in the textual syntax the construct denotation is indicated with a *keyword*, while in the graphical syntax it is indicated with a *symbol*.

Figure 3.1 shows an example of the text frame that defines a *constant* of Integer type with the name k and the value 5.

The SDL version SDL-2000 (published in 2000) distinguishes uppercase letters from lowercase ones; however, the keywords, such as the keyword SYNONYM in Fig. 3.1, may be written only with uppercase or only with lowercase letters (a keyword is a word that is always written in the prescribed form and determines the basic meaning of a construct or an operator). The version SDL-92 (specified in 1992) does not distinguish between uppercase and lowercase letters. In this book, all the keywords will be written only with uppercase letters (as was already done in Fig. 3.1), data types will be written with uppercase initials (as are written the data types that are built in SDL), and all the other names will be written only with lowercase letters, for the sake of better readability.

3.2.1 Building Blocks of a Specification

Structural building blocks of an SDL specification are systems, blocks and channels. Functional building blocks are processes and signal routes. Equally important are information carriers, such as signals and values.

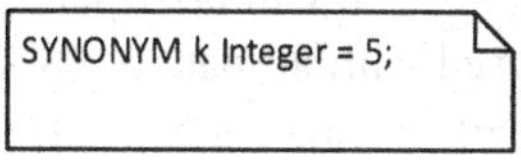

Fig. 3.1 Text frame

SYNONYM k Integer = 5;

The properties of a building block may be defined by defining a *type* of building blocks; thereafter the *instances* of a certain type can be used. An instance has the same properties as its underlying type. In this way several building blocks with the same structural, functional and information properties can simply be defined. However, a building block (a system, a block or a process) can directly be defined with its belonging properties, without a previous type definition; also in such a case, an appropriate type implicitly belongs to the element.

In SDL, a system description may be hierarchical; this means that a building block can be defined within another building block; in such a case the former building block is the child and the latter one is the parent. The *scope* of a building block (the domain in which it may be seen and used) is the building block in which it is defined and all of its descendants, but not its ancestors.

3.2.2   Values, Data Types and Variables

A *value* always has an associated data type. A *data type* defines the set a value can be drawn from and the operators which can be executed with the value as an operand or as a result; hence a value is a member of the set defined by the data type. In the documents defining the SDL language a data type is referred to as *sort*; however, the more frequently used term data type will be used here. One can calculate with values, using only those operators that are defined for the particular data type. Values may be stored (written) into *variables* from where they may also be retrieved (read); any variable also has an associated data type that determines which values can be stored in it. A variable must have a name. Variables model the memory, and a variable name indicates the address of the memory location where the value of a variable is stored. In the SDL language, only processes can use variables (see Sect. 3.4).

In SDL there are many predefined data types; using special generators, new data types can also be generated.

Some of the predefined simple data types are the following:

- Integer—is the set of integer values.
- Real—is the set of real values.
- Boolean—is the set of logic values true and false.
- Character—is the set of characters.
- Time—is the set of absolute time values.
- Duration—is the set of time differences.
- PId—is the set of process identifiers.

All of the above listed data types, except for the last three, are already known in all usual programming languages. The operators that may be used with the values of these types are known as well. If, in this book, an operator is used that is not so well known, it will be explained in time. Here, only two relational operators that may be used with all data types shall be mentioned: the operators » = « and » /= « mean »equal« and »not equal«, respectively; both operators yield a result of the type

```
SYNTYPE Index = Integer CONSTANTS 1:5 ENDSYNTYPE;
```

Fig. 3.2 Example of subtype definition

Fig. 3.3 Example
definition of array with five
logical values

```
SYNTYPE Index = Integer CONSTANTS 1:5 ENDSYNTYPE;
NEWTYPE Vector
    array(Index, Boolean)
ENDNEWTYPE;
```

Boolean. The data types Time and Duration are needed in the specification of systems that operate in real time. It is important to be aware that the unit which is used in SDL to measure time is simply the *time unit*; the relation between the SDL time unit and the standard unit to measure time (second) must be determined by a user of the language (a designer). In the real physical world, the absolute time without a defined reference point in time has no sense (only the time difference is meaningful); SDL therefore provides for the operator now which returns, when activated, the current value of the absolute time; all the other absolute times can be expressed with the operator now and a value of Duration data type. Here the data type PId must also be mentioned; this is the set of process identifiers (as will be told in the following, processes are the basic building blocks of functional specifications).

The construct SYNTYPE may be used to define a new data type that is a subtype of an already existing data type. Figure 3.2 shows the definition of the data type Index which is a subtype of the data type Integer, containing only the integer values 1–5, inclusively.

So-called *generators* may be used in SDL to define new data types. Using the generators string, array and powerset one can define new data types in the form of a *string*, an *array* or a *set*. Figure 3.3 demonstrates how to define an array of five logical values indexed with the indexes of the data type Index; the name of the new data type is Vector. If a variable a of the data type Vector is declared, $a(b)$ (where b is a variable of data type Index) denotes the element of the array a indexed with the value of variable b; $a(b)$ contains a value of data type Boolean. For example, $a(2)$ is the second element of the array a, as 2 is the second value of the type Index. One can use the assignment operator $a := (.\ false\ .)$ to assign the same logical value false to all the elements of the array a, which is very useful to initialise the values of an array.

As a variable represents a piece of memory reserved to store values, it is natural that a data type must be assigned to any variable; this assignment is expressed in a *variable declaration*. Variables may be used only by processes (which are functional components of a specification). A variable may therefore be declared only in a process; a variable is owned by the process where it is defined and can normally

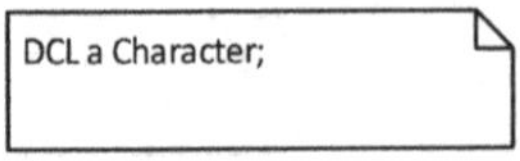

Fig. 3.4 Example of
variable declaration

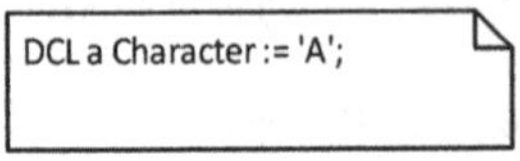

DCL a Character;

Fig. 3.5 Example
declaration of variable with
initialisation

DCL a Character := 'A';

not be seen from other system components.[1] The variable name and its associated
data type must be mentioned in a variable declaration. Variables may be declared
within a text frame contained in a process definition. The keyword DCL is followed
by declaration of one or more variables. Figure 3.4 shows an example of a variable
declaration, declaring the variable a of data type Character.

Before the value of a variable is used, a value must be assigned to this variable. A
variable may be initialised in the declaration statement, as can be seen in Fig. 3.5; in
this figure the variable a of the data type Character is declared and assigned the initial
value 'A' (the character A). The value of a variable may also be defined or modified
with assignment constructs that are used during a process initialisation (the transition
to the initial state) or during a state transition.

Figure 3.1 already showed the use of the SYNONYM construct to define a
constant with a specific name and a specific value.

3.2.3   Signals

Signals transport information between blocks and between processes. A process may
generate and transmit a signal, a process may also receive a signal. The transmission
of a signal is an output event of the process that transmits it, and an input event of the
process that receives it. The lifetime of a signal is short, only from its transmission to
its reception. A signal may bring zero or more values with it.

Any signal must be declared; this is done within a text frame following the
keyword SIGNAL. A *signal declaration* must define the signal name as well as
the number and the data types of the values that can be brought by the signal. At the
time a signal is transmitted, the values it brings must be defined; at the time a signal
is received, these values must immediately be stored into memory by the receiving
process as a signal ceases to exist immediately after it has been received.
(We mentioned already that signals enter processes through waiting queues; the
term »received« in the previous phrase means that the signal was retrieved from the
queue and »consumed«—the information it brought was used by the process.)

---

[1]To say the truth, SDL allows a variable to be seen in another process, too; however, this
mechanism will generally not be used in our book, as we will always specify a protocol entity in
a single process.

Fig. 3.6 Example of signals
declaration

```
SIGNAL s1;
SIGNAL s2 (Integer);
SIGNAL s3 (Boolean, Character);
```

Fig. 3.7 Example of signal
list declaration

```
SIGNAL a, b, c;
SIGNALLIST  abc = a, b, c;
```

Figure 3.6 shows an example of the declaration of three signals. The signal s1 brings no values with it; the signal s2 brings one value of type Integer; the signal s3 brings two values, the first one of Boolean type and the second one of Character type.

In SDL the values brought by a signal have no names; they may be distinguished between them only according to the order in which they are listed, e.g. the first one and the second one in case of the signal s3 in Fig. 3.6.

It is necessary to emphasise here that processes transmit signals to their environment and receive signals from their environment. Hence, signals must be visible in the process environment and must therefore be declared there; signals may only be declared in blocks that contain subblocks or processes.

A signal declaration actually declares a set of input and output events, as a process may transmit and receive a signal of the same type many times, possibly bringing different values.

In a text frame one may declare a *signal list* as well; a list of signals allows for an easier declaration of many signals that can be transported through a channel between processes or between blocks when the interprocess or interblock communication is described, as a whole list of signals can be referenced by a single name. Of course, the signals that are to be included into a list of signals must previously be declared. Figure 3.7 shows a declaration of the signal list named abc which consists of the signals a, b and c.

3.3   Blocks

A *block* is the basic structural element of a specification. A block instance is defined in a frame, such as the one shown in Fig. 3.8 (as was already told, a block type is implicitly associated with a block instance definition). In a block definition, its name is defined (block_name in Fig. 3.8). A block may be specified on more than one page (it can be too complex to fit on a single page); the page sequence number and the total number of pages in parenthesis are therefore shown in the upper right corner of the frame. In a text frame within the block frame one can declare signals to be used in this block and its descendants; constants, data subtypes and new data types may also be defined there.

Blocks may be hierarchically nested by defining one block within another. The block that is the highest in the hierarchy (i.e. the block which has no parent) is

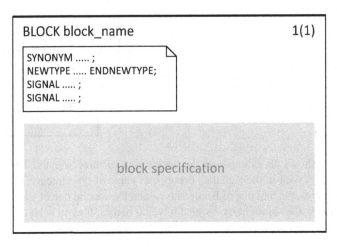

Fig. 3.8   Block definition

Fig. 3.9   Instance of block

Fig. 3.10   Instance of
process

referred to as *system*. A system is defined in the same way as a block with the only
exception that the keyword SYSTEM is used instead of the keyword BLOCK.

When a block is defined either its internal structure or its functionality may be
defined (but not both). The internal structure of a block is defined by specifying its
subblocks and their interconnections with channels; channels may also connect
subblocks with inputs/outputs of the block. Figure 3.9 shows a block instance
(named block_name) which can be used as a subblock in the specification of the
internal structure of another block. The functionality of a block is specified with
processes that run concurrently in the block; these processes communicate between
them and with the block environment via channels that interconnect them. Figure 3.10 shows a process instance (named process_name) which can be used in the
functionality specification of a block.

A *channel* is a building block which allows signals to be transferred between
blocks or between processes. Signals may also be transferred via channels that
interconnect a subblock or a process with the environment of the block which
contains the subblock or process. A channel may have a name and can be either
unidirectional or bidirectional. For each direction of information transfer a list of
signals which may be transferred via the channel in the given direction must be
shown in brackets; in this list a signal list name (which has previously been defined)

Fig. 3.11 Example of channel

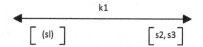

may also be cited in parentheses. Figure 3.11 shows an example of a bidirectional channel named k1; via this channel, the signals that form the signal list sl can be transferred in one direction, and the signals s2 and s3 can be transferred in the other direction. An SDL channel is lossless. The channel shown in Fig. 3.11 has no delay. SDL offers also the model of a channel with nondeterministic delay. Although all channels in the real world have delays as well as losses, all the channels to be used in the specifications in this book will have no losses and no delay.

The version SDL-92 of the language distinguished between channels to interconnect blocks and *signal routes* to interconnect processes. In the version SDL-2000 signal routes were withdrawn from the language definition, so channels may interconnect both blocks and processes. In this book only channels will be used.

3.4 Processes

A *process* is the basic building block of a functional specification. A process instance is defined within a frame that can be seen in Fig. 3.12. (As was already explained, a process type is implicitly associated with a process definition.) In a process definition the name of the process must be determined (process_name in the example in Fig. 3.12). Usually a process definition is too complex to fit onto a single page and must therefore be shown on several pages; the page sequence number and the total number of pages in parentheses must therefore be indicated in the right upper corner of the frame. Within a process definition frame a text frame may be used to define constants, data subtypes and new data types, and to declare variables and timers to be used in this process. The keyword DCL is used to declare variables, as was already explained in Sect. 3.2.2. The keyword TIMER is used to declare a *timer*; a timer declaration must specify the name of the timer. The timer is the only SDL mechanism which allows for the measurement of time. A timer may be activated (set); when a timer is set the time when it should expire must be determined. An active timer may be deactivated (stopped). The expiration of a timer is an input event for the process that owns it, just like the reception of a signal; when a process receives a timer expiration (i.e. when it retrieves it from the input queue), it must react to this event or remember it; just like in case of any input event, the timer expiration vanishes immediately after it has been consumed by the process; the timer itself, however, continues to exist and is prepared to be set again. The expression active(t) may be used at any time to test whether the timer t is active (is running) or not; this expression yields a result of the type Boolean (true or false). Signals may not be declared in a process, as they are received by the process from its environment, and they are transmitted to the environment.

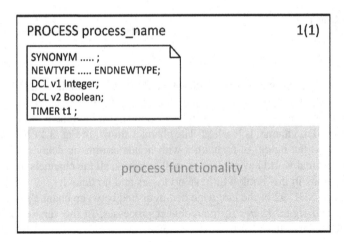

Fig. 3.12 Definition of process

The process functionality is specified as an extended finite state machine that was already discussed in Sect. 1.3.3. Variables and timers are declared in a text frame, as was told in the previous paragraph; in the area that is labelled »process functionality« in Fig. 3.12 one must therefore specify the operation of the extended finite state machine, or more specifically state transitions and state machine initialisation (the transition to the initial state).

The operation of an extended finite state machine is driven by input events. An input event of a process can be a signal reception or a timer expiration. An input *FIFO* (*first-in-first-out*) queue is implicitly associated with any process; all input events enter the process through this queue. When the automaton (finite state machine) retrieves an event from the queue it reacts to it and the event vanishes (does not exist any more). Hence the extended finite state machine must memorise the event and the eventual data associated with it as soon as this event has been received (retrieved from the queue). The memory of the extended finite state machine has two forms: the state and the variables. The automaton can therefore memorise an input event by changing its state, by modifying the values of its variables, or by both. Although the machine may execute several actions during a state transition, the duration time of a transition is theoretically zero (a transition is executed in an instant); hence the time in the automaton proceeds only in the moments when input events are received. An SDL process is therefore a typical discrete-event system.

A state is therefore the basic element of a finite state machine description. In SDL a state is represented with a *state symbol* as shown in Fig. 3.13; the name of the state must be written within a state symbol. All the state symbols within a process description that quote the same state name indicate the same state.

Fig. 3.13 State symbol

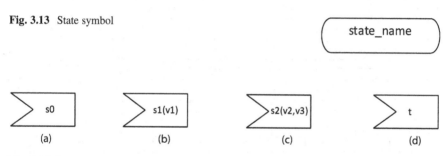

Fig. 3.14 Input symbol

In a state a process does nothing but waits for one of those input events which it may consume in that particular state. A state transition is conditioned by the reception of an event that is expected in the state of departure. In a process specification a state symbol is therefore in most cases followed by an input symbol that indicates the consumption of an input event. (The complete SDL language defines two additional conditions for state transition; they will, however, be neither explained nor used in this book.)

In Fig. 3.14 the *input symbol* is shown. This symbol represents the reception of an input event where the input event to be received is specified by a signal name or a timer name. Figure 3.14a shows the reception of the signal named s0 that brings no values with it; Fig. 3.14b shows the reception of the signal s1 that brings a single value which is stored by the process into the variable v1; Fig. 3.14c shows the reception of the signal s2 that brings two values—the first one is stored in the variable v2 and the second one is memorised in the variable v3. In this context v1, v2 and v3 represent memory addresses where the values brought by the signals are to be stored by the process, so they may only be variable names. Of course, the specifications (a)–(c) must be consistent with the signal declarations as well as with the variable declarations regarding the number and types of values. Figure 3.14 (d) shows the reception of the expiration of the timer named t.

In Fig. 3.15 an example of possible transitions from the state0 can be seen. If the process has received the signal named signal1 while in state0, it passes to state1; if it has received signal2, it goes over to state2; if the timer t has expired, it changes its state to state3; if, however, signal3 has been received, the process does not consume this reception and react to it, but retains the signal3 in its input queue to be able to possibly react to it later when it will be in some other, more appropriate state. If an input event is received for which an appropriate reaction is not specified (e.g. if the process has received signal4 in state0) such an input event is discarded (lost) and no reaction of the process takes place. The goal of the *save symbol* shown in Fig. 3.15 is therefore to allow a process not to react to an input event in a particular state while saving this event in the input queue; hence this event might be consumed by the process in some other state (capable to process such an event) and invoke a state transition in that case. In Fig. 3.15 signal3 is so saved in the input queue without invoking a transition from the state0. We have already told that a process does nothing while waiting in a state to receive an input event; however, it can execute

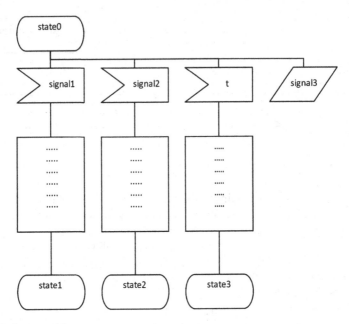

Fig. 3.15 State transitions

different actions (output signals, manage timers, calculate, modify values of variables. . .) during transitions between states; in Fig. 3.15 these actions are indicated with rectangles containing only dots. In the continuation of this chapter we will tell how to specify these operations.

Transitions from different states after reception of an input event may happen to be identical; in such a case a single state symbol at the start of the transition may contain several state names, separated by commas.

The new (final) state of a state transition may well be the same as the old (starting) state of the same transition. In such a case the new state may be indicated simply with the character »-«, meaning »remain in the old state«; this is especially convenient in a case described in the previous paragraph.

State transitions that are shown in Fig. 3.15 may also be specified separately, as shown in Fig. 3.16. Specifications in Fig. 3.15 and Fig. 3.16 are therefore equivalent.

The lifetime of a process is limited, a process is born and it dies. Most usually a process is born at the moment when the specified system is started (e.g. when its »on« button is pressed or batteries are inserted); however, a process may also be born later if it is generated by another process. In any case the specification of a process must define its initial state, including the initial values of its variables; the *process initialisation* hence determines how the process should start its operation. The process initialisation occurs unconditionally at the beginning of the process lifetime. In SDL specification a process initialisation is indicated with the *start symbol* as shown at the top of Fig. 3.17; the start symbol is never followed by a condition, such as an input symbol. During the initialisation a process may run different actions, just like during a state transition; after the initialisation the process enters a state which is

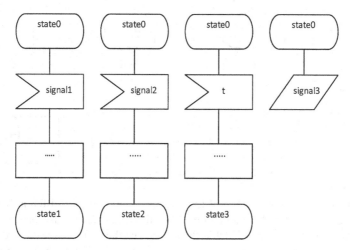

Fig. 3.16 State transitions

Fig. 3.17 Transition into
initial state (Initialisation)

Fig. 3.18 Stop symbol

referred to as its *initial state*. The start symbol (as seen in Fig. 3.17) is very similar to
the state symbol. The difference that is most evident is that a start symbol does not
contain any text while a state symbol always contains a state name; however, a closer
look reveals also that both symbols have slightly different shapes.

   A state transition may also be terminated with the *stop symbol* (which is shown in
Fig. 3.18) rather than with a state symbol; in such a case a process does not enter a
new state after the transition but stops and ceases to exist.

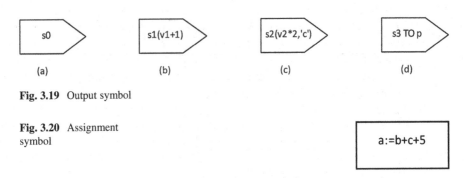

Fig. 3.19 Output symbol

Fig. 3.20 Assignment
symbol

As was already shown in Fig. 3.15, a process may execute different activities
during a state transition. Although all the activities of a single transition do theoretically occur at the same time (which is the reason that the system time advances only
at the moments when input events are received), the order in which these activities
are run may be important if these activities are interdependent. The activities which
will be described here are transmissions of signals, calculations and storing values
into variables, timer management and decisions.

In Fig. 3.19 the *output symbol* is shown that specifies a signal transmission. If a
signal brings one or more values with it, these values must be specified within the
output symbol in parentheses following the signal name. In parentheses one or more
expressions must be specified that are evaluated just before the signal transmission
and thus determine the values to be brought with the signal. The signal s0 in
Fig. 3.19a does not bring any values with it; the signal s1 in Fig. 3.19b brings a
value that equals the value of variable v1 at the time of execution increased by one;
the signal s2 in Fig. 3.19c transports two values: the first one is the double value of
the variable v2, and the second one is the character 'c'. The specification of values in
the output symbol must of course be consistent with the signal declaration both
regarding the number and the types of signal parameters. It may happen that it is not
clearly evident from blocks or processes interconnection which process should
receive the signal that is being output; in such a case the keyword TO may be
added to specify this; according to Fig. 3.19(d), the process named p should receive
the signal s3. Instead of a process name, a process identifier of the destination
process may also be used, i.e. a value of the PId data type.

The *assignment symbol* is shown in Fig. 3.20; when the assignment operation is
executed, the expression on the right-hand side of the assignment operator »:=« is
evaluated and the calculated value is stored into the variable cited on the left-hand
side of the assignment operator. In Fig. 3.20 an example expression is shown.

The two main operations of timer management are the timer activation and the
timer deactivation. A timer that is inactive may be activated using the *set symbol*. An
example set symbol is shown in Fig. 3.21a where the first parameter determines the
absolute time of the data type Time when the timer should expire, and the second
parameter is the name of the timer to be set. The absolute time when the timer should
expire is usually expressed as the sum of the current absolute time (retrieved with the

Fig. 3.21 Timer management

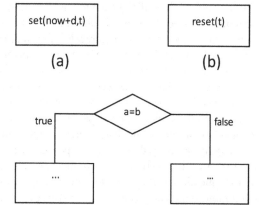

Fig. 3.22 Decision symbol

operator now) and a value of the data type Duration, called *timer expiration time*. An active timer may be stopped using the *reset symbol* shown in Fig. 3.21b; the argument of this operator is the name of the timer to be deactivated. If the operator set is executed on an active timer, this timer is stopped first and then immediately activated. If the operator reset is executed on a timer that is inactive, this has no effect. The expression active(t), where t is a timer name, tests the status of the timer t and returns the Boolean value true if the timer is active and false otherwise.

Figure 3.22 shows the *decision symbol*. A decision symbol offers more possibilities to continue the execution of the process. Which of the possibilities will actually be executed depends on the logical condition consisting of the expression written within the decision symbol and the values or value ranges that label the edges that exit the decision symbol. If one of the exiting edges is labelled with the keyword else, this edge will be executed if no other edge fulfils the condition. There may also be more than two exiting edges; however, all the conditions must exclude one another. Different exiting edges may even lead to different states. In the example shown in Fig. 3.22, the left edge will be executed if the values of *a* and *b* are equal; otherwise the right edge will be executed.

3.5   Simulation of SDL Specifications

The correctness of a specification must be verified; hence, one must verify that the specification specifies the same functionality as intended by the designer who conceived the specification. To do such a verification a simulation is most usually used. At first, specifications of processes must be transformed into simulation models; then simulation models are run in the environment provided for by the simulator. Computer-aided design tools that are based on the SDL language usually consist of SDL editor, automatic simulation model generator, and the simulator itself. However, one must be aware that a specification must be complete, entirely formal and syntactically correct to automatically transform it into the corresponding simulation model. This means that transmitting users that generate communication traffic, and receiving users that receive communication traffic and possibly verify the correctness of received messages, must be specified, in addition to protocol entities. Automatic verification of received messages is of course possible only if receiving users »know« what they must receive; in a real system this is of course a nonsense; if a receiver knew what it should receive, a communication system would not be needed at all! Usually special processes must also be specified to explicitly model channels with their realistic properties, such as losses and delays.

Usually not only the logical properties of a protocol are of interest, but also its performance in a network environment, e.g. its efficiency and delay. The nature of communication traffic is random in most cases, so a random values generator must be an important component of the traffic generator of a simulator. A random values generator is needed also for some other phenomena that occur randomly, such as channel errors and losses. In Chap. 6 we will explain that average values of performance measures are usually used; performance simulation results will therefore be the more accurate, the longer will be the simulation time and the more messages will be transferred during a simulation run.

Part II
Protocols and Their Properties

In the next four chapters a reader will come to know what protocols actually are, the environment in which they operate, how they interact with their environment, how they are structured and which are their properties. Concepts and terms that will be explained here will enable a reader to understand the topics of all the next chapters.

In Chap. 4 a protocol as well as its role and operation in a communication system will be introduced; a model of a simple communication system and a model of a protocol entity will be presented as well as the properties of channels that transport messages between protocol entities. The operation of a protocol and its interaction with the environment will be specified in the SDL language. In Chap. 5 the concept of a protocol stack will be introduced, i.e. how several protocols cooperate in a complex communication system to jointly implement the services of the complete system; protocol stacks will be presented as a method of reducing the complexity of a whole communication system. Three example protocol stacks will also be presented. Chapter 6 will acquaint a reader with the measures that can be used to assess the performance of a communication system; statistical methods that are used to describe the communication traffic will also be introduced. Protocol efficiency, protocol relative efficiency and message delay will be defined. In Chap. 7 the methods to assess the protocol performance using analysis, simulation and measurement will be shortly described.

Chapter 4
Communication Service, Communication Protocol

Abstract The aim of this chapter is to introduce the general concepts of communication protocols and the terminology used in the field. The notions of a communication service, protocol, service access point, primitives and protocol messages are first explained. Then the general constituents of protocol messages are described. The implementation of protocols with protocol entities and channels is discussed, as well as the interaction of protocol entities with users. The model of a simple communication system is shown. A more detailed model of a protocol entity is then presented. A simple communication system is formally specified in SDL, using both a hierarchical and a one-level specification technique, and the methodological difference between the two presentations is emphasised. At the end of the chapter, the transfer of a user message through a simple communication system, consisting of two users, two protocol entities and a channel, is specified in SDL, both to offer a simple example of SDL specification and to demonstrate how user messages are transferred through a communication system. This chapter has 18 figures.

In Sect. 2.1 we already told some basic and general facts about a communication protocol. In this chapter we will tell more about how a protocol operates, how it interacts with the users of a communication system and which properties a good protocol must have. The environment of a protocol operation will also be formally specified.

4.1 Service, Protocol, Service Access Point

The basic task of a communication system is to provide its users with some *service*. Hence, a communication system serves as a *service provider* for its users. A user and a service provider must somehow interact; for this purpose they use a contact which is referred to as *service access point* and is usually denoted with the acronym *SAP*. Each user interacts with its service provider using its own service access point. In this book a service access point will be considered a conceptual contact between a

Fig. 4.1 Service provider
and users

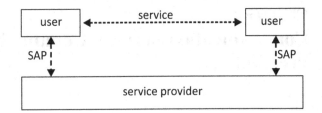

user and a service provider; in real life it can have very different shapes (e.g. a connector in a hardware implementation or a procedure call in a software implementation). A service provider and its users (there may also be more than two) are shown in Fig. 4.1.

A service provider consists of protocol entities and channels that interconnect them. Protocol entities exchange protocol messages via channels in accordance with a specific protocol; on the other hand, protocol entities interact with their users via service access points. In this way protocol entities implement services for their users; a protocol can therefore be considered an implementation of a service. The term telecommunications (or more briefly communications) means the information exchange between dislocated (distant) entities; hence users are separated by a physical distance. A service provider must therefore provide for the information (message) processing as well as for the information transfer across a physical distance. The former task is accomplished by protocol entities and the second task is done by a *communication channel*. The pair user–protocol entity and the service access point that interconnects them are assumed not to be physically separated; hence they are assumed to be located at the same place (e.g. in a same computer), while a channel has a nonzero physical size. Of course, the expression »at the same place« is a simplification (such simplifications are often used in technical sciences), as Archimedes found out already more than 2200 years ago that two physical bodies cannot exist at the same place! In our case, indeed, we assume that the sizes of a user, a service access point and a protocol entity are neglectable, while the physical distance« between them is also neglectable; the term »neglectable distance« in this context means that the delay of an electromagnetic wave propagating across this distance is neglectable, compared with other delays in the system.

In such a system users may communicate only using the services of a service provider; a message to be transferred from one user to another is actually transported from one user through a service access point to its associated protocol entity, then through the channel to another protocol entity, and finally through another service access point to the other user. Abstracting all the details we have just cited, one can, however, consider the two users to communicate directly through a *virtual channel* that directly interconnects them; this virtual channel is implemented by the service provider with the protocol entities and the channel. The structure of a service provider and the relation between the users and protocol entities are shown in Fig. 4.2.

Hence, a protocol is the implementation of a service that users are provided with by the service provider. This means that a certain service can be implemented by

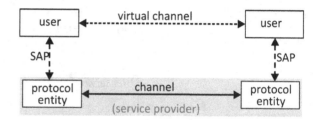

Fig. 4.2 Service and protocol

different protocols. In other words, this means that, in the course of a design process, one may decide to use a certain service without having a specific protocol in mind. A service may actually be viewed as the abstraction of the protocol.

A service can only be used if the users know exactly what this service offers them; it must therefore be precisely and unambiguously defined. The specification of a service must determine what this service offers its users, and also how a user can access the service.

A system designer who wishes to implement a service with a protocol must also precisely specify the protocol.

4.2 Primitives

Of course, a service provider will not execute a service on its own, but will execute it only upon the request of a user. The user that requests a service is referred to as *service initiator*, while the other one, there may also be more of them, which is to receive the service (each communication service concerns at least two users) is referred to as *service recipient*. The service provider must inform the service recipient about the service that was requested by the initiator. A service recipient may have (but not necessarily) the opportunity to receive the service or to decline it. The service provider may inform the service initiator about the success of the service realisation (such a service is called a *confirmed service*); this information may also not be provided (*unconfirmed service*). All these interactions between a service provider and its users are executed across the appropriate service access points.

The order in which the events of the interaction between users and the service provider occur is not arbitrary; this order is illustrated in the time-sequence diagram shown in Fig. 4.3. In this figure the space extends horizontally, while the time runs vertically (from top to bottom). There is a delay between the service request and the service indication, as well as between the response and the confirm; the reason for these delays is the physical size of the service provider which causes nonzero delays. On the other hand, a user and its associated protocol entity are assumed to be at the same place (as we have already explained in the previous section), so the delay needed to deliver a request or a response from a user to the protocol entity (and the delay in the opposite direction) is theoretically zero. There is also a delay between an

Fig. 4.3 Interaction
between service provider
and users

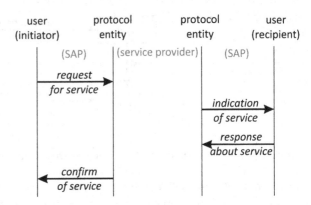

indication and its associated response; this delay is due to the processing that must be done by the recipient user and is referred to as *processing time*. A processing time is of course needed by all the entities, users as well as protocol entities.

In this section only the interaction of a service provider with its users is discussed; the space between the two protocol entities in Fig. 4.3 was therefore intentionally left void. The transfer of protocol messages between protocol entities will however be discussed in the greater part of this book.

The four types of the interaction between users and service providers that have just been discussed are sufficient to describe all possible interactions between users and service providers. These four basic types of interaction are therefore referred to as *primitives*. Standardised terms and their related abbreviations are used for the four primitives, as follows:

- *request—req.*
- *indication—ind.*
- *response—resp.*
- *confirm—conf.*

Primitives are normally used together with specifications determining what is being requested, indicated, responded or confirmed; primitives will therefore most often be written in the form *spec.primitive*, where *spec* will specify what is requested, indicated, responded or confirmed, while *primitive* will be one of the above-mentioned primitives (request, indication, response or confirm). For example, many different requests are possible: data.request requires a user message transfer, connect.request requires a connection establishment, release.request requires a connection release... A primitive may also contain different parameters which must be exchanged between a user and its associated protocol entity; e.g. a primitive requesting a user message transfer will certainly contain the user message itself as a parameter, and possibly also some other parameters (such as the address of the message recipient and the indication of the quality of service). A simple example of a primitive requesting user data transfer can therefore be shown as data.request (userdata).

All the four primitives are not always needed; only some of them may be used, depending on the service. The confirm primitive will be used only in the case of a confirmed service; otherwise it will not be needed. The most usual combinations of primitives are request–indication–response–confirm, request–indication–confirm and request–indication; in some cases a single primitive may even be used, such as the primitive indication (unfortunately, such a case may indicate that something goes wrong!).

The primitives which are actually used must always be used in the order in which they were listed above and which is also shown in Fig. 4.3. The primitives request and response are always generated by users to be received by a service provider, while the primitives indication and confirm are generated by a service provider to be received by users. The user that requires a service using the request primitive is the service initiator, while the service recipient receives the service by means of the indication primitive. Hence, the primitives request and confirm are always used at the initiator side, while the primitives indication and response are used at the recipient side.

4.3   Protocol Messages

Protocol entities are active elements of a communication system which implement the service as components of the service provider. Their activities must be coordinated; therefore they must exchange information between them in the form of messages. These messages must be in accordance with a protocol and are therefore referred to as *protocol messages* or *protocol data units* and usually designated with the acronym *PDU*.

As the information exchanged by users is transferred through a service provider, it is actually transferred through a channel between protocol entities as a part of protocol data units. A protocol message therefore contains the control information and possibly also the user information. The user information that is a part of a protocol message is referred to as *user message* or *service data unit* and is also indicated with the acronym *SDU*; the control information which controls the correct communication in accordance with a protocol is referred to as *protocol control information* and indicated with the acronym *PCI*. Hence, a protocol data unit (PDU) consists of a service data unit (SDU) and a protocol control information (PCI). As the transfer syntax of a protocol data unit is concerned, the protocol control information can be placed at the beginning of a protocol message (in this case it is called *header*), at its end (in this case it is called *tail*), or both (so a protocol message may contain both a header and a tail); it may even be distributed over the whole protocol data unit. A typical composition of a protocol data unit containing both a header and a tail is shown in Fig. 4.4.

Protocol data units that contain both user information and control information are referred to as *information protocol data units*; in this book they will sometimes be indicated with the acronym *iPDU*. However, some protocol messages contain only

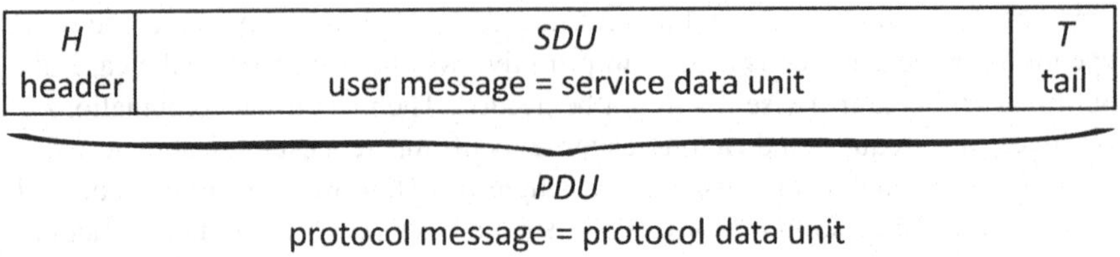

H header	SDU user message = service data unit	T tail

PDU
protocol message = protocol data unit

Fig. 4.4 Typical structure of protocol message

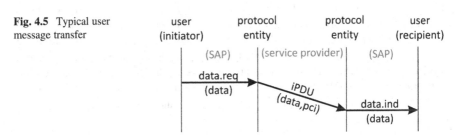

Fig. 4.5 Typical user message transfer

control information (without user information); they are called *control protocol data units*.

Both information and control protocol data units are important for the correct operation of a protocol; within a message, both user and control information are important as well. However, as users are primarily interested in the transfer of user information, the service data unit is often referred to as *payload* of a protocol message; on the other hand, the protocol control information is also indispensable as it guarantees the correct protocol operation, although it does not carry any user information, so it is often called *overhead*. As the transfer of control information requires some additional network resources, users do experience a lower transfer rate as provided by the channel, so a protocol requiring less overhead (though assuring the same quality of service) can be more efficient.

In Fig. 4.3 the interaction between users and protocol entities by means of primitives was shown; however, the space between both protocol entities was left void there, thus not showing how protocol entities interchange messages. Figure 4.5 shows a typical user message transfer from a user to another user; the user message is denoted with the word data in this figure. The procedure begins when a user (the service initiator) requests a user information transfer with the request primitive (data. req), passing the user message as the parameter of this primitive. The protocol entity (let us call it the transmitting protocol entity) composes the protocol data unit by adding the protocol control information to the user information and sends it through the channel to the other protocol entity (the receiving protocol entity). The receiving entity analyses the received protocol message and, if it finds it in accordance with the protocol, passes the user information contained in it to the recipient user with the primitive data.ind.

A reader must be aware that a user message is transferred from a user to another user, while a protocol control information is transferred only from a protocol entity to another protocol entity. The transfer of a protocol control information is a part of the service implementation and does not directly concern users; only the user information transfer represents the service!

On the other hand, the most important concern of protocol entities is the processing and analysis of protocol control information. Although the user information is transferred as a part of protocol messages, protocol entities do not process or analyse it; thus, a user information is somehow hidden within protocol data units. This phenomenon represents the use of the well-known methodological principle of *information hiding* which plays an extremely important role in the methodologies of complex systems design and is closely related to the abstraction principle. Of course, the protocol entities also could read and analyse the user information if they wanted to do so, unless users prevent them to do this (e.g. by ciphering the user information).

We have seen that a protocol control information is added to the user information before transmitting the protocol message. A user information is said to be encapsulated into a protocol message, the principle is referred to as *encapsulation*. At the receiving side, the receiving protocol entity extracts a user information from a protocol message before passing it to the user, it decapsulates it; this procedure is called *decapsulation*.

4.4 Specification of Service, Specification and Implementation of Protocol

The specification of a service must precisely and unambiguously determine:

- What does the service offer to users.
- Which is the *quality of service* (*QoS*).
- How can users access the service via service access points and which primitives can they use.

A protocol implements a communication service, using the services of the channel interconnecting protocol entities (see Fig. 4.2); the specification of a protocol must therefore be derived from

- The service to be implemented.
- The way protocol entities should interact with users.
- The properties of the channel that is to interconnect protocol entities (different protocols will implement the same service if channels with different properties are used).
- The way protocol entities should access the channel.

Hence the specifications of the service and the channel properties form the part of a protocol specification as they define which service the protocol can implement and with which channels it can be used. Additionally, a protocol specification must also define

Fig. 4.6 Transformation of
quality of service of channel

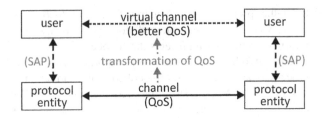

- The abstract syntax of protocol data units (the types of protocol messages as well as the number and types of the parameters brought by them).
- The transfer syntax of protocol data units (defining their structure).
- The set of rules specifying the behaviour of protocol entities in different situations in which they can find themselves (how they should react to different input events in different states); here one should keep in mind that protocol entities operate in real time and must therefore be able to measure time, as the rules of operation may be time-dependent.

Often the service implemented by a protocol is a message transfer with a better quality than is the quality of the channel interconnecting protocol entities. In such a case one must imagine a virtual channel to directly interconnect users, as was already shown in Fig. 4.2, while the protocol specification defines how a channel with a worse quality is transformed into a virtual channel with a better quality; this is illustrated in Fig. 4.6.

A protocol implementation is derived from the protocol specification; the protocol implementation actually defines the functionality of protocol entities. A protocol entity constructs protocol data units before transmitting them, a protocol entity interprets and analyses protocol data units it has received, a protocol entity also interacts with users. It is up to a protocol entity to decide what to do in specific situations and how to react to input events.

4.5 Properties of a Channel

The main topic of this book is protocols, hence how protocol entities should operate. As the functionality of a protocol can also be viewed as the transformation of a channel with a worse quality to a virtual channel with a better quality, as was discussed in the previous section, it is very important that a protocol designer be aware of the properties of channels that interconnect protocol entities. A protocol may strongly depend on these properties, as the task of many a protocol is to solve the problems posed by unfavourable properties of channels.

Communication channels pose numerous and diverse problems to communication systems. In this section only a few more frequent problems to be solved by protocols will be discussed.

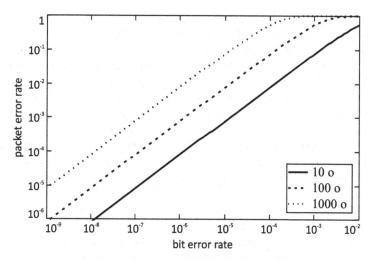

Fig. 4.7 Dependence of packet error rate on bit error rate

Due to the noise, disturbances and other physical phenomena, many a bit can be corrupted during the transfer through a channel; the consequence of such a corruption is that a receiver may interpret the received signal as a logical value that is different from the value which was transmitted. The usual measure to describe how frequently bit corruptions do occur is the *bit error rate*, most usually denoted with the acronym *ber*. The bit error rate is defined as the ratio of the number of bits that were corrupted during a transfer over the number of bits that were transmitted during the same time period. Errors do occur randomly, so the average value of bit error rate is usually of interest. If the only source of bit errors is the noise, the bit error rate does not change with time; in many cases, however, the bit error rate is time dependent, errors are said to occur in bursts.

If at least one bit in a message is corrupted, the whole message is corrupted. The *packet error rate* is defined as the ratio of the number of corrupted messages over the number of transmitted messages in the same time period. If the bit error rate is constant (not time dependent), the packet error rate *per* can be calculated from the formula

$$per = 1 - (1 - ber)^{L}, \tag{4.1}$$

where *ber* denotes the bit error rate and L is the length of a message in bits. As expected, the packet error rate is the higher, the higher is the bit error rate; the packet error rate is also the higher, the longer are messages (the longer is a message, the higher is the probability that at least one bit is corrupted). This fact can also be seen in Fig. 4.7 which was generated from the formula (4.1). In this figure the dependence of packet error rate on bit error rate for three different message lengths (10, 100 and 1000 octets, respectively) is shown.

During a transfer a message can be corrupted so severely that the receiver cannot recognise it as a message; such a message is lost. Furthermore, many receivers simply discard messages that they recognise as corrupted; in this way message errors produce message losses. A very important cause of message losses, in some networks by far the most important, is the lack of resources for information processing, storage and transfer in communication networks and communication devices; a network element that has no resources to process or store a message discards it. The measure usually used to describe losses is the *loss rate* which is defined as the ratio of the number of lost messages over the number of transmitted messages in the same time period. If the message corruption is the only source of message losses and receivers simply discard corrupted messages the loss rate equals the packet error rate.

In communication networks that are based on connectionless operation, the order of transferred messages may be modified during the transfer. This means that a message which was transmitted prior to another message is received later by the receiving entity. This phenomenon is referred to as *message reorder*.

A message may also happen to be received twice by the receiver. Such a message is referred to as *duplicate message*.

Special cases of duplicates which can in some cases be especially unpleasant are *floating corpses*. A packet may happen to be delayed unusually long in the network. The transmitter senses the loss and retransmits it; the receiver, unfortunately, receives and processes both the original packet and the retransmitted one. If the receiver is uncapable to recognise such a floating corpse as a much delayed copy of an already processed packet, it can react to it in a wrong way and the communication process enters an unforeseen and unpleasant trouble.

The *message delay* is one of the basic properties of a communication channel. A delay can of course be longer or shorter; this depends on the physical distance between communication entities as well as on processing times in intermediate network elements. A delay can either be constant and foreseeable, or it can be variable and random. Variable random delays are the consequence of the packet mode of information transfer; they can be very unpleasant with the transfer of some kinds of information, such as speech, especially if the delay variation is too large.

4.6 Protocol Design Process

A communication protocol design is similar to a system design as already described in Sect. 1.2.

Before the design of a communication protocol is begun, the specification of the services to be provided by the protocol and the properties of the channel interconnecting protocol entities must be known. Then the protocol itself is specified. Unfortunately, it may well happen that, due to a specification error, the protocol specification does not define the same functionality as was previewed and specified in the service specification; the equivalence of the protocol specification and the service specification must therefore be verified (which is most usually done by means of simulation). After that the implementation of protocol entities is defined in the form of software or hardware modules; the equivalence of the protocol implementation and the protocol and service specifications must also be verified which can be done with simulation already before the final implementation of protocol entities or with testing after the final implementation. The implementation step does not need to immediately follow the specification step, both steps also do not need to be done by the same designer, as an already existing standardised protocol specification is quite often used.

4.7   Properties of a Protocol and its Specification

This book is dedicated to communication protocols, their properties and how they can be specified. In this section the properties that a protocol specification must have will be discussed; of course, the properties of a protocol specification are closely related to the properties of protocols themselves.

The most important property of a protocol and its specification is the *logical correctness*. The logical correctness of a protocol means that the communication process as specified by the protocol always proceeds towards its goal, namely towards the realisation of the specified service; furthermore, this goal must be achieved in a finite number of steps and in a finite time. The communication process may not enter a state which it could no more leave; such a state is referred to as *deadlock*. The communication process also may not enter an infinite loop, referred to as *livelock*; a livelock is a sequence of states that are cyclically repeated but do not lead towards the final goal. Furthermore, any state must be reachable from the initial state. The initial state, the state in which a machine awakes when it is turned on, must be well defined (the process must be initialised).

The *nonambiguity* of a protocol specification means that this specification cannot be understood in different ways by different people. Natural languages have some nice properties, such as ambiguity and metaphor, which make them rich and allow writers to express themselves artistically; however, such properties are not desired in technical texts. Hence, a technical specification must be nonambiguous and self-consistent; different statements in it may not be contradictory.

The *completeness* of a specification means that the specification defines the reaction of an entity to any input event in any state (except, of course, if an event in a state is not possible). Hence a system cannot find itself in a situation where it would not know what to do. If, however, a system also has the property of *robustness*, it defines the behaviour even for a situation in which the system could not find itself in normal operation; such an abnormal situation can however occur due to a system malfunction or fault, or even due to a malicious action of an entity or a user. So, a robust specification will define the reaction even for a normally impossible combination of a state and an event.

An important property of communication protocols is the *standardisation*. Entities that are products of different producers and/or managed by different owners can cooperate only if they all operate in compliance with the same standardised specifications. Hence the standardisation of protocols allows for the *interoperability* of different communication entities; systems that are standardised so that they can interoperate are referred to as open systems. There are many organisations that develop and publish standards, such as *International Organisation for Standardisation—ISO, International Telecommunication Union—ITU, Institute of Electrical and Electronics Engineers—IEEE, European Telecommunications Standards Institute—ETSI* and *Internet Engineering Task Force—IETF*.

The *readability* of a specification means that an appropriately educated and experienced user (reader) can easily read and understand the specification. If a careless or unexperienced reader reads a hardly readable specification, the consequences are similar to the case of an ambiguous specification.

The *abstractness* of a specification means that the specification has a low implementation bias (it does not contain much information about possible implementations). Abstraction is an important methodological principle which allows designers to execute successive design steps as independently one of another as possible. This is especially important when designing complex systems.

The *machine readability* is usually also a desirable property of specifications. Only a machine readable specification can be automatically transformed into a hardware- or software-based implementation by means of dedicated computer-aided design tools.

4.8 Model of Protocol Entity

A protocol views a channel as a passive element with the only role to transfer a protocol message from one end to another; channels may have different properties, such as delay, errors and losses.

A protocol entity, on the other hand, is an active element which must be able to interact with users, transmit and receive protocol messages, and memorise protocol messages and other values (such as its state); it must make proper decisions and execute different activities according to the rules defined by the protocol. It uses a *processor* and a *memory* for these purposes.

A protocol entity, just like a communication system, operates in real time, so it must be able to measure time. For this purpose it uses a special device called a *timer* which can be activated or deactivated; after a timer has been active for a predefined time and has not been deactivated meanwhile it expires. A timer expiration is an event to which the entity can react. A usual purpose of a timer in a communication system is to warn an entity that an event the entity has waited for did not occur in due time, so the entity must react in some other appropriate way, as prescribed by the protocol.

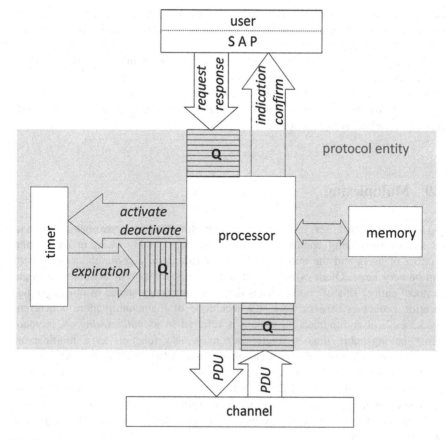

Fig. 4.8 Model of protocol entity

A protocol entity always has a limited number of processors (e.g. only one), so it can process a limited number of tasks at a time; all the requests for processing must therefore enter the entity through a specially organised memory module, referred to as *waiting queue* (or just a *queue*).

In Fig. 4.8 the model of a protocol entity is shown. In this figure it can be seen that receptions of protocol data units and primitives as well as timer expirations are possible input events of the protocol entity. All of these input events must access the processor through a single queue which is labelled with the letter Q in the figure (so all the elements labelled with Q in the figure form a single queue!). The protocol entity itself can read and modify the contents of its memory (including its state), transmit protocol data units and primitives and manage (activate or deactivate) its timer(s). A protocol entity is a reactive system, so it executes all its activities as reactions to input events.

The model of a protocol entity as shown in Fig. 4.8 is quite similar to the model of a computer operating in real time. Of course, a protocol entity can be implemented either as a piece of hardware (an electronic circuit) or as a software process; the difference between the two possible implementations will not make any difference for our discussions.

A careful reader will also notice the similarity between the model of a protocol entity that has just been presented and the model of an SDL process which was discussed in Sect. 3.4. This is one of the reasons why the SDL language is so appropriate for the specification of communication systems, whereas protocol entities are modelled as SDL processes.

4.9 Multiplexing

Up to now, only two users and two adjoining protocol entities were always shown in all figures illustrating communication systems (e.g. in Fig. 4.2); in reality, the number of users and the number of protocol entities in a communication system can be very large. Often even more than one users access the services of a single protocol entity. This of course means the concurrent coexistence of more communication processes in a network; the technique of maintaining more concurrent processes in a communication system is referred to as *multiplexing*. A protocol entity having more than a single user must also function as a multiplexor/demultiplexor. A protocol entity can distinguish between different users only if different users use different service access points to access the services of this protocol entity. The protocol entity understands different service access points as the addresses of different users. In Fig. 4.9 an example of multiple users can be seen, each one of them accessing the services of a common protocol entity via its own service access point.

As a protocol entity usually has a single processor (such a case was also shown in Fig. 4.8), it can process only a single task (e.g. a request or a response) at a time; the concurrent processing of multiple tasks is only virtual. The primitives of different users must therefore enter the protocol entity through a waiting queue, just like all the other input events. On the other hand, the protocol entity must pass each indication or confirm primitive to the appropriate destination user. Indication and confirm primitives must therefore be sent to users through a special module for distribution of primitives. This is only possible if the incoming protocol data units

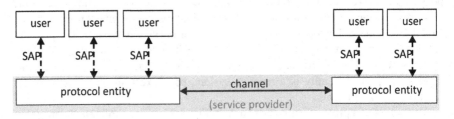

Fig. 4.9 Service access in multiplexing system

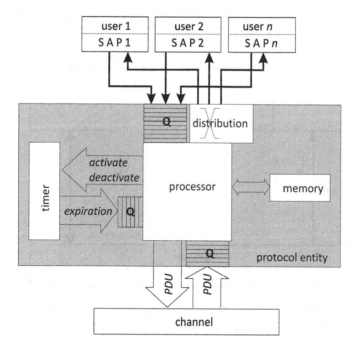

Fig. 4.10 Model of protocol entity in multiplexing system

that trigger these primitives contain addresses of destination users or identifiers of connections to which the primitives are associated. A protocol entity sees service access points as the addresses of its users. In Fig. 4.10 the model of a protocol entity providing multiple users with services is shown.

4.10   Hierarchical Specification of Users and Service Provider

The structure of a communication system that was already shown in Figs. 4.1 and 4.2 will still be specified in the SDL language in this section. This specification will be hierarchical; this means that a service provider will first be shown as a block with only its external structure, and after that its internal structure will also be specified.

In Fig. 4.11 the structure of a communication system with the name u_p is shown. The signals that are shown in this figure model primitives; from their placements in the system it is evident that the user ui is the service initiator, while the user ur is the service recipient. The block provider models the service provider.

In Fig. 4.12 the implementation of the block provider is shown; this implementation consists of the subblocks e1 and e2 interconnected by the channel named ch; evidently, the blocks e1 and e2 model protocol entities. The signals that can be

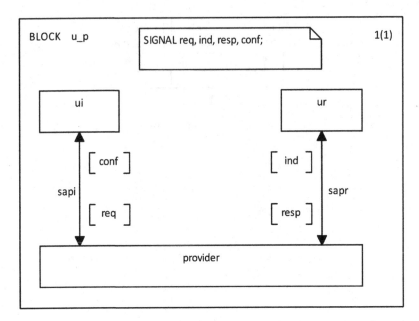

Fig. 4.11 Specification of users and service provider

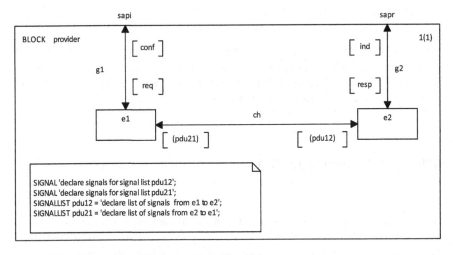

Fig. 4.12 Implementation of block provider in Fig. 4.11

transferred from e1 to e2 and from e2 to e1 are declared as signal lists pdu12 and pdu21, respectively. We wanted to remain sufficiently general when composing this specification (this specification refers to none of specific protocols), so the declarations of signals modelling protocol messages and signal lists are only informally pointed out. A very important detail that can be seen in Fig. 4.12 must also be emphasised. The signals req and conf can be transferred via the channel g1 inside the

Fig. 4.13 Functionality of
block e1

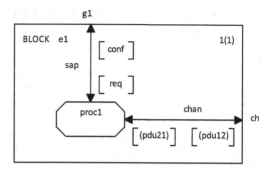

Fig. 4.14 Functionality of
block e2

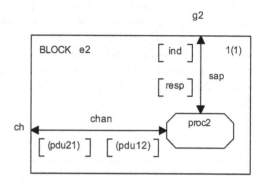

provider, but outside the provider (inside the block u_p) they can be transferred via
the channel sapi; it was therefore necessary to indicate outside the frame of the block
provider that the internal channel g1 of the block provider must be connected to the
external channel sapi in the environment of the block provider. The same is true of
the internal channel g2 and the external channel sapr. It is also important to
emphasise that the signals which model primitives must be visible both outside
and inside the block provider and were therefore declared within the outer block u_p
(because the block provider is a child of the block u_p, these signals are also visible
within the block provider); on the other hand, the signals that model protocol
messages (and were grouped into signal lists pdu12 and pdu21) need not be visible
within the outer block u_p, so they were declared within the inner block provider.
Syntactically and semantically the declaration of these signals in the outer block u_p
would be correct as well; methodologically, however, it is better to declare protocol
messages within the provider (so they cannot be seen outside the provider), as they
form a part of the provider implementation and should be of no interest to users
(which should be occupied only with services)!

In Figs. 4.13 and 4.14 the functionalities of the blocks e1 and e2 are specified.
These specifications are trivial, as a single process is running in each block; these
functionalities are also not complete, because the functionalities of both processes
are not specified. In these two figures the channels inside and outside the blocks e1
and e2, respectively, were also interconnected.

4.11 Single Level Specification of Users and Protocol Entities

The specification of a communication system as given in Figs. 4.11–4.14 is methodologically very sound as it separates the implementation of the service provider (its internal structure) from its specification (its external structure). This kind of hierarchical system specification and description is especially appropriate in case of complex systems as it allows for separate design of system components at separate levels of abstraction, which also allows for an easier mastering of the complexity.

This book is devoted to the treatment of communication protocols rather than to the treatment of complex systems; therefore a more simple system description will suffice which will not take the hierarchical structure into account. Figure 4.15 shows such a simplified (one-level) description of the same system as was hierarchically specified in Figs. 4.11–4.14. As can be seen in this figure, a communication system can only be specified with processes which model users and protocol entities.

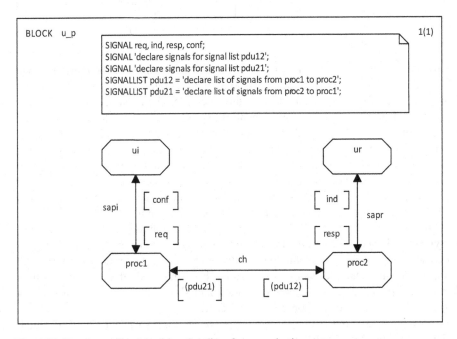

Fig. 4.15 Simple specification of functionality of communication system

4.12   Specification of User Message Transfer

In Section 4.3 and in Fig. 4.5 we already explained how a user message is actually transferred from one user to another one through two service access points, two protocol entities and the channel that interconnects them. In this section such a procedure will be specified in the SDL language.

In Fig. 4.16 the processes of both users and both protocol entities can be seen, interconnected by both service access points and the channel; the processes ut and ur model the transmitting and the receiving user, respectively, while the processes trans and rec represent the transmitting and the receiving protocol entity, respectively. If an unconfirmed service of user messages transfer is implemented, only the primitives req and ind are used. While both primitives bring only one parameter each, namely a user message, a protocol data unit brings a protocol control information (PCI) in addition to a service data unit (SDU). We wanted this specification to be general enough, so the definitions of data types Datatype (for user data) and Ohtype (for overhead) are only informally indicated. Let us also remark here that in SDL the names of channels are optional, so the channels in the specification of Fig. 4.16 have no names.

In Fig. 4.17 the operation of the transmitting protocol entity is specified. In the state wait this process waits to receive a request for the user message transfer. When it receives such a request it stores the user message into its memory (the variable sdu), determines the appropriate protocol control information and adds it to the user information, thus composing the protocol data unit which it then transmits as the signal pdu through the channel towards the receiving protocol entity. After this it is prepared again to receive a new request. In this very simple example the variable sdu is needed only because this is the only way in SDL to

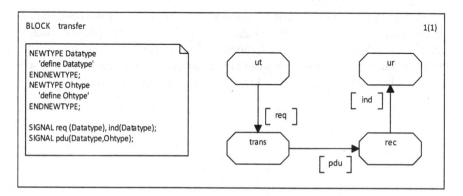

Fig. 4.16 Specification of user message transfer system

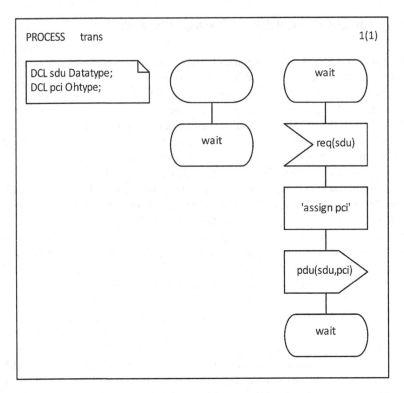

Fig. 4.17 Specification of functionality of transmitting protocol entity process

use a value brought by the signal (primitive) req when composing another signal—the protocol data unit pdu.

In Fig. 4.18 the specification of the functionality of the receiving protocol entity is shown. When it receives a protocol data unit pdu, it analyses the protocol control information it has saved in the pci variable. If no problem is found, it uses the primitive ind to pass the user message (without the control information, of course!) to the receiving user and then waits to receive a new protocol message.

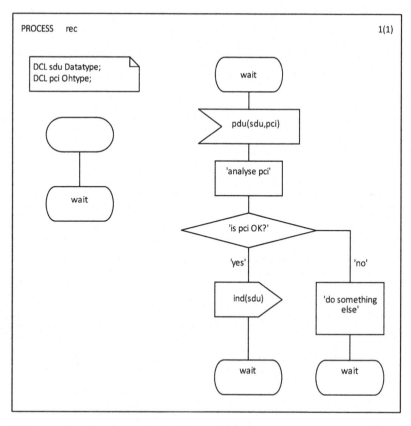

Fig. 4.18 Specification of functionality of receiving protocol entity process

Chapter 5
Protocol Stack

Abstract At first, the need for structuring a complex communication system to allow for an easier and a more reliable design is exposed. Then the concept of a protocol stack is explained and the top-down design of a complex communication system in the form of a protocol stack is described. The communication and interaction between the entities in a protocol stack are explained. The interaction of a protocol entity with the entities in the adjacent layers is also formally specified. The terminology that is usually employed to describe a protocol stack is explained. The two ways in which protocol stacks can be combined are also described, namely the communication planes and the concept of tunnelling. The OSI reference model, the TCP/IP protocol stack and the reference model for convergent networks are described as examples. The reference model for convergent networks is a peculiarity of this book and can usually not be found in other books. There are six figures and three tables in this chapter.

5.1 What Is Protocol Stack and Why Is It Needed

In Sect. 4.1 and in Fig. 4.2 the relation between a communication service and a communication protocol was explained as well as the roles of protocol entities, channels and service access points. A protocol was said to implement a service and a service may be viewed as a virtual channel that directly interconnects users. The virtual channel interconnecting users should have better quality than the channel that interconnects protocol entities or should it provide services that are not available with the channel between protocol entities; otherwise the protocol (which consumes some system resources for its operation) would have no sense. Hence a protocol implements a kind of the transformation of a channel with less services or a worse quality of service into a channel with more services or a better quality of service, as already stated in Sect. 4.4 and in Fig. 4.6. It is easy to imagine that a protocol will be the more complex, the bigger will be the difference between the services of both channels; however, a higher complexity of a protocol implies a higher probability of design errors when specifying and implementing the protocol. A protocol design

D. Hercog, *Communication Protocols*, https://doi.org/10.1007/978-3-030-50405-2_5

must therefore also follow the guidelines of an appropriate protocol design methodology based on similar principles as already discussed in Chap. 1, hence on the
principles of structuring and abstraction. The structure of a communication system
that is based on such design principles is referred to as *protocol stack* or *protocol
suite*.

A protocol, designed in accordance with Fig. 4.2, which used a real, physically
realisable channel connecting protocol entities and implemented services applicable
to real-life distributed applications, would in most cases be too complicated to be
designed in a single step as a single component. The functionality of a whole
communication system must therefore be partitioned into components that are
separately manageable and can be treated and designed mostly independently one
from another.

In Fig. 5.1 the design of a protocol stack will be explained as a top-down
sequence of design steps; from this procedure the significance of designing a
communication system in the form of a protocol stack will also be clearly seen.
Let us begin this procedure with two users and a communication service represented
as a channel between these two users which makes the operation of a distributed
application feasible; this is shown in Fig. 5.1a. Unfortunately, such a channel is in
most cases not available, so it must be implemented with the protocol n, running
between protocol entities n interconnected by a channel; the channel between the
users thus becomes a virtual channel (see Fig. 5.1b). Protocol entities n must
however be interconnected by a channel with such services and such a quality of
service that the protocol n is not too complex, so it can be designed and implemented
without excessive problems. If such a channel is also not available, it is implemented
with the protocol $n-1$ running between the protocol entities $n-1$ across the channel
interconnecting them, as can be seen in Fig. 5.1c; so the channel interconnecting
protocol entities n, too, becomes a virtual channel. The channel interconnecting
protocol entities $n-1$ must also provide proper services and sufficient quality, so that
the protocol $n-1$ can be designed without excessive problems and risk of errors.
This channel, too, must therefore be implemented with a lower-layer protocol. This
procedure continues until the protocol 1 is designed that runs between protocol
entities 1 interconnected by a real channel which is somehow already available (see
Fig. 5.1d).

As can be seen in Fig. 5.1d, the communication system is partitioned into n
layers; in each layer there are two (or more) protocol entities which communicate
according to the appropriate *protocol of the layer* by exchanging protocol data units
through the channel that interconnects them. Communication processes in all layers
are running concurrently, so there are n protocols running concurrently in the
system. Each one among the n protocols implements its share of the functionality
of the whole system; they all together collectively implement the whole functionality
of the communication system. The information interchange between protocol entities in the system is however restricted: a protocol entity of the i-th layer may
exchange protocol data units only with its *peer entity* (or entities), or *peer*(s) for
short, in the same layer, according to the protocol of this layer; these message
exchanges are done in all layers via virtual channels, except for the lowest layer

Fig. 5.1 Protocol stack design

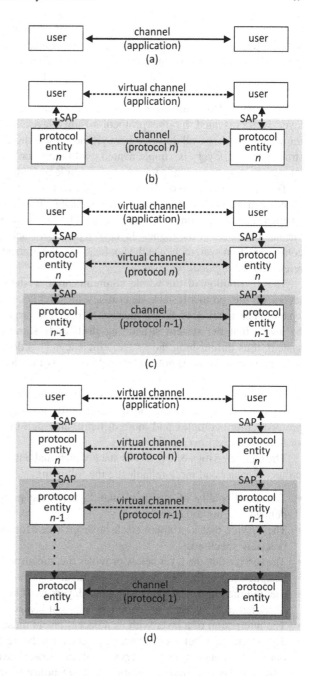

where real channels are used. Additionally, a protocol entity may directly interact by means of primitives across service access points with protocol entities that are running in the adjacent layers directly above and below it. The protocol of a layer implements services for the adjacent layer immediately above it, while at the same time using the channel (the services) implemented by the protocol of the adjacent immediately lower layer. Although a protocol of a specific layer directly uses only the services of the immediately lower layer, it indirectly uses the services of all the lower layers. The services a layer implements to the immediately higher layer directly and all the higher layers indirectly represent the abstraction of this layer and indirectly also of all lower layers. On the other hand, the implementation of a layer is represented by the protocol of this layer and indirectly also by the implementations of all lower layers. Hence, the whole system is hierarchically structured: a layer is implemented with protocol entities and a channel interconnecting them, while this channel is implemented with a lower-layer protocol.

The protocol entities of the lowest layer, named the *physical layer* and denoted with the number 1 in Fig. 5.1d, communicate via a real, also called a physical channel (hence the name physical layer), while the channels of all higher layers are virtual. The users of the highest layer of a protocol stack are the users of the whole communication system, hence the applications; the highest layer is therefore usually referred to as *application layer*.

The distribution of the whole communication system functionality among the layers of the protocol stack determines the architecture of the communication system. The number of layers must be neither too large nor too small. One must be aware that the system structuring itself adds some complexity to the whole system, thus diminishing the system efficiency. This does not pose much problem if the number of layers is not too large; this slightly increased complexity can even be viewed as the price that must be payed for an easier and more reliable design of individual protocols. On the other hand, if the number of layers were too low, the complexity of layers would be too high, thus making the design of individual protocols more complex and more difficult, although the complexity of the whole system would be lower in this case.

The characteristics of a protocol stack which make the design of individual layer protocols easier and more reliable are the following:

- The functionality of any particular layer of the protocol stack must be well defined and not too complex.
- The service that a protocol stack layer provides for the layer above it must be precisely specified.
- The service that the layer below provides (i.e. the virtual channel to be used by the designed layer) must be precisely specified.
- The interaction of any layer with other layers must be as simple as possible and restricted to the interaction with only adjacent layers.
- The interaction of a layer with its adjacent layers must be precisely specified.
- The interaction between adjacent layers must be restricted to only the basic operations concerning the services of the lower layer, represented with the primitives request (req), indication (ind), response (resp) and confirm (conf).

When designing a protocol stack the architecture of the whole communication system is first designed and only then the design of protocols is tackled; the order in which particular protocols are designed is not important. In the architecture design

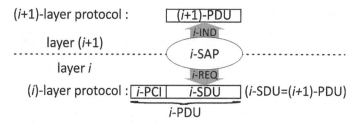

Fig. 5.2 Adjacent layers in protocol stack

phase the functionality of the whole system is partitioned into the functionalities of particular layers; when designing the protocol of a particular layer the specification of the services of the layer and the properties of the channel of this layer (hence the services of the next lower layer) are followed. Different protocols can be combined in different layers, provided the protocol of any layer implements the services the higher layer requires, and the lower layer implements the services this layer expects.

In some modern communication systems the efficiency and the complexity of a system are of utmost importance, so designers sometimes intentionally violate some rules for the design of protocol stacks that are posed by design methodologies. Thus, they may merge the functionalities of stack layers that are usually separately designed, or the adjacent layers interact more closely than is usually done with the four basic types of primitives. Sometimes even the layers which do not lie one immediately one above the other (are not adjacent) interact directly. Such design methods are referred to as *cross-layer design*.

Now let us explain still some basic concepts and technical terms which are usually used when speaking about protocol stacks. These concepts and terms are illustrated in Fig. 5.2 where two adjacent protocol stack layers are shown, namely the layer i and the layer $(i + 1)$ immediately above it, as well as the service access point in between. The protocol that is running in the i-th layer is referred to as *i-layer protocol*; this protocol operates between two peer entities in the layer, so it can also be called *peer-to-peer protocol*. The layers i and $(i + 1)$ can interact by means of primitives that are used across the *i-layer service access point*, also denoted as *i-SAP*. The specifications of primitives are often added prefixes which denote the services of which layer are requested, indicated, responded or confirmed. The *i-protocol data unit* (*i-PDU*) is the protocol message which is transferred within the i-th layer and consists of the *i-protocol control information* (*i-PCI*) and possibly the *i-service data unit* (i-*SDU*). The layer $(i + 1)$ is the user of the layer i, hence the $(i + 1)$-PDU is the user message within the protocol message i-PDU; in other words, the $(i + 1)$-PDU equals the i-SDU. Now let us consider how the protocol message $(i + 1)$-PDU is transferred from one protocol entity of the $(i + 1)$-th layer through the virtual channel to another protocol entity of the same layer. As the channel is virtual, the message must actually be transferred through the next lower layer, namely through the layer i which implements the virtual channel of the $(i + 1)$-th layer. The protocol entity of the $(i + 1)$-th layer requests the data transfer from its adjacent lower layer i by means of a primitive of the request type, such as i-data.req, with the message to be

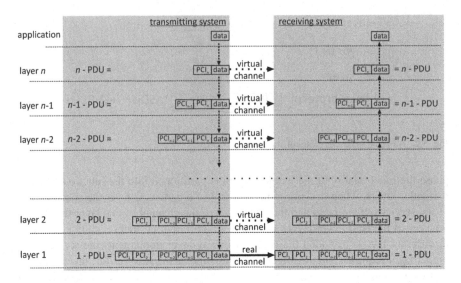

Fig. 5.3 Application data transfer through protocol stack

transferred as the parameter. The protocol entity of the i-th layer adds the control information i-PCI to the message $(i + 1)$-PDU, which thus becomes i-SDU, and so composes the protocol message i-PDU which it sends to its peer protocol entity in the i-th layer through the (virtual) channel of the i-th layer. As the control information can be added both before and after the user information, this procedure is referred to as *encapsulation*. The i-th layer protocol entity at the receiving side analyses the control information i-PCI after it has received the message i-PDU; if everything seems OK, it extracts the user message i-SDU from the protocol message i-PDU (this is referred to as *decapsulation*) and passes it to the protocol entity of the layer $(i + 1)$ as the parameter of the primitive indication, e.g. i-data.ind; this represents the reception of the protocol message $(i + 1)$-PDU for the protocol entity in the $(i + 1)$-th layer. The specification of this procedure in the SDL language is almost equal to the specification given in Figs. 4.16–4.18; only the processes ut and ur are substituted by the transmitting and receiving entities of the higher layer, and the processes trans and rec are substituted by the transmitting and receiving entities of the lower layer, respectively.

The data transfer between two applications that employ an n-layer protocol stack for information exchange is illustrated in Fig. 5.3. The initiator side application uses a request type primitive to ask the n-th layer protocol entity to transfer data to the application at the recipient side. The n-th layer protocol entity adds its control information to the data and passes the n-PDU message to the protocol entity immediately below it (in the layer $n-1$), again using a request primitive. The $(n-1)$th layer entity executes a similar action... Data that were generated by the application at the initiator side are thus passed down from one layer to the next lower layer, while each layer adds its own control information to the data. When the data, together with the control information added by all the layers, reach the physical layer, everything is transferred through the physical channel to the recipient side. Here data pass upwards from one layer to the next higher layer; the protocol entity in each layer analyses the control information which was generated by its peer at the transmitting side and, if everything goes well, passes the information it sees as the user data, to the entity in the next higher layer, using the primitive of the type indication; this procedure is repeated until the data originally generated by the initiating application arrives to the application at the recipient side.

5.2   Model of Protocol Entity within a Layer of Protocol Stack

In Sect. 4.8 in Fig. 4.8 the model was shown of a protocol entity that communicates with its peer entity via a channel and thus implements a communication service for a user. The model of a protocol entity running in the i-th layer of a protocol stack is of course not much different and is shown in Fig. 5.4.

In Fig. 5.5 the specification of the external structure of a protocol entity running in the i-th layer of a protocol stack is shown, together with protocol entities in both adjacent layers. It must also be mentioned here that most usually the kind of the service to which a primitive refers is specified in addition to the primitive type and the layer for which services are provided. A typical example of such a specification is i-data.req which is used by a protocol entity in the $i + 1$-th layer to ask the adjacent lower layer to transfer user data.

5.3   Combining Protocol Stacks

The organisation of a communication system is quite flexible as a protocol stack structure determines only the partitioning of the system functionality into the functionalities of particular layers, and also the interactions between adjacent layers. Within the frame of this architecture, one can employ different protocols. However, different protocol stacks are suitable for different kinds of communication systems and different kinds of applications. Quite often a brand new protocol stack must therefore be developed for a new networking technology, or an existing protocol stack must be adapted. Often it may also happen that a combination of two or more protocol stacks is used to implement some services of a communication system.

The two most usually used modes of combining protocol stacks are communication planes and tunnelling.

Sometimes different kinds of information must be transferred in order to provide for some communication services; to transfer different kinds of information, however, different protocol stacks may be suitable. In such cases several protocol stacks may be employed simultaneously to transfer different kinds of information. The

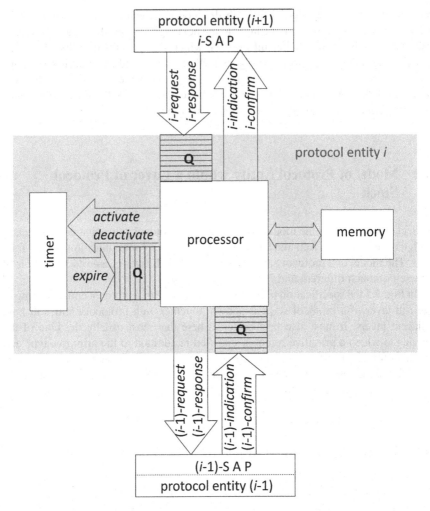

Fig. 5.4 Model of protocol entity in i-th layer of protocol stack

communication is said to be running in different *communication planes*. The most usually used communication planes are the *user plane*, in which the user information is transferred, the *control plane*, in which the data needed to manage communication processes (such as connections) are transferred, and the *management plane*, in which the data needed to manage the communication network itself are transferred. A typical example of a system using communication planes is a telephone network where the speech is transferred in the user plane, telephone connections are managed (set up and released) in the control plane, and the telephone network operation is supervised and managed in the management plane.

The *tunnelling* is a technique of combining two protocol stacks so that the upper part of one of them (the tunnelled stack) is put above the lower part of the other one

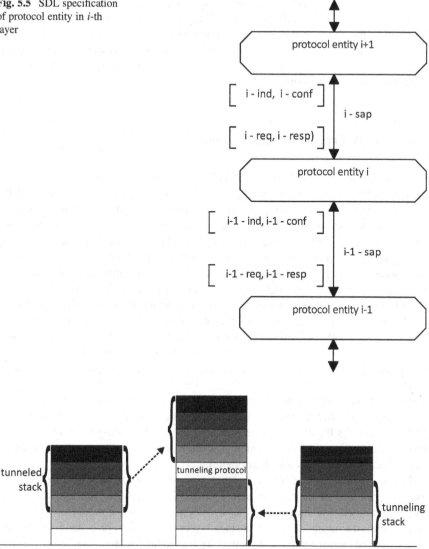

Fig. 5.5 SDL specification of protocol entity in i-th layer

Fig. 5.6 Principle of tunnelling

(the tunnelling stack), whereas the lowest layer of the tunnelled stack is normally lower or at the same level in a protocol stack as the highest layer of the tunnelling stack. Between the tunnelling and the tunnelled stack an additional layer may be inserted in which the *tunnelling protocol* is running. The principle of tunnelling is illustrated in Fig. 5.6. The term tunnelling follows the usual practice that a protocol entity does not inspect and interpret the contents of the user information it transfers, so the contents transferred in the tunnelled stack are normally hidden from the protocol entities operating in the tunnelling stack; however, if the entities of the tunnelling stack are not trusted, the tunnelling protocol can provide for the confidentiality of the transferred information. The tunnelling is usually used to provide for secure information transfer, for connection of more networks of the same kind across a network using another technology and another protocol stack, and to provide for easier mobility management.

5.4   A Few Protocol Stack Examples

In recent decades a number of protocol stacks were developed, as well as the guidelines for protocol stack design. Partly this development was influenced by the development of communication technologies that concurrently ran in different directions. On one side these technologies were developed by private companies and were consequently used in their private non-standardised networks; such systems, developed by different companies, were of course unable to cooperate. On the other side, some protocol stacks were published as the standards that were publicly accessible; in cases where such standards were successful, the systems developed according to them were able to successfully interoperate, even if developed and used by different organisations. Systems that communicate according to standardised protocol stacks and standardised protocols are usually referred to as *open systems*. Due to continuous development of new communication technologies and systems, new protocol stacks are also being developed and existing ones are adapted to new needs.

Protocol stacks determine the architecture of communication systems operation and at the same time the frame within which protocols operate. Understanding of protocol stacks is therefore indispensable for the understanding of communication networks operation, and at the same time indispensable for the understanding of communication protocols and their operation.

In this section three different examples of protocol stacks will be presented. Indeed, only one of them is an actual protocol stack, while the two others are reference models for the protocol stack design. A *reference model* of a protocol stack defines the number of protocol stack layers, the services provided by particular layers, and the mode of interaction between adjacent layers. Hence a reference model does not determine particular protocols to be used in particular layers; it only defines the specifications from which protocols are to be chosen or designed.

5.4.1   OSI Reference Model

The reference model for the interconnection of open systems (*OSI—Open Systems Interconnection*) was standardised by the International Organisation for Standardisation (ISO) long ago, in 1984; with this standard the principle of complex

Table 5.1 OSI reference model

Number	Designation	Name of the layer
7	A	Application layer
6	P	Presentation layer
5	S	Session layer
4	T	Transport layer
3	N	Network layer
2	DL	Data-link layer
1	Phy	Physical layer

communication systems design in the form of protocol stacks was standardised, partitioning the system functionality into seven layers and defining interfaces between these layers. At the very beginning of its existence, OSI was criticised by some experts; it was pretended to be too complicated, to have too many layers and too many options. In all these years, OSI also has become obsolete due to the rapid development of communication technologies. Above all, the protocol stack TCP/IP (we will discuss it in the next section) has become more successful in practice. However, this does not mean that OSI is completely dead today. The designers of the OSI model followed important and clearly defined methodological principles which we have discussed in this chapter; the notions of a service, a protocol and a service access point were thus clearly defined and delimited. Moreover, the OSI standard introduced technical notions and terms which are still used today. This model is therefore used still today to teach communication systems and protocols at universities. Modern protocols and protocol stacks that are being used nowadays are often compared to the OSI reference model.

As we have already told, the OSI reference model has seven layers. The layers are numbered bottom-up; each layer also has a name and a designation that is the abbreviation of the layer name. Designations are also used to indicate service access points and primitives (e.g. DL-SAP and DL-data.req indicate the point of access to the services of the layer DL, and the primitive to request the data transfer service from the layer DL, respectively). In Table 5.1 the designations and the names of layers are shown.

The basic functionalities of the layers are the following:

1. Phy—The *physical layer* provides for the transport of binary values between neighbouring network elements.
2. DL—The *data-link layer* provides for the formation of protocol data units at the transmitting side and their recognition (synchronisation) at the receiving side. A protocol data unit of the data-link layer is usually referred to as *frame*, so the formation of frames is often called *framing*. The data-link layer can also provide for the error detection and recovery, as well as for the flow control. A protocol of this layer always runs between two neighbouring network elements. The message switching is also executed in this layer.
3. N—The basic task of the *network layer* is the message routing; hence, it is the duty of this layer that any message finds its way from the source through the

network to its destination. The protocol data units of this layer are often referred to as *packets*.

4. T—The *transport layer* provides for the communication between network terminals directly (without the intervention of intermediate network elements); it can also provide for an appropriate quality of service of data transport between terminals. Protocol entities of this layer are only implemented in network terminals, but not in intermediate network elements.

5. S—The *session layer* provides for the session management. Sessions are frameworks for the interchange of data between applications.

6. P—The *presentation layer* provides for the transformation of the format of data which are interchanged between applications, regarding the internal formats used by terminals; hence all data are interpreted as exactly the same values by all applications, even if the formats of their storage in different computers are different.

7. A—The *application layer* provides for the direct support to different kinds of applications. Hence the application layer protocols execute different transactions which are necessary for the data exchange between applications.

Only the basic functionalities of layers were mentioned here; particular layers may also provide for some additional functionalities.

In local area networks an additional problem appeared which had not been previewed by the original developers of the model; this problem is the access of terminals to a common medium for information transfer. To resolve this problem, the data-link layer was divided into two sublayers for the use in local area networks. The lower sublayer is referred to as *medium access control* sublayer and is usually denoted with the abbreviation *MAC*; it provides for the coordination of the access of different terminals to the common resources of the medium. The upper sublayer is called *logical link control* sublayer and usually denoted as *LLC*; its task is to transform the services provided by the MAC sublayer to the standard services that the network layer expects from the data link layer, regardless of which medium access technology is used.

5.4.2 Protocol Stack TCP/IP

TCP/IP is the protocol stack that is used in the so-called *IP networks*; the operation of IP networks is essentially based and dependent on the IP protocol. Although there are several kinds of IP networks in existence today, the most used and the best known IP network nowadays is still the public data network called *Internet*. The protocol stack TCP/IP was developed in the 70s of the twentieth century; in 1983 the US Department of Defence adopted this protocol stack for the use in the military network *Arpanet*, the predecessor of what is the Internet nowadays. Today the IP technology is becoming the prevailing networking technology.

Table 5.2 TCP/IP protocol stack

Layer name	Protocols
Application layer	SSH, FTP, HTTP, SMTP, POP3, SNMP, SIP
Transport layer	TCP, UDP, SCTP
Internet layer	IPv4, IPv6, ICMP
Network access layer	MAC, ARP, DSL, PPP

While the OSI reference model was a result of the theoretical development in international standards organisations and industry, the TCP/IP protocol stack was developed at American universities and financed by the American department of defence. This difference resulted in some differences in the networking philosophy. The design philosophy in TCP/IP networks is much more practically oriented: already in the design phase, the developers of the TCP/IP stack had in mind some practical protocols which had already been developed, implemented and tested. The protocol stack TCP/IP is by far not as theoretically and methodologically clean as is the OSI reference model. In some way it violates the severe design rules as enforced by OSI and presented in this chapter; however, it is simpler and therefore also more efficient. In practice it has proved to be very successful; presently it seems that all future networks will be based on this technology.

While in one of TCP/IP layers the same protocol, namely IP, is always used (actually there are two protocols IP used, version 4 and version 6, both of them being quite different and not compatible), different protocols may be used in other layers. Hence, even the TCP/IP protocol stack has some characteristics of a reference model.

Hence the operation of an IP network is based on the protocol IP. The second most frequently used protocol in this protocol stack was (and probably still is) the protocol   TCP.   Therefore   the   whole   protocol   stack   was   named TCP/IP. Unfortunately, there are some people speaking about the »protocol TCP/IP« which is erroneous! TCP and IP are two separate protocols, although most usually used in the protocol stack TCP/IP.

Formally the TCP/IP protocol stack has four layers; we said formally because in the lowest layer multiple protocols can run concurrently in several separate layers. We can speak about one layer (namely network access layer) because the protocols running in this layer are not standardised by standards *RFC* (*Request For Comment*), developed and published by the *IETF* (*Internet Engineering Task Force*), which standardises the other protocols used in the TCP/IP protocol stack. In Table 5.2 the names of TCP/IP layers are listed, along with some most usually used protocols in these layers.

Let us now shortly describe the functionalities of these layers.

- The most important layer is the *internet layer* in which the protocol IP is running. Currently two versions of this protocol are used, namely IPv4 and IPv6, which are so different and incompatible that many people consider them to be two separate protocols, not two versions of the same protocol. Together with the protocol IP, the protocol ICMP is always running (also having two versions, ICMP and ICMP6). The IP network properly (as can be seen in the internet layer) consists of hosts (network terminals) and routers. The protocol IP is implemented in all hosts and all routers; although IP packets are processed in all hosts and routers, they are always transferred (with minor modifications) from a source host to a destination host. IP packets are routed from one host to another one in the internet layer.

- A host and a router or two routers can be interconnected by means of various communication technologies (e.g. LAN, DSL, leased lines...); information transfer in these connections can run differently, depending on the particular interconnecting technology, hence according to different network access protocols which are running in the *network access layer*.
- Protocols of the *transport layer* are implemented only in network terminals (hosts); transport layer protocols are therefore always running directly between terminals, they are said to be *end-to-end protocols*. The most usually used transport layer protocols are TCP (providing for a reliable data transport) and UDP (not providing for a reliable data transport). Because a very reliable data transport is more and more needed, a newer protocol SCTP, which is even more reliable than TCP, is also becoming more and more important.
- In the *application layer* many different protocols supporting different distributed applications can be used. Because the protocol stack TCP/IP does not contain session and presentation layers (as the OSI reference model does), application layer protocols in the TCP/IP protocol stack must also provide for session management and data format transformation. The application layer protocols are also end-to-end protocols.

5.4.3   Reference Model for Protocol Stacks of Convergent Networks

While telephony was by far the most important service provided by communication networks for a century, the data transfer has been growing in recent decades to such a degree that the amount of transferred data is equal or even greater than the amount of transferred speech nowadays, which is in a large part due to the rapid development and employment of computer technologies. Consequently, the importance of so-called *convergent networks* has been growing due to both economic and technical reasons; the characteristic of convergent networks is the convergence of both services and technologies, hence the concurrent transfer of different kinds of information (e.g. speech and data) over the same communication network infrastructure. However, different kinds of information require different modes of transfer and different protocols, so some problems have to be solved to design convergent systems.

Table 5.3 Reference model
for protocol stacks of convergent networks

Layer name
Adaptation layer
Relay layer
Transmission layer

The characteristics of protocol stacks that are used in convergent networks can be summarised in the *convergent networks reference model* shown in Table 5.3.

The important characteristics of the layers of the convergent networks reference model are as follows:

- The *transmission layer* provides for framing and information transfer between neighbouring network elements.
- The *relay layer* provides for multiplexing information flows and relaying information from sources to destinations. For efficiency reasons, the information transfer in this layer is packet oriented. The relay layer must also provide for some level of the quality of service.
- In the *adaptation layer* various protocols are used that transform the quality of service provided by the relay layer to the quality of service required by various applications. Evidently, these protocols essentially depend on the kind of information to be transferred.

As this model is strictly information transfer oriented, additional higher layers, supporting distributed applications, such as an application layer, must of course be used above the adaptation layer.

Two typical examples of protocol stacks in convergent networks are the ATM protocol stack (ATM was the first technology which really provided for the convergence of services and adaptable quality of service), and the protocol stack TCP/IP which is adapted for the transfer of both data and speech.

Chapter 6
Communication Protocol Performance

Abstract The basics of the traffic theory, as related to communication protocols, are presented; the traffic intensity, offered load, throughput and delay are defined. The usual performance measures the protocol efficiency and the delay are defined and explained. Additionally, the relative efficiency of a protocol is defined and described. While the protocol efficiency can usefully be employed to assess the protocol performance in a nonmultiplexed communication system, the relative efficiency is shown to be an especially useful measure to assess the performance of protocols in multiplexed systems. The discussion of the relative efficiency is a special feature of this book; this performance measure was proposed by the author of this book and is not presented in the other books on protocols and their performance. The performance of a protocol stack is also discussed and the nominal transfer rate of a whole communication system is assessed. There are six figures in this chapter.

The most important property of any protocol is of course its logical correctness; a protocol that is not logically correct is useless and has therefore no sense. It is not however unimportant how many resources a protocol consumes to implement a service, or which services and which quality of services it can provide using a certain amount of resources. This is especially important when several protocols or several versions of a protocol are available to implement a service, or when the resource consumption depends on the parameters of a protocol. A decision on which protocol is most appropriate to implement a service in a certain environment or which protocol parameters are most appropriate may well be based on the consumption of resources.

Communication systems use various resources to implement communication services; some of these resources are time, bandwidth and transmission rate of channels, as well as processing power and memory of protocol entities. (It is well known that the transmission rate and the bandwidth are closely related, the ratio between them, usually referred to as spectral efficiency, being strongly dependent on the modulation and other procedures, often carried out in the physical layer; in this book the protocols that run in higher layers, from the data-link layer upwards, will be of main interest, so we will only think about the transmission rate in the rest of this

D. Hercog, *Communication Protocols*, https://doi.org/10.1007/978-3-030-50405-2_6

text, rather than about the bandwidth.) Depending on which resources are most important or which resources are more important than the others in certain circumstances, different measures can be defined and used to assess the performance of protocols. In general, one cannot pretend one measure to be more appropriate than another, this being dependent on the requirements of users and communication network managers and on the environment in which a protocol operates. It is however important to use the same performance measure when comparing two or more protocols.

The resources that will be of main interest in this book are the transmission rate and the time; from time to time the processing power and the memory will also be mentioned. As the resources that are to be used are concerned, three different performance measures will be discussed in this chapter, namely the efficiency of a protocol, the relative efficiency of a protocol, and the delay of messages.

The findings of this chapter will be referenced later in this book when performance of various protocols will be compared or assessed.

6.1 Traffic Intensity

Let us imagine we have two protocol entities interconnected by a channel providing the transmission of information[1] with the rate R. This rate is called *nominal channel rate*. The unit to measure the transfer rate is of course b/s (bits per second). One must be aware that the nominal rate denotes the maximum rate that can be achieved in a channel. In a physical channel the information is transferred with the nominal rate only if protocol messages are transferred through the channel one immediately after another, without non-active intervals between them; the nominal rates of virtual channels in higher layers of a protocol stack depend on the properties of the channels and the protocols in lower layers of a protocol stack. Most usually the information is not transferred through a channel with the maximum possible rate, either because sometimes there is nothing to be transferred (e.g. there is no user information) or because the transfer is sometimes not allowed by protocols. The actual *transfer rate* r is therefore lower than the nominal rate, $r < R$. The actual transfer rate can be expressed with the formula

$$r = \lambda \cdot L, \tag{6.1}$$

where λ denotes the rate of transfer of messages (hence the number of packets transferred in a time unit) and L denotes the length of messages. Because the nature of communication traffic is most usually random, the quantities r, λ and L also

---

[1]In this text the rate of information transfer means the number of bits that are transferred in a time unit (bit rate); considering the definition of the amount of information, as understood by the information theory, this is not correct, as the transferred bits usually include redundant bits, too.

Fig. 6.1 Lossless channel

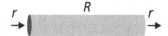

change randomly; in such cases Eq. (6.1) relates the average values of the quantities r, λ and L.

The actual transfer rate normalised with the nominal transfer rate is referred to as *traffic intensity* and denoted as y:

$$y = \frac{r}{R}.$$

(6.2)

Evidently the traffic intensity is unitless,[2] and its maximum possible value is 1. Eq. (6.2) most usually denotes the average value of the traffic intensity. One can clearly see from Eq. (6.2) that the traffic intensity actually means the usage of the nominal transfer rate of a channel. If the expression (6.1) is inserted into Eq. (6.2) one gets the formula

$$y = \frac{\lambda \cdot L}{R}.$$

(6.3)

The ratio

$$T_t = \frac{L}{R}$$

(6.4)

denotes the time that is needed to transmit a packet with the length L. If the channel is observed during the time period T while N packets are transferred through it, then

$$\lambda = \frac{N}{T}$$

(6.5)

and also

$$y = \frac{N \cdot T_t}{T} = \frac{t}{T},$$

(6.6)

hence the traffic intensity is also the ratio of the actually used time t in period T over the total time T and therefore means the usage of time.

Let us now emphasise once more that all the equations we have shown most usually relate the average values of random variables.

Figure 6.1 shows a channel with the nominal transfer rate R; the information enters it and leaves it with equal transfer rates, namely r. Such a channel is called *lossless channel*.

---

[2]In the traffic theory the nonphysical unit erl (Erlang) is often used for the traffic intensity.

Fig. 6.2 Lossy channel

Unfortunately, a portion of transferred information can be lost for various reasons (e.g. a packet may be corrupted to the degree to be unrecognisable, it can be discarded by a network element due to the lack of processing or memory resources, or a receiver recognises it as unusable and discards it). Furthermore, from the users' point of view only information data units (bearing the user information with them) are useful. For these reasons the input transfer rate r_g and the output transfer rate r_s will be distinguished. While the *input transfer rate* comprehends all the protocol messages that are transmitted into the channel, the *output transfer rate* comprises only those protocol messages that are successfully received and their user contents forwarded to a user with an indication primitive. A channel where the output transfer rate is lower than the input transfer rate, $r_s < r_g$, is referred to as *lossy channel*. A lossy channel is shown in Fig. 6.2. As a channel by itself is incapable to generate information, the output rate cannot be higher than the input rate. If a channel is lossless, as described in the previous paragraph, $r_s = r_g$ holds.

The rates r_g and r_s normalised with the nominal rate of a channel yield the *offered load G*,

$$G = \frac{r_g}{R},$$
(6.7)

and the *throughput S*,

$$S = \frac{r_s}{R},$$
(6.8)

respectively. Of course, the offered load and the throughput also cannot be greater than 1; $S \leq G$ is also true, where the equality holds only in the case of a lossless channel. If the quantities r_g and r_s represent the average values of the input and output rate, then the expressions (6.7) and (6.8) also represent the average values of the offered load and the throughput, respectively.

6.2 Efficiency of Protocol

Let us assume we have a channel, as shown in Fig. 6.2; let the information be transferred through this channel according to a specific protocol. As we have already stated, both the offered load and the throughput in the channel are bounded; furthermore, the throughput depends on the offered load. However, the throughput does not depend only on the offered load, but also on the protocol which is used to govern the transport, on the protocol parameters, as well as on the channel properties. Because both traffic intensities (the offered load and the throughput) are bounded, there exists some throughput that is maximal. The maximum throughput through a channel is referred to as *protocol efficiency*:

$$\eta = S_{\text{max}}. \tag{6.9}$$

The protocol efficiency depends not only on the protocol itself and its parameters but also on the properties of a channel. In practice, the throughput is maximal (hence equal to the protocol efficiency) if the protocol entity which transmits user information, hence information protocol data units, has always user information to be transmitted at its disposal, or, in other words, if the input queue between the transmitting user and the transmitting protocol entity is never empty. This means that the throughput is only limited by the properties of the protocol and the channel.

The protocol efficiency hence determines the maximum throughput and consequently also the maximum possible rate of information transfer between protocol entities that is useful for the users. In other words, the protocol efficiency tells which share of the nominal transfer rate of the channel the protocol is capable to use to transfer user information. The protocol efficiency is therefore a performance measure which is interesting for users at the first place, because users want to use as much system resources as possible and achieve as high a useful transfer rate as possible. However, this performance measure is important in cases where all the resources of a channel are available to the user who wants to use the highest possible share of them. The protocol efficiency is therefore the performance measure that assesses the performance of a protocol from the users' viewpoint in cases when multiplexing is not used in the channel.

6.3   Efficiency Experienced by Users and Nominal Transfer Rate of Virtual Channels

We already have stated in Sects. 4.3 and 4.12 that a protocol data unit in general contains the protocol control information, exchanged between protocol entities in order to control and regulate the correctness of a communication process, and the service data unit (user message) which is transferred between users. More information is therefore transferred through the channel interconnecting protocol entities (see Figs. 4.2 and 4.16) than through the virtual channel interconnecting users directly. Though only the transfer of user information is directly interesting for users, the overhead must also be transferred through the channel, thus consuming its resources. Users therefore see a lower usage of the system resources than protocol entities; the ratio of the *efficiency experienced by users* and the efficiency experienced by protocol entities is referred to as *overhead factor k_r* which is given by the expression

$$k_r = \frac{L_{sdu}}{L_{pdu}} = \frac{L_{sdu}}{L_{sdu} + L_{pci}} \; ; \qquad (6.10)$$

in this formula, the quantities L_{pdu}, L_{sdu} and L_{pci} mean the lengths of a protocol data unit, a user message and a protocol control information, respectively. Hence, the efficiency that is experienced by users equals

$$\eta_u = k_r \cdot \eta. \qquad (6.11)$$

Because the efficiency that is seen by users equals the ratio of the maximum possible transfer rate of user information between users through the virtual channel over the nominal transfer rate of the channel interconnecting protocol entities, the maximum transfer rate of user information between users R_u can be written as

$$R_u = \eta_u \cdot R = \eta \cdot k_r \cdot R. \qquad (6.12)$$

R_u in Eq. (6.12) can be called *virtual channel nominal rate*.

Hence the overhead factor has an important impact: if two protocols are available which implement a same service with equal quality and are also equally efficient, but require different amounts of protocol control information, then the protocol that requires more overhead offers its users a lower transfer rate.

6.4 Relative Efficiency of Protocols

In Fig. 6.3, a multiplexed channel with the nominal rate R is shown; n concurrent communication processes are established through it, the i-th process having the input rate r_{gi} and the output rate r_{si}, or the offered load $G_i = r_{gi}/R$ and the throughput $S_i = r_{si}/R$, respectively. The total offered load and throughput of the multiplexed channel are of course $G^M = \sum G_i$ and $S^M = \sum S_i$, respectively. A statistic or packet multiplexing is assumed, where a multiplexer dynamically assigns communication resources to individual channels, with respect to their current needs. The maximum throughput of the i-th subchannel $(S_i)_{max}$ is not important in such a case because a subchannel does not have the total nominal rate R at its disposal but only a part of it; the share it is assigned depends on the traffic flows of all subchannels. The protocol efficiency is therefore not a very appropriate measure to asses a protocol

Fig. 6.3 Multiplexed channel

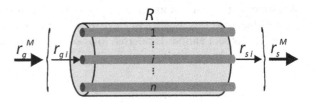

performance in the case of multiplexed channels. From the viewpoint of a system and a multiplexed channel, however, the maximum total throughput through the multiplexed channel is important which will therefore be referred to as the system efficiency,

$$\eta^M = \left(S^M\right)_{max}. \tag{6.13}$$

In the case of a statistic or packet multiplexing, it is important that the burden of a protocol which is used in a subchannel on the multiplexed channel common resources is as low as possible, so as much resources as possible remain to be used by other communication processes running in other subchannels. An important measure to assess the performance of a protocol running in a multiplexed environment is therefore the ratio of the throughput over the offered load. This ratio depends only on the protocol and its parameters, as well as on the properties of the channel, but not on the resources that the protocol has at its disposal. This ratio is hence independent of whether the protocol is used in a multiplexed or a nonmultiplexed environment.

The ratio

$$\sigma = \frac{r_s}{r_g} = \frac{S}{G} \tag{6.14}$$

is called *relative protocol efficiency* and is an appropriate measure to assess the performance of a protocol that is used in a multiplexed environment.

The system efficiency (6.13) of a multiplexed channel is not dependent on the protocol efficiencies of the protocols used in subchannels, but rather on their relative efficiencies (6.14). It can be shown that in the special case where the same protocol is used in all subchannels and all subchannels are loaded with the same traffic intensity ($G_i = G_j$), one can write

$$\eta^M = \left(S^M\right)_{max} = \sigma \cdot \left(G^M\right)_{max}, \tag{6.15}$$

where $(G^M)_{max}$ is the maximum possible offered load that can be achieved in the multiplexed channel. If the number of subchannels is large $(G^M)_{max} = 1$ is usually (but not always) true; in such a case

$$\eta^M = \sigma \tag{6.16}$$

also holds. Hence, in many cases the system efficiency equals the relative efficiency of protocols.

6.5   Delay

The *delay* is also a very important measure for assessing the performance of communication protocols.

The delay is a time difference between two events; it can be defined in different ways, according to the events that are of interest. In this book a delay will be considered as the time difference between the transmission of a message into a channel and the reception of the same message from the same channel; however, even with this restriction in mind, a delay can be defined in different ways, according to where a channel is considered to begin and where it is considered to end.

Delays can be constant during the duration of a communication process, or they may randomly change from message to message. In the latter case, the average value and the variation of a delay are of primary interest, sometimes also the maximum value or some other statistical measure; a delay variation can be given as the standard deviation. In a real channel the delay is constant and foreseeable, while in a virtual channel it is variable and random, hence not foreseeable—its average value can only be assessed based on the statistical analysis of past communication processes with the aid of the traffic theory.

A delay is most easily analysed in the simple case of the message transfer through a real channel. The basic properties of a real channel are as follows. The transmitting entity transmits a message into the channel with the nominal rate of the channel; this means that the *transmit time* T_t needed to transmit a message is proportional to the length L of the message and inversely proportional to the channel nominal rate R, hence $T_t = L/R$, as already told in Eq. (6.4). In case of the message transfer through a real channel, the transmitter and the receiver must be synchronised as the transmit/ receive rate is concerned; hence the transmit time and the receive time of a message must be exactly equal. Furthermore, the bearer of information through a real channel is an electromagnetic wave whose propagation speed v is known, so the *propagation time* T_p of information through a real channel is also known and foreseeable: it is proportional to the physical length of the channel D and inversely proportional to the propagation speed v:

$$T_p = \frac{D}{v}, \qquad\qquad (6.17)$$

where the electromagnetic wave propagation speed depends on the transmission channel properties, its upper bound being the propagation speed of the light in vacuum, $c = 3 \times 10^8$ m/s. In Fig. 6.4 the transfer of a message through a real

Fig. 6.4 Message transfer through real channel

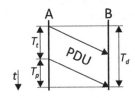

channel is shown as a timing diagram. In this figure the entities A and B are the transmitting and the receiving protocol entities, respectively. The channel that interconnects both entities has a nonzero physical length; the reception of the beginning of a message, the reception of its end, and also the reception of all other parts of it are therefore delayed by the propagation time T_p; the arrows that indicate the transfer of the beginning and the end of the message in the diagram are therefore drawn inclined. The transmit time needed at the transmitting end T_t equals the receive time needed at the receiving end; the arrows indicating the transfer of the beginning and the end of the message are therefore parallel.

If the message delay is defined as the time difference between the moment when the transmitting protocol entity begins transmitting the message, and the moment when the receiving entity entirely receives the message (see Fig. 6.4), then the delay, as can easily be seen from the figure, equals

$$T_d = T_t + T_p = \frac{L}{R} + \frac{D}{v}. \tag{6.18}$$

Let us now consider an interesting case where the delay of messages transferred through a virtual channel between the users UA and UB randomly changes; we will see how one can find the average delay of messages transferred between the users, based on a simple statistical analysis of what is going on in the channel between the protocol entities. Let the protocol entities A and B, interconnected by a real channel, communicate according to a protocol that implements the virtual channel between the users UA and UB. Let us assume that information protocol data units PDU which bear user messages can be lost in the real channel between the protocol entities; the receiving protocol entity B must therefore acknowledge all correctly received messages with special acknowledgments ACK sent to the transmitter. If the transmitting protocol entity A does not receive the acknowledgment, it retransmits the information protocol data unit PDU. (Much more about protocols of this kind will be explained later in this book.) In Fig. 6.5 three scenarios of the transfer of a message PDU are shown: in the scenario (a) the message is received by the receiver B in the first attempt, in the second scenario two attempts are necessary and in the third scenario even three (the messages which are crossed out in Fig. 6.5 have been lost). It must be noted here that the transmitter A retransmits a message only after it senses it has not received an acknowledgment in time, hence when it would receive the acknowledgment were the message not lost (the mechanism that allows the transmitter to recognise the absence of the acknowledgment will also be presented later). As the virtual channel that is of interest interconnects both users the message is considered transmitted when the user UA passes it to the protocol entity A with the request primitive, and it is considered received when the user UB receives it from the protocol entity B with the indication primitive; the message delay is therefore the time difference between the two primitives. A primitive is assumed to be executed in zero time because the service access point is assumed to have zero size; the two primitives are therefore shown as horizontal arrows. It can be clearly seen from Fig. 6.5 that the delay is different in the three cases. Let us assume that, based on the

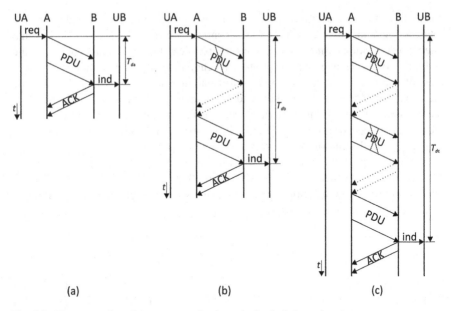

Fig. 6.5 Three scenarios of message transfer through physical channel

statistical analysis of past communication processes, it was stated that the scenario
(a) occurs with the probability P_a, the scenario (b) with the probability P_b and the
scenario (c) with the probability P_c where $P_a + P_b + P_c = 1$ must of course be true
(it is sure one of the three scenarios happens). The average delay T_d can be calculated
as the sum of the delays in the three scenarios, weighted with their respective
probabilities: $T_d = P_a \cdot T_{da} + P_b \cdot T_{db} + P_c \cdot T_{dc}$. Readers may themselves develop
the formula for the average delay if the transmit times of information protocol data
units and acknowledgments, as well as the propagation time in the channel are
known; doing this, they may refer to Fig. 6.4.

The two users UA and UB in Fig. 6.5 are not interconnected directly but
indirectly through the two processors (protocol entities) A and B which process
messages, thus implementing the virtual channel between the two users; this is the
reason why the channel interconnecting the users is virtual. The example we have
just discussed was simplified by neglecting the times of message processing in users
and protocol entities.

In the example shown in Fig. 6.5, we have shown what is happening both in the
channel interconnecting protocol entities and in the virtual channel interconnecting
users which is implemented by the protocol, because we knew the protocol running
between the protocol entities A and B. Often, however, the implementation of a
service is unknown (or intentionally abstracted). In such cases a message transfer is
shown more abstractly, as illustrated in Fig. 6.6a or even in Fig. 6.6b.

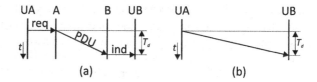

Fig. 6.6 Message transfer through virtual channel

6.6 Performance of Protocol Stacks

In Fig. 5.3 it was shown that the protocols of all protocol stack layers cooperate and are concurrently running in order to transfer application data through a communication system. Each of these protocols impacts the transfer rate experienced by the users of the whole system with its efficiency and overhead. If Eq. (6.12) is used recursively for all the layers the nominal rate which is experienced by the applications, hence the users of the entire communication system, is

$$R_a = \prod (\eta_i \cdot k_{r_i}) \cdot R, \tag{6.19}$$

where R is the nominal transfer rate of the physical channel in the lowest layer, R_a is the nominal rate of the whole system (hence the maximum rate that is available to the applications and the users of the whole communication system), and η_i and k_{r_i} are the protocol efficiency and the overhead factor, respectively, of the protocol running in the i-th layer of the protocol stack. As is usual in mathematical notation, the operator $\prod f_i$ denotes the product of all factors f_i (in our case the index i is running over all layers of the protocol stack).

Both Eq. (6.19) and Fig. 5.3 show that a lot of overhead is being transferred through the physical channel of a communication system. It is true that all this overhead lowers the transmission rate which is available to the applications (all the factors η_i and k_{r_i} are less than 1); one must however be aware that all this overhead is the price which must be payed in order to have an adequate service with an adequate quality of service available to the users of a communication system.

The least possible delay of a message in a communication system is determined by the properties of the real channel in the physical layer through which any message has to be transferred; this minimum delay is determined by Eq. (6.18). The physical path through which a message must be transferred from one terminal through the network to another terminal may of course consist of several physical channels between pairs of network elements; the least possible delay of a message to be transferred from the source terminal to the destination terminal is in this case the sum of the delays in all constituent physical channels.

In any layer of a protocol stack the message delay is increased for various reasons. One of these reasons may be due to the protocol itself which does not allow protocol data units to be transmitted at any time, or even requires multiple transmissions of a message; the transmission and transfer of control protocol data units also consumes some time; messages must also be processed in both terminals as well as in intermediate network elements (which are a part of a virtual channel implementation), which also takes some time; a very important, often even the most important, part of a delay is the time a message spends in waiting queues which are associated with all processing elements in a network. The delay of a message transferred between applications can therefore be much bigger than the minimum delay imposed by the physical layer. The delay added by all the layers of a protocol stack must also be seen as the price that must be payed in order to receive more and better services.

An important task of a designer of a protocol stack and its protocols is to assure as many services as possible with the best possible quality while using the network resources which are available.

Chapter 7
Protocol Analysis, Simulation and Measurement

Abstract The importance of the verification of the logical correctness and performance of a communication protocol is first exposed. Then the three methods that can be used to verify the logical correctness and performance of protocols are described and compared, namely analysis, simulation and measurement. Their usefulness within the frame of a design process and within different phases of a protocol lifecycle is assessed. The complexity of analytic and simulation models and consequently the need for simplified models of real-life protocols are pointed out. The difference between the logical simulation and the performance simulation of a protocol is emphasised. The role of the protocol measurement in the communication system management is described. Protocol analysers are also shortly presented.

When a protocol is designed, the correctness of any design step must be verified. Although the logical correctness of a protocol is most important, care must also be taken of its performance. Both correctness and performance can be verified by means of analysis, simulation or measurement.

During the protocol design phase the analysis and the simulation are most helpful, as the protocol is not yet implemented at this time, hence there is nothing yet to be measured.

After a protocol has already been implemented and let to operate, either for testing purposes or to offer useful services in an operational network, its logical correctness and performance can be verified with measurements. Special measurement systems, referred to as protocol analysers, are used for this purpose.

System management is usually related to system measurement. A system manager inspects the results of system measurements, compares them to the results which are considered the best or normal, and then tunes the system parameters and consequently the system operation with the aim to improve the system performance. A protocol management is very important, as it can prevent some unwanted problems in a network operation, due to which a network could even cease to work. A protocol management is therefore sometimes built into the protocol itself. Such a kind of protocol management is the flow/congestion control which will be discussed in a separate chapter of this book, due to its utmost importance for an effective protocol

D. Hercog, *Communication Protocols*, https://doi.org/10.1007/978-3-030-50405-2_7

performance in various network conditions. In a communication system a separate network management system can also be used where special management elements communicate with managed network elements according to special management protocols (running in the management plane) in order to control and manage them.

7.1 Analysis of Protocols

The protocol *analysis* is always based on a logical or performance model of the communication process; such a model can more or less reflect the actual operation of the protocol. The more closely a model mimics the modelled system (protocol), the better and the more accurate are the results of the analysis; unfortunately, a model that is more accurate is usually also more complex and therefore more difficult to develop and use.

The logical correctness of a protocol can be analysed by systematically considering its operation which in any case is based on the informal or formal specification of the protocol. For a formal analysis or even correctness, proving formal specifications based on the process algebra (see Sect. 1.3.3) is most appropriate.

The analysis of the protocol performance is based on mathematical models (equations) which describe the traffic intensity, efficiency, relative efficiency and delays. An introduction to performance analysis of protocols was already given in Chap. 6; based on the knowledge acquired in that chapter some simple performance models will be developed in later chapters which will allow us to have some better and more precise ideas about the protocol performance.

In general, analytic logic or performance models of communication protocols are the more complex, the more they are accurate, which makes the analysis of protocols complicated and difficult. In practice, simplified models are therefore usually used where those effects which are not essential for the basic protocol operation are deliberately neglected; such simplified models are easier to analyse, but the accuracy of results is limited.

In any case, simplified models are very appropriate to study, explain and understand the basic functionality and performance of protocols. To this end simplified logic and performance models will also be used in this book.

7.2 Simulation of Protocols

The simulation of a protocol is based on simulation models of protocol entities; simulation models are run under the control and coordination of a simulation program. The method used to simulate communication protocols and other discrete systems is the *discrete-event simulation*. Simulation models of protocol entities are procedures called by the simulation program when an input event occurs at an input of a protocol entity or a channel.

Standardised protocols can be simulated using simulators with built-in models of protocol entities. However, if the model of the protocol we want to simulate is not available, a simulation model can be developed either in a general-purpose discrete-event simulation-oriented modelling language or in a general-purpose programming language, such as C or C++. When developing a simulation model, one always follows a protocol entity specification.

The accuracy of simulation models of protocol entities and channels is limited, too; the more accurate is a model, the better and the more accurate are simulation results. However, more accurate simulation models require more engineering effort to be developed; in general, the simulators that yield more accurate results also run more slowly.[1]

The timing accuracy of a simulator which is used for a logical correctness verification must be such that simulation results yield a correct sequence of events in the simulated system, as the logical (in)correctness is determined by the sequence of input and output events of protocol entities and channels. However, if the goal of a simulator is to assess the performance of protocols, the timing accuracy must be such that simulation results yield correct results for the traffic intensity according to Eq. (6.6) and along with this correct results for the efficiency and the relative efficiency; of course, the accuracy of results for delays does also depend on the timing accuracy of simulation models. A simulator of communication protocols performance must also have built-in models of random processes to be able to generate input traffic, errors, losses, etc. As a performance simulator usually yields the average values of results, a simulation must be running so long that a sufficient quantity of results is gathered to be able to calculate the average value with a sufficient accuracy.

In most cases the development of a simulation model of a protocol requires less time and less money than the development and implementation of the protocol itself; in the development phase it is therefore much more wise and economic to verify the correctness of a protocol by means of simulations first, before implementing and testing it.

7.3 Protocol Analysers

A *protocol analyser* is an element of a communication network which receives protocol data units from the network, decodes them and finds out according to which protocol they are composed; it also can verify if these messages are transferred in accordance with the rules of that particular protocol, it even can analyse the performance of the transferred traffic. This kind of protocol analyser operation is

---

[1]The speed of a simulation program does of course not depend only on the accuracy of simulation models, but also on the system that is simulated, the quality of software that implements the simulator, and, very importantly, on the speed of the computer on which a simulation runs.

referred to as *passive mode* of operation; a protocol analyser also can work in the *active mode* if it itself generates communication traffic, transmits it into the network and finds out the network response to this traffic.

Protocol analysers are intended to test the correctness, efficiency and safety of the operation of a network that has just been installed or to supervise all of these properties during the normal operation of a network, as well as to search for errors in case of an erroneous or temporarily erroneous operation of a network and its components.

A protocol analyser can be implemented as a hardware or as a software, or as a combination of both. However, as it must have a physical contact with a network, at least the physical layer must be implemented as hardware, which is often also true for the data-link layer. The physical and possibly also the data-link layer are implemented as a *network interface card.* This card must have a very special property, different from the usual network interface cards in computers. If multiplexing is used in data-link layer, addressing must also be used there; if the data-link layer entity is running in the normal mode of operation, it only processes protocol data units which are addressed to it and discards all the others, just like any other network interface card. A protocol analyser must however be able to recognise and process all protocol data units, regardless of their destination addresses; the network interface card of a protocol analyser must therefore be able to run in the *promiscuous mode* in which it receives and processes all protocol messages. Naturally, a protocol analyser can analyse only those protocols it knows (the knowledge about which is built into the analyser).

In a communication network huge quantities of messages can be transferred, while only some of them are of interest to the user of a protocol analyser; a protocol analyser therefore contains filters that determine which protocol data units are to be processed and which of them are to be discarded (with regard to protocols, protocol stacks, source and destination addresses, etc.); these filters can be configured by a user of the protocol analyser. Furthermore, an analyser may contain various triggers and timers which determine the conditions under which the analyser starts or stops collecting the results of protocol analysis.

Users of a protocol analyser must be acquainted with the protocols they want to analyse with the analyser, in order to be able to interpret the results of a protocol analysis and decide for further actions if necessary. However, some of this knowledge can be built into a protocol analyser itself which allows it to warn users about the anomalies in the network operation; this kind of protocol analysis is referred to as *expert analysis.*

Part III
Methods of Message Transfer

The chapters that follow constitute the longest and the most important part of this book. Some of the most important communication methods on which the operation of many specific protocols is based will be explained here. The lifetime of a general communication method is longer than the lifetime of a specific communication protocol; furthermore, many specific protocols may be based on a single communication method; last but not least, an engineer who understands a communication method will have no problems to understand the specific protocols that are based on that method. All of these are the reasons why this book focuses on communication methods rather than on specific protocols.

A communication method determines the procedure (algorithm) of solving a communication problem. When specifying a method, not all the details of the procedure need be specified; a single method can therefore be implemented in different ways with different specific protocols which are not necessarily interoperable. In the discussions of communication methods that follow, only abstract syntaxes of protocol data units will be explained; even these will be shown simplified in many cases (e.g. the set of all nonnegative integers will be used to count protocol messages, which is however not practically possible); the message parameters that are not essential for the explanation of a method will even not be mentioned. The concrete syntax will mostly not be discussed.

In each of the chapters that follow in this text, several methods to solve a specific communication problem will be explained. Thus, we will be able to focus on a specific problem in that chapter; this way of explanations should be easier for a reader to grasp.

When discussing a method, the services it can provide will be explained as well as the environment in which it can be used; often, a method will also be formally specified in the SDL language. Some of the methods will be additionally illustrated with one or more characteristic communication scenarios; time-sequence diagrams, timing diagrams and MSC diagrams will be used to this end.

In Chap. 8 the formatting of protocol data units in accordance with their transfer syntaxes will be explained. In Chap. 9 it will be shown what a protocol can do if some user messages are too long to be transferred through the channel. The addressing of protocol entities and transformations between different formats of addresses indicating a same entity will be discussed in Chap. 10. In Chap. 11 the management of states in which a communication process can be will be described; in that context, the connection management will especially be considered, a connection being the most important kind of a communication system state. Chapter 12 is the longest in this book, also having more subchapters than the others; it might also be the most difficult to be understood by a reader. In this chapter we will explain how protocols solve problems that are due to nonideal properties of communication channels. Along with providing a reliable information transfer through a communication network, many protocols also try to avoid congestions in the network and its terminals or to recover from them should have they already occurred; Chap. 13 will explain how this can be done. Chapter 14 is the last one of this part of the book; in this chapter it will be shown how terminals of a local area network can communicate although all of them use the same communication medium for information transfer.

Chapter 8
Coding and Decoding of Protocol Data Units

Abstract In this chapter the general principles of coding/decoding protocol messages and the different styles of transfer syntaxes of protocol data units are explained. At first, the need for the message synchronisation is discussed, and also how this can be done and where it must be done. Then the notion of the transparent data transfer and the methods for its provision are presented; the difference between this provision in the respective cases of synchronous or asynchronous data transfer is pointed out. The purposes and functionalities of various fields, which are necessary to assure a reliable message synchronisation and decoding, even if the protocol is developed and implemented in several versions, are explained. The importance of the flexibility and simplicity of the transfer syntax of a protocol is also pointed out. Then the need for the transfer of values, rather than just bit sequences, between communication endpoints is emphasised. Various styles of building protocol messages are then described more in detail, namely character, binary, time-length-value, matched tag, MIME and ASN.1 coding techniques; their advantages and disadvantages are listed.

In Sect. 2.2 we explained what the abstract syntax is and what is the transfer syntax; in Sect. 2.4.2 we told also that, as far as the transfer syntax is concerned, a protocol data unit consists of constituent parts, referred to as fields, which can even be hierarchically organised (there can be fields, subfields...). Hence a protocol data unit has a structure (fields, subfields...) that is determined by the transfer syntax for this PDU type; furthermore, an interpretation is always associated with any field that defines the meaning of that field in the sense of a value specified as a message parameter in the abstract syntax of the message type. Both the structure of a message and the interpretation of its fields must of course be standardised. A transmitting protocol entity must compose a message so that a receiving protocol entity can decompose the received message into fields and correctly interpret the values contained in them in accordance with the values specified in the abstract syntax. Composing and formatting messages at the transmitting side is called *message coding* or *protocol coding*, while recognising and analysing messages at the receiving side is referred to as *message decoding* or *protocol decoding*.

D. Hercog, *Communication Protocols*, https://doi.org/10.1007/978-3-030-50405-2_8

The service implemented by the physical layer is the transfer of a sequence of bits. Protocol data units of the data-link layer (frames) must therefore be formatted so that a data-link receiving protocol entity can easily find out where in a received bit stream a frame begins and where it ends; the formation of protocol data units at the transmitting side is called *framing*, while finding the beginning and the end of a message at the receiving side is referred to as *message synchronisation*. The framing and synchronisation may be the task of a layer higher than the data-link layer as well. As the examples of such procedures in higher layers, the application layer protocols of the TCP/IP protocol stack which use the services of the TCP protocol in the transport layer can be mentioned; the service provided by the TCP protocol being the reliable transport of a sequence of octets, an application layer message may not be transferred through the network as one packet, but can be transferred as parts of more TCP protocol data units, or a single TCP protocol data unit can contain several application layer messages. In many cases, however, the protocol entities in higher layers (above the data-link layer) receive from their lower layer entities protocol data units in their integrity, so the synchronisation in such layers is not needed. Sometimes it may also happen that a transmitting protocol entity must add to a protocol data unit some additional bits, referred to as *padding*, in order to fulfil some necessary condition, e.g. a minimum packet length or a packet length that is a multiple of some basic length, such as an octet; in such a case the transmitting protocol entity must somehow indicate the length of the »true« message (without the padding).

The beginning of a frame is most usually indicated with a special condition, often with a *synchronisation sequence* of bits or octets which must be found and recognised by the receiving entity, so it can know where in the incoming bit or octet sequence a message begins. The end of a message can be separately marked as well, or the length of the message can be written somewhere within the message; in this latter case the end of a message need not be separately indicated. There are also protocols where all the protocol data units have a standardised length; in such cases the end of a message also does not need to be marked.

Protocol data units can contain the user information; however, a protocol may not prescribe which contents a user message may contain and which it may not. It may therefore well happen that a user message contains a sequence which could be understood as the synchronisation sequence by a receiving protocol entity. It is the duty of a protocol that such a sequence in a user message cannot disturb the operation of the protocol, hence that such a sequence would not be understood as the end of message indication by the receiver; a protocol that provides this is said to provide a *transparent transfer* of user information (which means that the protocol is capable to transfer any user information without restrictions). When messages are transferred asynchronously, additional bits or characters are inserted into a bit or character sequence when needed to prevent the misinterpretation of a synchronisation sequence in a user message; such inserted bits or characters are removed by the receiver before passing the user message to a user. When messages are transferred synchronously, a receiver must, after it has detected a supposed beginning of a message, continue to check the start of the message condition also in some following messages before concluding that it really has found the message delimiter and achieved the synchronisation.

After a protocol has been developed and standardised, it begins to be used; only then some of its deficiencies can be seen, especially because of new information transfer techniques which are emerging all the time. A communication protocol is therefore being further developed even after it has been released to be used. After some considerable modifications have been made, a new *protocol version* is issued. Transfer syntaxes of different versions of a protocol may be slightly different; the protocol version must therefore be indicated at a standard location within a protocol data unit in order to allow a receiver to correctly decode the message. Often transfer syntaxes of new protocols also preview one or more unused fields within messages to allow later additions of new functionalities to the protocol.

In Sect. 2.2 where the abstract syntax of a protocol was explained, we told already that protocol data units of various types can be used and most often they actually are used. Usually, different transfer syntaxes are associated with different message types. A receiving protocol entity must therefore know the type of the message it has received in order to be able to decode it. Hence the message type must be clearly indicated within a message in a way and at a place that are independent of the message type. After a receiver has received a message and synchronised with it (if the synchronisation is necessary), it must therefore find out the type of the message first and only then decode the rest of it.

Often the abstract as well as the transfer syntax define a number of values that can be brought by a protocol data unit; however, not all of these values are always necessary to correctly control a communication process. Some of the parameters of a message (and the transfer syntax fields associated with them) are therefore mandatory while the others are optional; optional fields may be present in a message only when they are needed. A field of a message may have always the same length or the field length may change from message to message. Furthermore, the order in which fields are placed within a message may be always the same or it may change from one message to another. The structure of a protocol data unit can therefore be fixed or variable. Often, however, a fixed and a variable structure are combined in a single message type.

The flexibility of the syntax of protocol messages is therefore one of the important properties of a protocol on which the quality and the usability of a protocol depend; a protocol that is flexible may simply, easily and effectively be used in different communication network environments. A structure that is simple is easier to be understood by humans and easier to be processed (encoded and decoded) by machines, i.e. protocol entities, and therefore requires less processing power. The transfer of protocol data units through a communication system consumes system resources, so a protocol that uses shorter protocol messages uses less resources and is therefore more efficient.

A value can be coded differently in different types of computers; when a value is transferred between applications, however, the transferred value (the transferred sequence of bits) must be correctly and equally interpreted by both applications, regardless of its encoding in terminal computers. Hence, a value must be transferred between applications, rather than a sequence of bits. The task of the protocols in the presentation layer of the OSI reference model (see Sect. 5.4.1) is therefore to transform the local value encoding of the source computer into the transfer syntax encoding (this is the reason why it is so called!) at the transmitting side, and to transform the transfer syntax of a message to the format used locally by the computer at the receiving side.

In what the transfer syntax of protocol data units is concerned, there are bit-oriented protocols and character-oriented protocols.

The rules for encoding protocol data units can be specific for any individual protocol, or they determine how messages can be formatted in a systematic way. The systematic methods to be explained in this chapter are the TLV method and the matched tag coding method.

8.1   Character Coding

Character-oriented protocols employ the *character encoding* of protocol data units. This means that all fields of a message can be interpreted as sequences of characters, coded in accordance with some character coding rule. The *ASCII* (*American Standard Code for Information Interchange*) coding rule was most in favour for many years; this rule allows for coding of uppercase and lowercase letters, numbers, punctuation marks and control characters. Basic ASCII coding table, also referred to as US-ASCII, is a 7-bit code and allows only the coding of the letters of the English alphabet, so many extensions of the basic coding table are also used to allow coding of other alphabets. In recent years the more universal code *UTF-8* (*Unicode Transformation Format—8-bit*) has become more popular; this code includes the ASCII code as its subset.

Character-oriented protocols allow for the transparent transfer of user information using the procedure called *character stuffing*; when needed, a transmitting protocol entity inserts a special character into the character stream in order to prevent the receiver to misinterpret the synchronisation sequence within the user information as an end-of-message indication, while the receiver recognises the true nature of the added character and removes it.

The character coding is quite wasteful; to encode a field in the ASCII format, too many bits are needed that must of course be transferred through a channel. Let us assume that one octet (eight bits) is needed to encode a character and that a field in a PDU is encoded as a sequence of n characters; hence $n \cdot 8$ bits are needed to code such a field, which allows one to encode 35^n combinations of character sequences (assuming that only 25 letters and 10 numerals are used). On the other hand, if the binary coding is used, $2^{n \cdot 8}$ combinations can be encoded with $n \cdot 8$ bits. If $n = 1$, the binary coding makes 7 times more combinations possible as the character coding; if $n = 2$, this ratio is 53, and if $n = 3$, this ratio is already 391... However, the character coding is simple and easily readable by a human.

Decades ago the character coding was much in favour, due to its simplicity and also because the efficiency did not yet seem to be so important. Special control characters were used as message and field delimiters, such as SYN (synchronous idle), SOH (start of header), STX (start of text) and ETX (end of text). In order to assure the transparent transfer of user information, the control character DLE (data link escape) was inserted before a sequence equal to the synchronisation character within a user message. Later, this mode of coding was mostly abandoned in favour of the more efficient binary encoding. However, still nowadays many character-oriented protocols are used in the application layer of the TCP/IP protocol stack (e.g. FTP, SMTP, POP3, HTTP, SIP...); these protocols employ the matched tag coding method. In the lower layers of protocol stacks, the binary coding prevails.

8.2   Binary Coding

Bit-oriented protocols employ *binary encoding*. This means that the values of all fields are coded as binary sequences which in general cannot be interpreted as character sequences. This also means that the lengths of message fields are not necessarily multiples of a character length.

As was already told, the binary coding is more efficient than the character coding, as binary coded messages are in general shorter than character coded messages. The binary coding is therefore much more used nowadays, except for numerous application layer protocols which use the character coding due to its simplicity, better readability and better understandability.

In bit-oriented protocols the transparent transfer of user information is implemented with the procedure called *bit stuffing*, where a transmitting protocol entity inserts additional bits into user information before a bit sequence that might be understood by a receiver as the synchronisation sequence; consequently, a transmitter transmits the synchronisation sequence only where it is really needed to make the synchronisation possible; a receiver detects the inserted bits and removes them.

The protocol HDLC (see Chapter 15) and most of its descendants use a so-called *flag* as the synchronisation sequence; this is the binary sequence 01111110. Any HDLC frame begins and ends with a flag. In order to implement a transparent transfer of user information, a transmitting protocol entity inserts a zero after any sequence of five consecutive ones (of course, except in a flag itself); a sequence 01111110 can therefore be transmitted only as a real flag. A receiving protocol entity, after it has removed flags from a received message, removes the final zero of a received sequence 1111110.

8.3   TLV Coding

The *TLV coding* (*type-length-value coding*) is the most flexible method of coding protocol data units; it is, however, not always the most efficient one, as the message lengths are mostly longer as with the coding methods that are not flexible. In case of the TLV coding a message field consists of three subfields: the type subfield indicates the type of the field, related to the parameter type in the abstract syntax; the length subfield determines the length of the third subfield in bits or octets; in the third subfield a value that relates to a parameter in the abstract syntax is written. The length of the third subfield can be variable, while the first two subfields must either have a fixed length or be coded so that a receiver can recognise their lengths when decoding them.

If the TLV coding is used, the number and the enlistment of values contained in a message need not be fixed; optional fields can easily be included. The order in which fields are written is also not fixed. According to the types of fields, a receiver can easily recognise which fields are actually present and in which order.

Decoding of protocol data units that are TLV encoded is simple and fast; a receiver must decode only those fields that are actually present and are of interest. After it has decoded a field, it can easily calculate, using the length subfield, where the next field begins. If some field doesn't interest it, it can easily skip it. Such a technique has already been known and used for a long time in data structure design and is referred to as *linked list* technique.

The TLV coding is better used in the binary coding mode.

The TLV coding technique is most usually used to encode optional fields of protocol data units.

8.4   Matched Tag Coding

If the *matched tag coding* technique is used, the beginning and the end of each field of a message are marked with special *tags*.

The matched tag coding is used by many character-oriented protocols in the application layer of the TCP/IP protocol stack (such as SMTP, POP3, HTTP or SIP). In SMTP (and many other) protocol the beginning of a field is marked with a *keyword* which indicates the type of a field while the end of a field is marked with the end-of-line indication—the sequence of ASCII characters '*CR*' (*carriage return*) and '*LF*' (*line feed*). The header of a message and the user message are delimited with an empty line (hence with a sequence 'CRLFCRLF'). A user information is terminated with a single period in a line, hence with a sequence 'CRLF.CRLF'; if such a sequence is contained in a user message, the transmitting protocol entity adds an additional period to the existing period, while the receiving protocol entity removes a period from the received sequence 'CRLF..' and thus implements the transparent transfer of user information.

The matched tag coding method is used also in several *markup languages* which are used to describe various documents. Two well-known and intensively used markup languages are *HTML* (*Hypertext markup language*) and *XML* (*Extensible markup language*).

8.5 MIME Coding

It was already told that numerous application layer protocols of the TCP/IP protocol stack employ the character coding. At first these protocols (such as the protocol SMTP, used to transfer electronic mail) were used only to transfer text-based information. However, in last decades, the need to transfer also nontextual information, such as images, audio, video, data and texts encoded with non-ASCII codes, has also been growing more and more. A special method of coding was therefore developed and standardised that allows for the transfer of nontextual information with character-oriented protocols. At the beginning, this kind of coding was intended to transfer nontextual information with electronic mail, so it was named *MIME encoding* (*Multipurpose internet mail extensions*). Nowadays, MIME encoding is heavily used also in other applications, such as to transfer different kinds of information in the World Wide Web with the HTTP protocol.

A MIME-encoded message may consist of several parts, each of them being additionally coded with a method that is suitable to transfer that kind of information with a character-oriented protocol. (The word »additionally« in the previous phrase means that a message which has previously already been coded in some way is additionally MIME-encoded to allow it to be transferred with a character-oriented protocol.)

A MIME message consists of a *header* and a *body*. A body may either contain only user information or consist of several parts, each of them again containing a header and a body. Hence, the structure of a MIME message can be hierarchical.

A MIME header contains the information about the type of information in the body (Content-Type), e.g. text/plain for a plain ASCII text, text/enriched for a formatted text, image/jpeg for an image encoded in jpeg format, audio/basic for an audio encoded according to the ISDN standard, application/octet-stream for an arbitrary sequence of octets, application/postscript for a postscript-encoded text, or multipart/mixed for a document, itself consisting of several parts containing different kinds of information. A header may also contain the information on the method of additional encoding (Content-Transfer-Encoding) and some other data.

We have already told that binary contents or a non-ASCII text can be additionally coded so that it can easily be transferred with a protocol that is normally used to transfer ASCII texts. The method that is usually used for such additional encoding is *base64 encoding*. A binary sequence is partitioned into subsequences, each of them 6 bits long, which are further encoded with the aid of the so-called base64 table into ASCII characters from a 64-character subset of the standard ASCII table (containing uppercase and lowercase letters, as well as the characters '+' and '/').

8.6   ASN.1 Coding

In Sect. 2.5.1 the language *ASN.1 (Abstract Syntax Notation One)* that is used to describe the abstract syntaxes of protocol data units was shortly presented. A standardised transfer syntax is associated to the messages whose abstract syntax is defined in ASN.1. This transfer syntax is based on the TLV coding.

8.7   Examples of Protocol Data Unit Transfer Syntaxes

General principles of coding and decoding protocol data units were explained in this chapter. How these principles are used in practice will be shown in the last part of the book where some specific protocols will be presented, including the transfer syntaxes of some of them.

Chapter 9
Segmentation and Reassembly of Protocol Data Units

Abstract Most communication networks determine an upper bound on the length of protocol messages that are transferred through them; the segmentation and reassembly of messages is therefore the task of many protocols. The need for this to be done without the users even being aware of it is emphasised. The general principles of segmentation and reassembly procedures are first explained. Then the methods for segmentation and reassembly of protocol data units are described with more details for both connection-oriented and connectionless transfer of messages. The special problem of segmentation and reassembly in heterogeneous connectionless networks is also discussed, where several strategies are possible for where the messages should be segmented and reassembled. The chapter has four figures, one of them displaying the principle of segmentation and reassembly, and the others showing three different scenarios of segmentation and reassembly in a heterogeneous network.

For various reasons numerous protocols restrict the length of protocol data units. Hence, it may well happen that a protocol entity receives a user message from its user that is too long to be transferred in a single protocol message. In such a case the user message must be segmented (fragmented) into more shorter segments; to each of these segments is added its own protocol control information and is transferred through the channel as a separate protocol data unit. This procedure is called *segmentation* or *fragmentation* of messages. The segmentation must be transparent for users of the protocol and for the higher layer in the protocol stack; hence the receiving user must receive exactly the same message as was sent by the transmitting user; both users may even not be aware that segmentation has been done in the lower layer. The receiving protocol entity must therefore reassemble all the segments into the original message (if this can for some reason not be done the whole message must be discarded and is therefore lost); this procedure is referred to as *reassembly* of messages. Hence, segmentation and reassembly are carried out by a layer in a protocol stack without the knowledge of the higher layer. The procedures of segmentation and reassembly of a message are illustrated in Fig. 9.1.

D. Hercog, *Communication Protocols*, https://doi.org/10.1007/978-3-030-50405-2_9

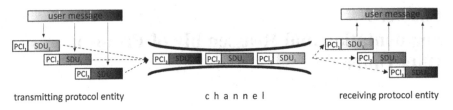

transmitting protocol entity   c h a n n e l   receiving protocol entity

Fig. 9.1 Segmentation and reassembly of messages

All protocols do not support the message segmentation and reassembly. If a protocol entity which is incapable of segmentation receives a message that is too long, such a message must be discarded.

From what has already been told as well as from Fig. 9.1 it is clear that the message segmentation increases the amount of overhead which must be added to user information and transferred through a network, as protocol control information is added to each segment separately, rather than to the message as a whole. Hence, segmentation decreases the efficiency of user information transfer. Segmentation and reassembly is therefore used only when it is really needed; furthermore, a message should be segmented into the longest possible segments that can be transferred through a network, in order to minimise the number of segments and the overhead increase.

A reverse procedure may also be executed by some protocols. At the transmitting side several short user messages may be merged and transferred within a single protocol data unit; thus the overhead can be decreased because a single protocol control information is added to the whole protocol data unit, rather than to each original user message. This increases the efficiency of information transfer. At the receiving side the merged messages must of course be separated into the original user messages.

The segmentation and reassembly procedure that can be used in connection-oriented protocols which assure a reliable information transfer is different from the procedure that must be used in connectionless protocols which are incapable to provide a reliable information transfer.

9.1   Segmentation and Reassembly—Connection-Oriented Protocols Case

The segmentation and reassembly procedure is most simple to be carried out in case a connection-oriented protocol is used which assures a reliable transfer of user information. As will be seen in Chapter 12, a reliable transfer means that no message is lost, no message is duplicated, and the order in which messages are transmitted is preserved at the receiving side, too. A transmitting protocol entity partitions a user message into so many segments as is needed. Usually all of the segments (except for the last one) have the maximum permitted length. To each segment is added the protocol control information in which it is clearly indicated whether it is the last segment of the original user message or not; if all segments (except for the last one) have the same length, the one with a shorter length is known to be the last one and therefore does not need to be separately marked as the last one (if the length of the original user message is divisible by the maximum segment length an additional segment with zero length can be added). Protocol data units containing segments are then transmitted one at a time into the channel. The receiving protocol entity concatenates the received segments, until it receives the last one and adds it to the message which is being reassembled; only then the complete reassembled user message can be passed to the user with an appropriate primitive. If the segmentation is not needed, there is of course a single segment which is also the last one.

9.2 Segmentation and Reassembly—Connectionless Protocols Case

Connectionless protocols do not assure a reliable transfer of user information; hence some messages may be lost, they may also be reordered (possibly they are received in an order that is different from the order in which they were transmitted). In spite of this the receiving protocol entity must correctly reassemble all the segments into the original user message if possible; if this is not possible (e.g. because one or more segments were lost), all received segments are discarded and the user message is lost. The following procedure is necessary. The transmitting protocol entity partitions a user message into so many segments as necessary. To each segment is added the protocol control information in which its place in the original user message is indicated (to assure a correct reassembly even if segments were reordered in the channel); the last segment is also marked, so the receiver can know when it has received all segments (the message partitioning and marking of the last segment is done the same way as described in the previous section). The protocol control information of all the segments of a same original message must also contain a unique message identifier (so the receiver can also distinguish between segments of different original messages, as these could also be mixed up). All of this allows the receiving protocol entity to correctly assemble segments of each original message separately, regardless of the order (correct or incorrect) in which they were received. After it has successfully reassembled a user message it passes it to the user with an appropriate primitive; if, however, one or more segments are lost, the original message cannot be reassembled and all of its segments are discarded. The receiver can wait for the segments of an original message for a limited time, measuring this time with a timer; if the timer expires before the message is successfully reassembled, this means that the message cannot be reassembled and the received segments are discarded in order to release the memory occupied by them.

9.3   Segmentation and Reassembly—Connectionless Transfer through Heterogeneous Networks

In heterogeneous networks (such as the Internet) a message can be transferred through different kinds of constituent subnetworks on its way from the source to the destination, where constituent subnetworks may impose different restrictions on the protocol data unit lengths. The question arises about the necessary segment lengths and where the segments should be reassembled! If a connection-oriented communication is used, the answer is simple: when the connection is set up, the shortest maximum permitted length on the whole communication path through the network can be found out and that segment length is used for the segmentation in the source network element; further segmentation is then not needed in interme- diate network elements and the segments are reassembled in the destination network element. This is possible because in case of a connection-oriented communication all the messages belonging to a connection follow the same transfer path through the network. However, in case of a connectionless communication, the path to be followed by messages is not known in advance, because different messages may well follow different paths; therefore a message may need to be segmented again in one or more intermediate gateways between different subnetworks. In such cases the question about where and how to reassemble segments is very relevant.

The simplest possibility is to partition a message in the source network element into segments so short that they will need to be resegmented in none of the constituent subnetworks, regardless of the path they will follow through the network. With this strategy, the segments may be too short in most situations, which will have a harmful impact on the transfer efficiency.

The other possibility is that the source element which knows only the require- ments of the subnetwork to which it is directly connected segments a message according to the requirements of that subnetwork. In such a case it may well happen that message segments must be further segmented in one or more gateways on the message path through the network. This method, too, has several variants, according to where and how segments are reassembled into the original message.

The simplest possibility is that segments can only be further segmented in gateways between subnetworks, according to the requirements of individual sub- networks, while they can be reassembled into the original message only in the destination network element. Such a case is shown in Fig. 9.2; in this figure a user message 1080 bits long is transferred through three subnetworks with their maximum protocol data unit lengths 512, 256 and 1024 bits, respectively, while the length of protocol control information is 40 bits in all of them. In this figure the lengths of user information parts are indicated in all user messages (in the original user message and service data units of all segments), while the protocol control information is filled with grey colour.

If segments are reassembled only in the destination network node, a great number of quite short segments may be transferred through the network, along with a large amount of overhead; this can be quite inefficient, especially in subnetworks that allow the transfer of longer messages (e.g. in the network 3 in Fig. 9.2). Better

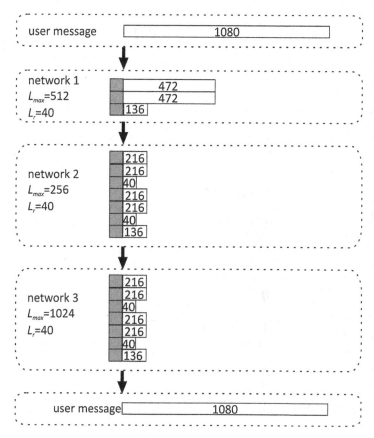

Fig. 9.2 Message reassembly at receiving end

results can be achieved with the next strategy where segments can be partially reassembled in gateways between subnetworks. In a gateway from a subnetwork which permits shorter messages to a subnetwork that allows longer messages so many segments of the original message, transferred through the previous subnetwork, as can fit into a protocol data unit of the next subnetwork, are reassembled. Such a case is shown in Fig. 9.3 where longer messages can be transferred through the network 3 which results in a better efficiency. However, one must be aware that more processing power is consumed in the gateway for this purpose.

If one looks attentively at Fig. 9.3, they may see that the whole original message could be transferred through the subnetwork 2 in only five segments instead of in seven segments. Segments with maximum lengths can be transferred through all subnetworks if any gateway reassembles the complete original message and then segments it according to the requirements of the next subnetwork. This strategy can be seen in Fig. 9.4. Unfortunately, this strategy requires not only more processing power in intermediate gateways but also more memory for the buffers in which messages are reassembled.

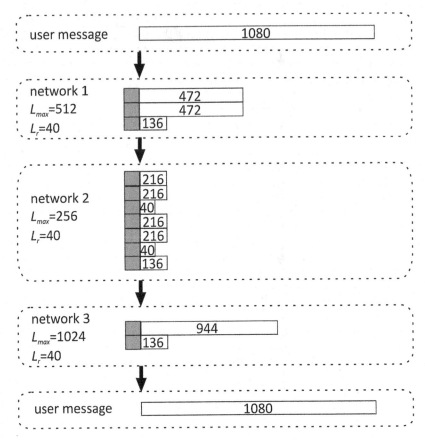

Fig. 9.3 Partial message reassembly in gateways between networks

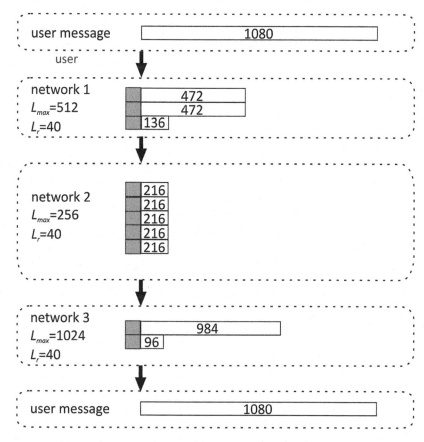

Fig. 9.4 Complete message reassembly in gateways between networks

Chapter 10
Addressing

Abstract The notions of addresses and names of network elements, and also that of the address space, are introduced. Different kinds of addresses are discussed. The scope of addresses is also explained. The addressing of applications using a protocol stack for communication is also described. From the fact that a single network element can have several addresses and/or names at the same time stems the need for the translation between different addresses of the same network element. Special protocols are used for this task, which, however, differ in local area and wide area networks. One protocol for the address translation in local area networks and two protocols for the address translation in wide area networks are described and formally specified in the SDL language. There are 13 figures in this chapter, ten of them specifying the address translation protocols in SDL, and three of them showing the examples of address translation procedures in the form of MSC diagrams.

In a communication network there are many network devices exchanging information. Each of them is connected to the network via one or more interfaces; furthermore, there can be several processes running in each of the network devices that transmit and receive information. It is of course very important that any message, sent by a source device/process, reaches the proper destination device/process to which it was sent. Any network device, any interface and also any process must therefore be uniquely recognisable within the network; hence it must possess a unique *address*. The term *addressing* is used to indicate a set of procedures concerning formatting addresses, the assignment of addresses to communicating entities, transporting them as important parameters of protocol data units, and also their use in routing information through a network. The addressing system is determined by a protocol which uses and transports these addresses. The concept of addressing is therefore closely related to communication protocols; the concept of addressing is also very closely related to the procedures of *routing* (finding communication paths through networks). In this book the addressing will be treated from the point of view of protocols, routing being more the matter of communication network theory.

D. Hercog, *Communication Protocols*, https://doi.org/10.1007/978-3-030-50405-2_10

In Sect. 4.9 we already told that a service access point can be seen by a service provider as the address of the user using this service access point.

10.1 Addresses and Names

As any value that can be stored in a computer memory or transferred as a parameter of a protocol message, an address is written and stored as a binary sequence as well. The set of all possible addresses an addressing system allows to be used is referred to as *address space*, and the number of all possible addresses is called *address space size*.

Addresses are not processed and memorised only by computers but must often also be remembered by humans. As binary sequences are difficult to be memorised by humans, addresses are often written and interpreted also as sequences of characters. Addresses which can be interpreted as character sequences are referred to as *names*.

An address may have a structure; hence it can consist of several components which are recognisable within the address structure and have some meaning associated with them. The structure of addresses is often hierarchical; a hierarchical structure is related to the hierarchical structure of the communication network itself, thus allowing better recognisability of network elements within the network structure and easier routing.

Sometimes more than just one addressing system is used in a communication system at the same time; a single communication entity can therefore have more than just one address or name at the same time. Furthermore, a communication entity may also have an alternative name, sometimes referred to as *alias*; an alias may tell more about the entity to human users than an official address or name. Whenever a communication entity has more than just one address or name or alias, a special mechanism for the translation between different entity denotations is needed.

An address in most cases designates a single well-defined communication entity; such an address is called a *unicast address*. A *multicast address* denotes a well-defined group of communication entities, while a *broadcast address* designates all communication entities in a network or a subnetwork. In recent years so-called *anycast addresses* have more and more come into use which denote any one entity from a well-defined group of communication entities.

Any address or name has a *scope*; the scope of an address is a network or a part of it where the address is known and uniquely recognisable. An address must be unique within its scope which means that an address must indicate a single communication entity within the scope (of course, if this is a unicast address). The addresses that are known only within a part of a network (e.g. within a subnetwork) are *local addresses*, while the addresses which are known within the whole network are *global addresses*. Assigning addresses to individual communicating entities is a special problem which is especially tough in large networks with large numbers of network elements, as the uniqueness of all addresses must be assured. The address assignment can be manual or automatic; a manual assignment is carried out by network managers, while an automatic assignment is done by a special server. A hierarchical addressing system makes the assignment problem easier as the address assignment at one level of hierarchy can be done independently of the assignment at other levels; furthermore, addresses can be assigned in one subnetwork independently of the assignment in other subnetworks at the same level.

A communication network has a physical topology which tells how communication elements are physically interconnected in the network. Accordingly, a physical address of a communication element tells where in a network the element is physically located. This information is very important, without it routing would not be possible. However, a network may have a logical topology as well which can differ from the physical topology; a logical topology describes the organisation of a network from the viewpoint of managers and users, rather than its physical topology. A network element can therefore have both a physical address that defines its physical location in the network and a logical address which defines its placement in the organisational topology. Quite often logical addresses are names. The routing is always based on the physical topology, so logical addresses of network elements must be translated into their physical address equivalents before they are used in a routing process.

The address of a communication entity can be permanent or temporary. Temporary addresses are used in two cases. In the first case a communication element is assigned a temporary address if it communicates only periodically. Whenever it wants to communicate it is assigned a temporary address by the network; when it terminates the communication and does not need an address any more, the network releases the address so it can assign it to another communication entity. In this way the network can save addresses for those who really need them, if there is a lack of addresses. Temporary addresses are used also in mobile networks. A communication element that is mobile must have a permanent address which is used by other elements to reach it; the permanent address does not depend on the element's current location. A mobile entity also has a temporary address which depends on its current location. Information is routed towards a mobile element via a router which is indicated by the permanent address of the element and the router that is related to its temporary location and address. Although other users know only the permanent address of the element to be contacted, the network must know also its temporary address which is currently assigned to it. In this case, too, address translators are needed which translate permanent addresses into temporary addresses.

10.2 Addressing in a Protocol Stack

In Sect. 5.1 it was explained what is a protocol stack and also how protocol entities are running in a layer of a protocol stack, communicating according to the protocol of this layer. A set of protocol entities of a layer can be seen as a virtual network operating in this layer. If the multiplexing is used in this layer, the protocol of this layer must define and use the addressing, too; without the addressing protocol entities could not be distinguished between them and consequently multiplexing would not be possible.

From what has already been told it is clear that the addressing can be concurrently used in several layers of a protocol stack. An application that uses a protocol stack for its operation is directly or indirectly supported by all the layers of the protocol stack; the application is therefore uniquely determined within the network with the addresses of the protocol entities in all those layers of the protocol stack that use the addressing. In other words, this application is identified with the sequence of the addresses of the protocol entities in all layers that use multiplexing. Such a sequence of addresses is referred to as *address concatenation*.

We told in Section 10.1 that several addressing systems can concurrently be used in a communication network. Hence any protocol entity in a layer can have several addresses of different kind at the same time; in such cases some mechanism of address translation must be used. It also may happen that a protocol entity and its user in a higher layer have each of them its own address. Both entities provide the same service, directly or indirectly. Both addresses do therefore denote the same service; however, for the communication in one layer one address is used, while for the communication in the other layer the other address is used. In such cases, too, the address translation between both types of addresses is usually used.

10.3   Address Translation

Several times we have already mentioned the need for the translation between addresses of different kinds, either between addresses in a same protocol stack layer or between addresses in different protocol stack layers; in all cases, however, all the addresses denote a same communicating entity.

Basically, the *address translation* is done using translation tables; such a table is usually called *directory*. A directory contains a list of addresses according to one addressing system and a list of equivalent addresses according to another addressing system. In most cases a communicating entity that employs the address translation maintains a local translation table containing those addresses that are frequently used, because a search in a local table is carried out much quicker than a search for a translation across the network. Any record in a local translation table must however have a limited lifetime; if an entity has not used a certain translation for some time, it removes it from the table and thus limits the size of the table and the memory usage. Hence a search for a translation across the network is needed only if it has not been found in the local table.

In wide area networks dedicated servers are used for the address translation that maintain dedicated databases containing address translations; such servers are referred to as *name servers*. Because the number of addresses in large networks is huge, these databases are distributed; this means that different addresses can be stored on different servers which exchange data between them when needed. If a name server does not know the translation which it is asked for, it must know how to find it in the distributed system of name servers. Hence name servers and translations stored on them must be organised systematically to simplify as much as possible translation queries in the distributed database. In hierarchically organised networks using hierarchically structured addresses name servers are hierarchically organised as well.

Dedicated protocols are used for translation queries. In this section three examples of such protocols will be presented; while one of them can be used for queries in local area networks, the other two are appropriate for the use in wide area networks with a distributed system of name servers.

10.3.1   Address Translation: Local Area Network Case

The basic characteristic of a local area network is a small number of communication entities and consequently a small number of addresses (therefore, name servers are not needed for the address translation). Communication devices in a classical local area network communicate via a common transfer medium in the broadcast mode of operation; whatever is transmitted by any communicating entity all the other entities can hear. In many modern local area networks, such as in most ethernet networks, switches are used for the message routing, so the mode of information transfer is not broadcast; however, in these networks, too, the broadcast mode of information transfer is possible if broadcast addresses are used.

Often, a local area network (LAN) is a part of a wide area network, such as the Internet (usually a LAN serves as an access network). While the network addresses (e.g. IP addresses) are used for the packet routing in wide area networks, the communication in local area networks is based on the MAC addresses (which are data-link layer addresses); any communicating element in a LAN must therefore have both addresses. Whenever a packet arrives from the wide area network to the LAN, its network layer destination address must be translated into its MAC equivalent, so it can reach its destination within the local area network.

Any communication element in a LAN of course knows its own addresses (this is a part of its configuration); besides this it also maintains a local translation table with the addresses of some other communicating elements and their translations. If an element needs the translation of an address, it first checks its local directory to find it. If it cannot find it there, it broadcasts a query for this translation to all the other network elements (hence it transmits a query with a broadcast destination address); if one of the network elements recognises the queried address as its own address, it responds with the translation which the originator of the query writes into its local translation table. Of course the size of the table is limited; an excessively long table would consume too much memory and a search through it would consume too much time. For the sake of simplicity, the lifetime of directory records will not be shown in our specification.

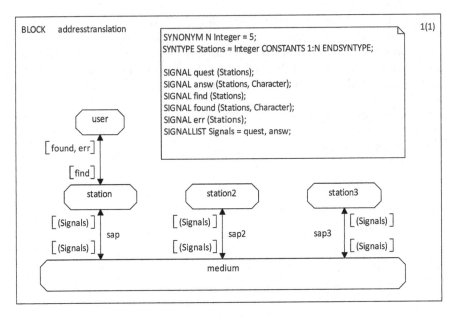

Fig. 10.1 Address translation in local area networks

In Fig. 10.1 and Fig. 10.2, respectively, an example of a simple local area network with three communicating entities (stations) is shown and a protocol for the translation of addresses into names of communicating stations in this network is specified. In this simple example the addresses of stations are values of data type Stations (integer numbers between 1 and 5), while station names are single characters (values of data type Character). The SDL specification language is used in both figures.

In Fig. 10.1 the communicating entities are shown as the processes station, station2 and station3, while the common communication medium is shown as the process medium. The process medium will not be specified here; let us tell only that any signal that has been received by the medium is immediately forwarded to all stations. Any communicating entity has an address of data type Stations and a name of data type Character. The process user in Fig. 10.1 models a user of the address translation protocol (its specification will also not be shown here) which may use the signal find (with an address as its parameter) to request the translation of the address given as the parameter; it receives the answer with the signal found. Of course, all stations may have a user each of them.

The processes station, station2 and station3 have exactly equal functionalities, only their addresses and names are different. In Fig. 10.2 the SDL specification of the process station is shown. During the initialisation this process is configured with the address 1 and the name 'A'. The array nametable of data type Names is the translation table (local directory); all the elements of this table are initialised with the special ASCII value nul, only the element indexed with the address 1 is assigned the value 'A'. (The ASCII character nul is coded with the decimal value 0; it is

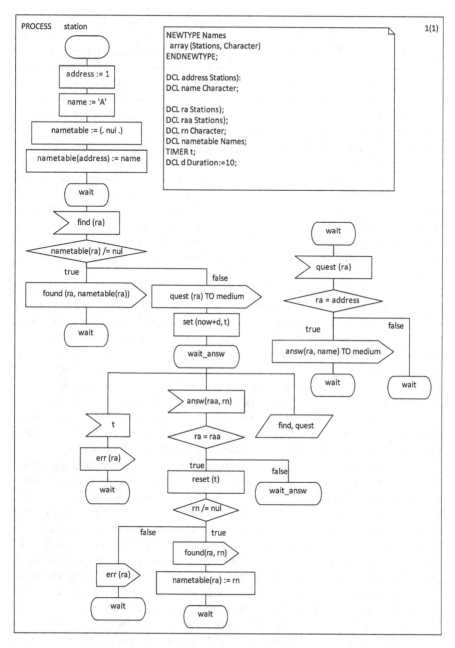

Fig. 10.2 Address translation protocol for local area networks

assumed that the name of no network element can be assigned this value.) If the process station receives a request for a translation with a signal find it first executes a local table lookup; if the translation is found there, it immediately replies the query with the signal found. If the translation is not found (the »name« corresponding to the queried address is nul), it sends a query (with the signal quest) to all other stations via the common medium; when it receives the answer with the signal answ, it forwards it to the user with the signal found. In the local area network many queries can be present concurrently; the station must therefore verify that it has received an answer to its question. It also writes the name corresponding to the queried address into its translation table, so later it would quickly find it locally if needed.

If the process station receives with the signal quest a query for the translation of its own address into its name, it replies with the signal answ via the common medium; however, it discards all queries that do not refer to its own address. As the mode of communication in a local area network is broadcast it is sure that a translation query will be heard by the element whose address the query refers to, and will therefore be replied (unless an element with the queried address does not exist or is currently not working).

In the specification shown in Fig. 10.2 one can see that a user may be informed about an error. If an element with the queried address does not exist or if it currently does not work or malfunctions or if the common medium does not work, the station does not receive an answer, it times out and sends the signal err to the user. Furthermore, a station sends the signal err also if it receives an answer with the character nul as the name (this should not happen in a normal, error-free operation). The use of a timer and the signal err increases the robustness of the protocol. The expiration time of the timer is important, of course; it depends on the network size, the number of network elements and the traffic intensity in it. (In our specification the delays in common medium were not modelled, so the value of the timer expiration time is not relevant and was deliberately set to 10 time units.)

In a local area network with a large number of network elements the length of the translation table can change dynamically, for the sake of memory saving. When a network element acquires a translation, it saves it in its translation table; however, if it does not need it for some time, it deletes it from the table, thus shortens the table and decreases the memory consumption. In this way only those translations can stay in the table for a longer time which are frequently used. One can expect that the translation table in practice is not as simple as is shown in our specification in Fig. 10.2; in this figure the amount of memory needed for the table is proportional to the maximum number of network elements. In a real system such a table is usually implemented as a linked list, its length being proportional to the number of translations currently present in the table. However, such details will not be further discussed here, this book being dedicated mainly to principles, rather than implementations, of the protocol design.

In Fig. 10.3 an MSC diagram shows an example address translation in the local area network specified in Figs. 10.1 and 10.2. In this example the entities station, station2 and station3 have the addresses 1, 2 and 3, and names 'A', 'B' and 'C', respectively. The address 4 is not assigned to any station. The user requests the

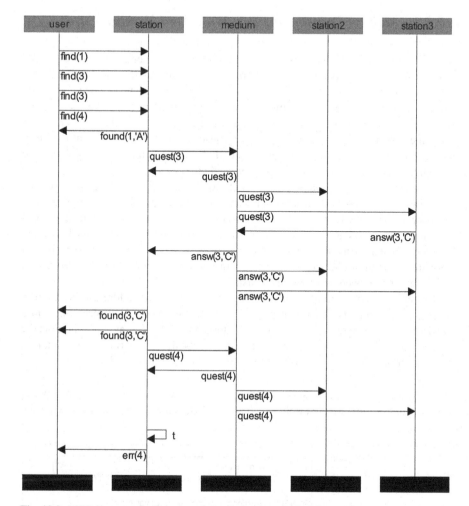

Fig. 10.3 MSC diagram of address translation in local area network

translation of addresses 1, 3 (two times) and 4. One can see that the process station, after it has been asked the translation of the address 3 the second time, does not query the translation in the network, as it has found it in its local directory. When it queries the translation of the address 4 it receives no answer, times out and sends the signal err to the user. The arrow labelled t that originates in the process station and closes in the same process in the MSC diagram indicates the expiration of the timer t in the process station. It must still be mentioned that the process station can process a next user request only after it has finished the processing of the previous request, due to the save symbol used in the state wait_answ of the process station.

10.3.2   Address Translation: Two Wide Area Network Cases

In a wide area network the number of communication elements may be huge, so a broadcast mode of operation is not possible. Distributed databases (distributed directories) containing address translations are therefore used in such networks. Addresses and their translations must be reasonably distributed over various servers so that database queries are as simple and as fast as possible. In spite of the existence of name servers each communicating entity also maintains its own local directory containing those addresses which have recently been used, in order to accelerate the searches for the translations of frequently used addresses.

Any communication element that wants to use the address translation service must know the address of at least one name server which will be referred to as its primary server here. If an entity that needs a translation does not have it in its local directory, it queries its primary server. If the primary server does not know the queried translation as well, it can either acquire it from other name servers and forward the result to the communication entity or it sends the entity the address of a name server it expects to know the translation.

In Fig. 10.4 an example of a very simple system for the address translation is shown which consists of a client and only two name servers (a primary one and a secondary one), where the primary server returns the address of another name server to the client if it does not know the translation itself. This variant of the address translation protocol will be referred to as the variant 1 in this section.

In Figs. 10.5–10.7 the specifications of the client (process client) as well as the primary and the secondary servers (processes p_server and s_server, respectively) are shown. For the sake of simplicity, the mechanisms that are not mandatory for the

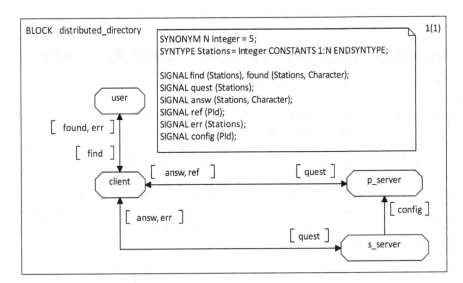

Fig. 10.4 Address translation in wide area networks, variant 1

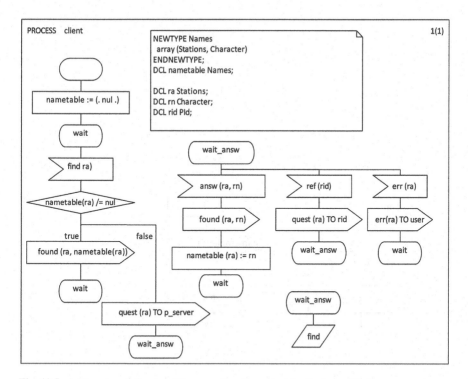

Fig. 10.5 Address translation in wide area networks, variant 1, client specification

basic operation, such as the mechanisms to enhance the robustness, are omitted. However, some peculiarities must be pointed out.

Before the system of name servers can be used it must of course be configured and »filled« with necessary data (in a real system this is being done all the time during the operation, as the system is alive, so its configuration may be changing all the time). How the configuration is done depends on the organisation and distribution of data in the distributed database (i.e. the system of name servers). In our case this configuration is simplified as much as possible: during the process initialisation, the name related to the address 1 is written into the table of the primary server and the name related to the address 2 is written into the table of the secondary server; all the other names are assigned the value nul. Furthermore, during the initialisation, the secondary server sends to the primary server its own address as a value of the data type PId (see Sect. 3.2.2), using the signal config—the SDL operator self returns the identifier of the process which executes this operator.

When a client receives from its user a request for an address translation it first verifies if that translation is already present in its local directory; if it is not it queries its primary server. If the primary server does not know the translation, it sends the address (process identifier) of the secondary server to the client with the signal ref. If the secondary server also does not know the translation, it sends the signal err (with the queried address as parameter) to the client. If the client receives the queried

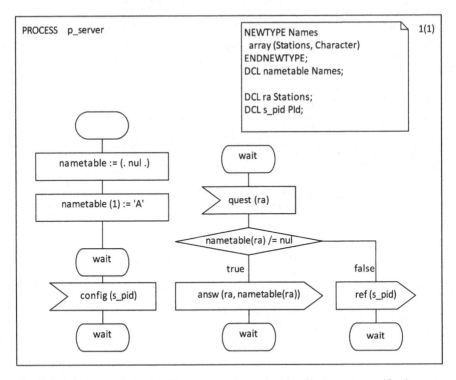

Fig. 10.6 Address translation in wide area networks, variant 1, primary server specification

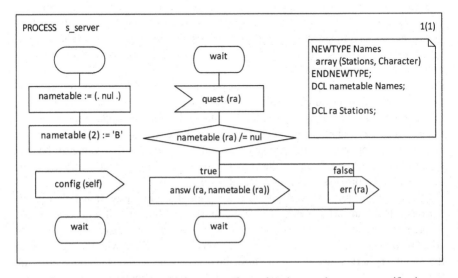

Fig. 10.7 Address translation in wide area networks, variant 1, secondary server specification

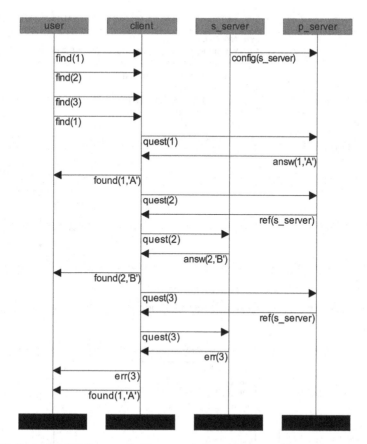

Fig. 10.8 Example MSC diagram of address translation with name servers, variant 1

translation from either the primary or the secondary server, it registers it into its local directory and forwards it to its user.

In a real system there are of course many name servers. When a primary server receives a request for an address translation it first analyses it to verify if the queried address is legal; if it is and it does not know the translation it determines which among the other servers should know it. Furthermore, the protocol must also provide for a robust operation for the case when the network or name servers do not work correctly. As the implementation of directories is concerned, the same remarks as were explained in the previous section already hold.

In Fig. 10.8 an example of the address translation using two name servers and the variant 1 of the protocol, according to the specifications in Figs. 10.4–10.7, is shown as an MSC diagram. The user requests the translation of addresses 1, 2, 3, and then 1 once more. While the translations of the addresses 1 and 2 are stored in the primary and secondary server, respectively, no name corresponds to the address 3. When the user requests the translation of the address 1 for the second time, the client finds it in its local directory and can therefore immediately reply to the user.

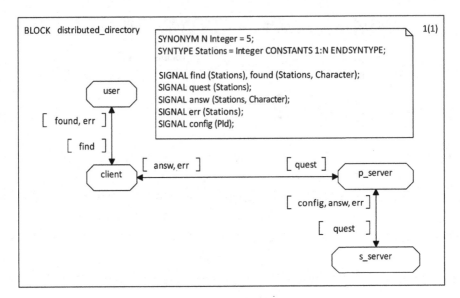

Fig. 10.9 Address translation in wide area networks, variant 2

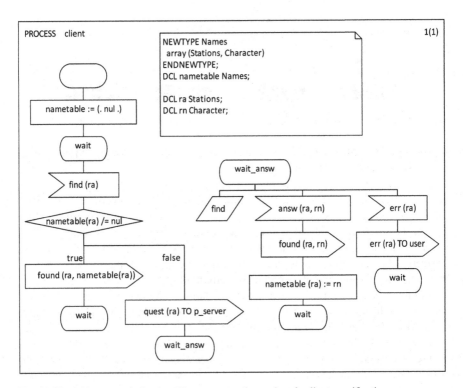

Fig. 10.10 Address translation in wide area networks, variant 2, client specification

In Fig. 10.9 another simple system for the address translation in wide area networks is shown; besides the client, there are only a primary server and a secondary server. In this variant of the protocol which will be referred to as the variant 2 in this section, the primary server tries to acquire the translation from the secondary server if it itself does not know it, and forwards the result to the client.

In Fig. 10.10, Fig. 10.11 and Fig. 10.12 the specifications of the client, the primary and the secondary server, respectively, are shown. In these specifications, too, only those activities that are indispensable for the basic operation of the protocol are shown.

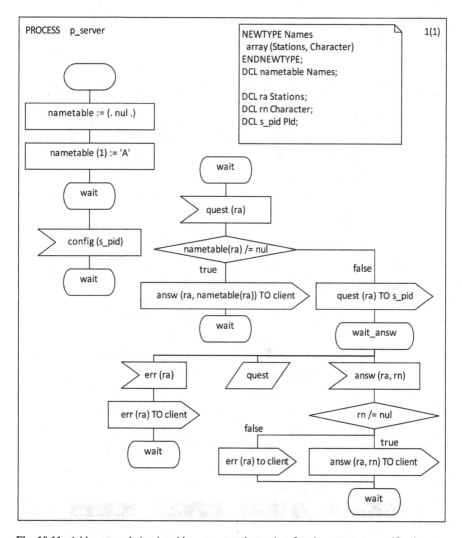

Fig. 10.11 Address translation in wide area networks, variant 2, primary server specification

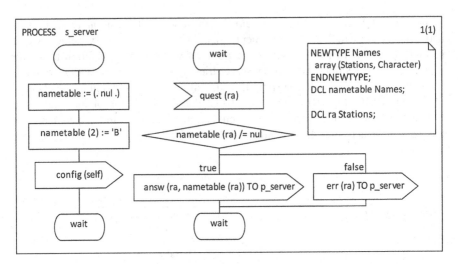

Fig. 10.12 Address translation in wide area networks, variant 2, secondary server specification

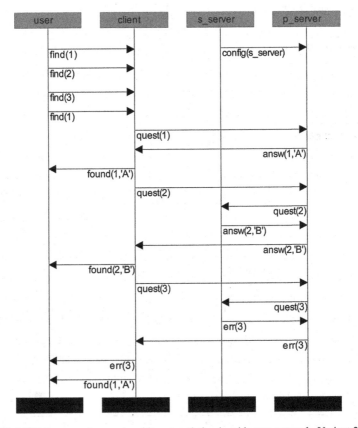

Fig. 10.13 MSC diagram of example address translation in wide area network, Variant 2

It is the opinion of the author that a reader who has carefully studied the specifications of the address translation in local area networks (Figs. 10.1 and 10.2) and especially in wide area networks, variant 1 (Figs. 10.4–10.7), will have no problems to understand the specifications of the variant 2 of the address translation in wide area networks, as shown in Figs. 10.9–10.12. We will therefore give no further comments and explanations here.

In Fig. 10.13 an example of the address translation in a wide area network according to the variant 2 of the protocol, as specified in Fig. 10.9–10.12, is shown as an MSC diagram. The configuration of the translation tables in both servers as well as the basic scenario of the queries are the same as in Fig. 10.8 where an address translation scenario was shown for the variant 1 of the protocol used in wide area networks.

Chapter 11
Distributed State Management

Abstract The notion of a distributed state is first explained and the need for its management discussed. Because two protocol entities must coordinate their actions in order to change their common distributed state, special protocols are needed to do this. The insufficient knowledge of the system state by any particular communication entity is pointed out as one of the basic communication problems. The distributed state management is described and formally specified first for the simplest case where the protocol entities are interconnected with a lossless channel, then for the case of a lossy channel, and finally also for the environments where floating corpses can occur. The methods to increase the robustness of the state management procedures are also discussed. Then the special case of the connection management is discussed, including the strategies for resolving collisions, the connectivity testing and the negotiation of connection parameters. There are 18 figures in this chapter, eleven of them specifying the distributed state management protocols in SDL, and seven of them showing characteristic scenarios in the form of MSC diagrams.

Usually, two or more communication entities have to agree upon the change of a common state; such a common state is referred to as *distributed state*. Hence, a distributed state consists of the states of two or more entities; these entity states must be correlated so that the entities can successfully communicate. An entity must possess some knowledge about the states of other entities that constitute the distributed state. Unfortunately, this knowledge is limited. The entities adjust their common state by exchanging messages; however, any information, sent by one entity, needs some time, called delay, to reach another entity which is the reason that an entity can only know the state of the other one which was in effect some time ago. This very uncertainty is one of the basic problems of telecommunications.

The most common cases of a distributed state are a *connection* and a *session*. There is no essential difference between a connection and a session; while the term connection is usually used in the lower layers of a protocol stack (up to the transport layer), the word session is normally employed in the upper layers (namely, the session and the application layers). Hence, the role of a connection is to organise the information transfer, and the role of a session is the organisation

D. Hercog, *Communication Protocols*, https://doi.org/10.1007/978-3-030-50405-2_11

of the application support. In both cases, there is a distributed state in which protocol entities can interoperate and exchange information.

In order to manage distributed states, special procedures are used because the change of a distributed state always necessitates the agreement of several involved entities and hence the information exchange between them. Usually these procedures are not special protocols, but are rather components of more general protocols that also provide for other services in addition to the state management.

In principle, the distributed state management could be a very simple process if the information transfer through a channel did not involve problems, such as errors, losses and floating corpses. Therefore, the basic procedure for distributed state changes will be considered first which can only be used if the information is transferred across a reliable channel. Only then, we will study how to overcome the problems of losses and floating corpses. The case of message errors will not be considered here, as a receiver will be assumed to detect errors and discard corrupted messages; in this way, corrupted messages are transformed into lost messages.

11.1   Distributed State Change, Basic Procedure

In Fig. 11.1, a simple system consisting of two users and two protocol entities is shown. ui and pi are the user and the protocol entity, respectively, on the initiating side, the former requesting the service of the state change and the latter initiating the procedure for changing the state. ur and pr are the user and the protocol entity, respectively, on the recipient side, the former receiving the service of the state change and the latter receiving from its peer the proposal for changing the state. If a confirmed service is to be provided, the conf primitive is used, otherwise it is not needed. If the possibility of an unsuccessful state change exists, the service must be confirmed; the change of the distributed state of a system implies the change of the state of both users, too, so they must both be made aware of the (un)success of the procedure. This is important because in the new state new services can be provided for which may not be available in the previous state.

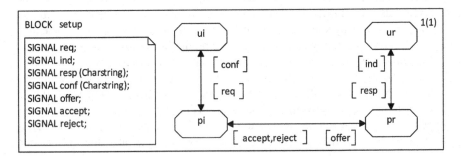

Fig. 11.1 Distributed state change

In the specifications shown in this section, the protocol entity pr, after it has received a proposal for the state change, may inquire if the recipient user agrees to change the state, using the ind primitive; the success of the procedure in such a case depends on the response from the user, so the primitive resp is needed. Of course it would also be possible that the protocol entity on the recipient side itself declined the change (maybe because it had not sufficient resources). The recipient entity could also just inform the recipient user about the change, without asking it; in such a case the resp primitive would not be needed.

Figures 11.2 and 11.3 specify the procedure for changing the state for the above-mentioned case that the recipient user may accept or decline the service. In both entities, the old state is simply called old and the new state is named new. It must be emphasised that the procedure shown in Figs. 11.2 and 11.3 is logically correct only if the channel interconnecting both protocol entities is reliable (i.e. no errors, losses or floating corpses can occur).

In Fig. 11.4 the scenario of a successful state change according to the specification in Figs. 11.1, 11.2, and 11.3 is shown. The scenario of an unsuccessful procedure is similar with the signals resp('Yes'), accept and conf('Success') substituted by the signals resp('No'), reject and conf('Fail').

If the recipient user is not allowed to accept or not the service the procedure is very similar to that shown in Figs. 11.1, 11.2, 11.3, and 11.4, only more simple. The most evident difference is that the resp primitive is not needed in this case. The ind primitive, however, is mandatory to make the recipient user aware of the state change.

As can easily be seen in the scenario of Fig. 11.4, two protocol data units are exchanged between the two protocol entities in case of the basic procedure for the state change. Such a procedure is therefore referred to as *two-way handshake*.

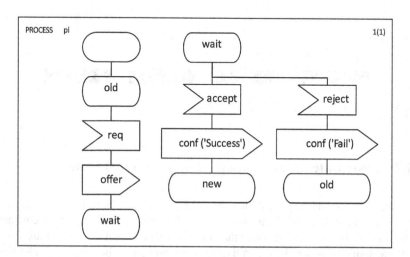

Fig. 11.2 Basic procedure for state change—initiator side

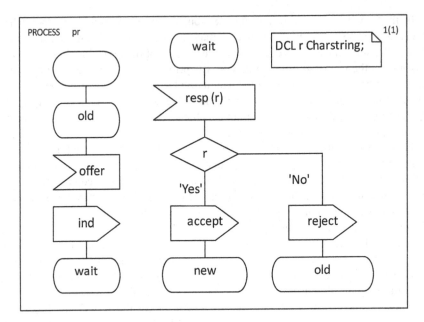

Fig. 11.3 Basic procedure for state change—recipient side

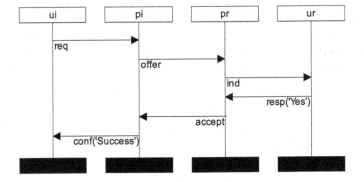

Fig. 11.4 Basic procedure for successful state change

11.2   Distributed State Change, Lossy Channel Case

The two-way handshake may also be used to change the state if the channel interconnecting two protocol entities is lossy. However, the initiating protocol entity must use a timer to guard the offer that has already been sent but not yet answered.

If the signal offer sent by the initiating entity happens to be lost, the sender must detect this loss. Therefore it sets the timer immediately when transmitting the offer; if it does not receive the response in time the timer expires (the timer expiration time must be set properly), the initiating entity retransmits the offer and waits again for either the response or another timer expiration. Of course, it may well happen that the response has been lost rather than the offer; in any case, the offer is retransmitted, as the sender has no way to know what has happened. The receiver pr must respond even if it has received both the original and the retransmitted offer; indeed, it cannot know whether it has received a repeated or a brand new offer. If it is already in the new state it sends an accept; in the other case it asks the user again (which is correct in both cases, whether it has received a repeated or a new offer). It is very important to be aware that the reply of the entity pr is always necessary; if the previous reply was lost and pr did not reply to the repeated offer, the timer in pi would expire again and again and the system would enter a livelock. When the sender pi receives the answer, it must immediately stop the timer to prevent an unnecessary timer expiration.

The *timer expiration time* must be slightly greater than the time that is needed to receive the answer to an offer (of course in the case that neither the offer nor the response to it was lost); this time is usually referred to as *round trip time* and shall be denoted as T_{rt} in this book. If the processing time (i.e. the time that is needed by both protocol entities to make decisions and process protocol data units) is disregarded, the round trip time can be written as

$$T_{rt} = T_{off} + T_{resp}, \qquad (11.1)$$

where T_{off} and T_{resp} are the delays of the message offer and the response to it (either accept or reject), respectively.

Here, it is necessary to explain why the timer expiration time must be greater than the round trip time, rather than equal to it. The first reason for this is that the equality does not exist in engineering; system and device designers must therefore always take tolerances into account, regarding either timing or some other characteristics of systems and devices. By the way, with these tolerances some effects that were disregarded during the design phase, either because they were not known or they were considered neglectable in comparison with some other effects (e.g. processing times), can be covered. In the unlikely case that the two quantities were exactly equal and the two events (e.g. the reception of the response and the timer expiration) occurred at precisely the same time, the further operation of the system would be unpredictable because nobody could know to which event the system would react first (this would be implementation dependent).

Let the communication system structure be the same as in the previous section, hence the one that is shown in Fig. 11.1. In Figs. 11.5 and 11.6 the specifications of both protocol entities are shown. Here, too, the old and new states in both processes are referred to simply as old and new. In the specification, the timer expiration time was deliberately set to 10 time units; however, in a real system it must of course be set according to Eq. (11.1). Otherwise, a reader is expected to be able to understand the specification without any further explanation.

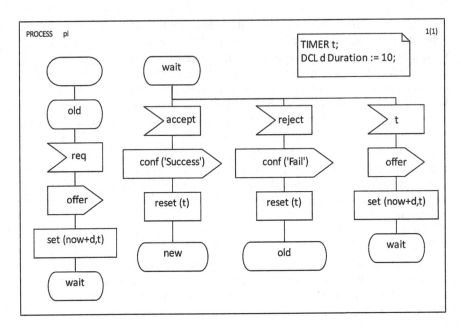

Fig. 11.5 Procedure for state change—initiator side, lossy channel case

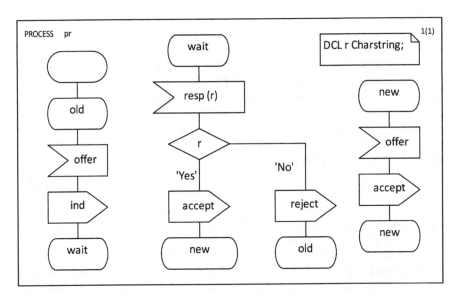

Fig. 11.6 Procedure for state change—recipient side, lossy channel case

In Figs. 11.7, 11.8, and 11.9 the state change scenarios are shown for the cases where the signals offer, accept and reject, respectively, are lost.

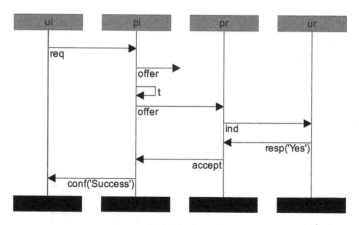

Fig. 11.7 State change in case signal offer was lost

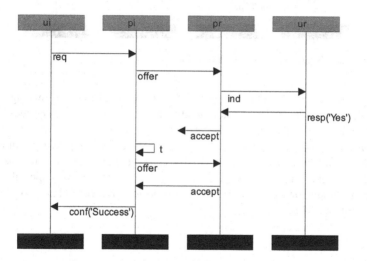

Fig. 11.8 State change in case signal accept was lost

At the beginning of this chapter it was mentioned that the knowledge a protocol entity possesses about its peer entity is limited; however, this limitation is not only due to the message delay, it can also be due to the message loss. This is clearly evident in the state change specification that has just been given in this section. When the protocol entity pr sends the signal accept to its peer it changes its state to new, although it cannot be sure that the entity pi will receive the message and itself change the state, as the message can well be lost on the way. Well, if pr does not receive another offer within the round trip time T_{rt}, it could assume that the accept was received and the state changed also at the initiating side, but what if the retransmitted offer was also lost!? One might expect this uncertainty could be cleared by the entity pi by sending the acknowledgment of the acknowledgment;

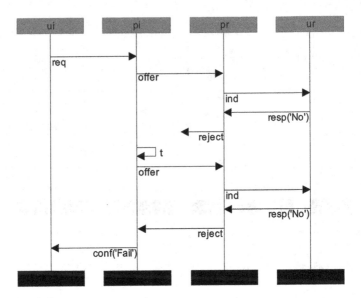

Fig. 11.9 State change in case signal reject was lost

unfortunately, pi also could not know whether pr received this acknowledgment of acknowledgment. Evidently, it is not possible that any of the protocol entities be sure that the other one really changed the state. This problem is known as *two generals problem* and is in general not soluble; using sufficiently complex procedures the probability that both parties have actually changed the state can only be made sufficiently high, but never 100%. The two generals problem is a special case of the *byzantine generals problem* which has to do with the state change of possibly more than two entities.

11.3   Distributed State Change, Channel with Losses and Floating Corpses

The two-way handshake which was treated in the previous section is simple and yet can master message losses in the channel. However, it can fail if a floating corpse appears in the channel which is not unusual in the case a connectionless network implements the channel between the protocol entities. If the two-way handshake is used, the protocol data units do not contain any parameters; the protocol entities are therefore incapable to distinguish the messages belonging to an old, already given up process (floating corpses) from the similar messages of the ongoing state change procedure.

In cases where floating corpses are possible, it is therefore preferable to use the procedure called *three-way handshake*. In this procedure, three protocol data units are exchanged that contain two parameters which are independent one of the other and are usually chosen randomly or pseudorandomly. If a floating corpse happens to appear during a process of changing the state using a three-way handshake, it is highly improbable that its parameters coincide with the parameters used in the ongoing process.

The procedure for the state change using a three-way handshake is shown in Figs. 11.10, 11.11, and 11.12. A simple case is shown where the recipient user ur has no chance to either accept or decline the service, it is just informed about the change. Only three primitives (req, ind and conf) are therefore used. As previously, the old and new states are called old and new in both protocol entities. The three protocol data units that constitute the three-way handshake and hence carry out the state change are offer, accept and check. If the entity pi discovers a floating corpse, it informs about it its peer pr with the signal invalid, in order to prevent an unnecessary timer expiration at the recipient side. Although the two parameters used in the three-way handshake can be of any type, the Integer type is used here for simplicity. While the signal offer carries only one parameter, the other three signals carry both of them; that is because one of the parameters is chosen by the initiator entity, while the other one is chosen by the recipient entity.

As the initiating entity pi sends a state change offer to its peer it adds its reference value to the message. The recipient entity pr replies to the offer with a signal accept and includes its own reference value to it, in addition to the value related to the initiator's reference (in our case simply equal). The reference value of the recipient is independent of the reference value of the initiator. When the initiator receives the signal accept, it replies to it with the signal check which again contains two values that are related to both reference values (in our specification simply equal to them). In this book only principles are discussed; in a specific protocol, these messages can be named differently. In the specification shown in Figs. 11.11 and 11.12 the reference values of 0 and 1 were chosen (some specific values had to be chosen in order to be able to run simulations). However, in most cases these values are chosen randomly or pseudorandomly and are determined for each state change instance separately; they can of course be different or equal.

The protocol entities use each its own timer. The initiator pi guards the unanswered signal offer with its timer, while the recipient pr's timer is used to guard the unacknowledged message accept. If the timer of the initiator pi expires,

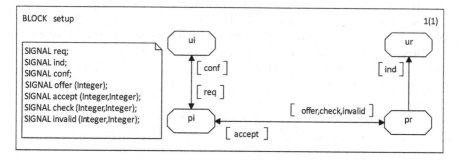

Fig. 11.10 State change, three-way handshake

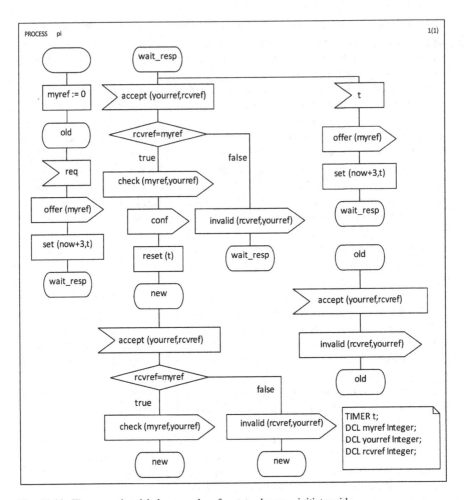

Fig. 11.11 Three-way handshake procedure for state change—initiator side

this indicates the loss of either the signal offer or the signal accept. If the timer of the recipient pr expires, this indicates that either the signal accept or the signal check was lost. Whichever entity detects the loss it retransmit its recent message.

The timer expiration times of both timers are of course very important. Both of them must be slightly longer than the respective round trip times, where one must keep in mind that pi can resend the offer only after its timer has expired. Let the timer expiration time T_{toi} of the initiator pi be longer than its round-trip time by some small tolerance δ,

$$T_{toi} = T_o + T_a + \delta, \tag{11.2}$$

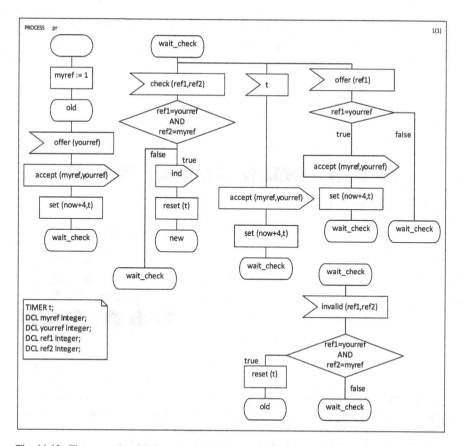

Fig. 11.12 Three-way handshake procedure for state change—recipient side

while the tolerance of the recipient must be still longer than the pi's tolerance by an amount ε; hence the pr's timer expiration time T_{tor} must be

$$T_{\text{tor}} = T_a + T_c + \delta + \varepsilon, \qquad (11.3)$$

where T_o, T_a and T_c are the delays of the signals offer, accept and check, respectively, while δ in ε are some appropriately small constants.

If the equality of all the three message delays is assumed, hence $T_d = T_o = T_a = T_c$ (which is often, but not necessarily, true, as all the three messages are control messages), the timer expiration time of the initiator pi can be written

$$T_{\text{toi}} = 2 \cdot T_d + \delta, \qquad (11.4)$$

and the timer expiration time of the recipient entity pr is

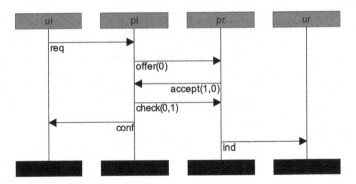

Fig. 11.13 Three-way handshake scenario with no floating corpses

Fig. 11.14 Three-way handshake scenario in case a delayed signal offer appeared

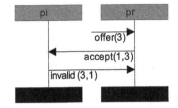

$$T_{\text{tor}} = 2 \cdot T_d + \delta + \varepsilon. \tag{11.5}$$

In practice, standards usually prescribe abundant values for both timer expiration times; as state changes occur rather unfrequently, the efficiency (which is somehow affected by excessively large values) is not important in these cases.

In the specification shown in Figs. 11.11 and 11.12 the delays of all signals as well as the values of constants δ in ε were all assumed to be simply 1 time unit.

The initiating protocol entity can detect a floating corpse in two cases. In the first case, it receives the message accept before it has even initiated the state change procedure, hence when it is still in the old state. In the second case, it receives the accept signal with an improper reference value. In both cases, it sends its peer the signal invalid to prevent an unnecessary timer expiration at the recipient side (the entity pr might be expecting the check signal).

The recipient entity discards all messages check that are either received in an improper state or with wrong parameters. After it has received the message invalid which is the reply to its own signal accept, it returns to the old state; if it receives the signal check which is not the answer to its message accept, it discards it and keeps waiting for the right answer.

In Figs. 11.13, 11.14, and 11.15 the MSC diagrams of three possible scenarios of the three-way handshake state change that was specified in this section are shown.

Figure 11.13 shows the scenario without any losses or floating corpses. The state change is successful.

Fig. 11.15 Three-way
handshake scenario with
two floating corpses (offer
and check)

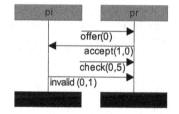

In the scenario shown in Fig. 11.14 a delayed message offer appears that has not
been sent by the entity pi (at least not recently). As the recipient pr does not know
this it replies with an accept which the entity pi discards, sends the signal invalid and
remains in the old state; after receiving the message invalid, pr also returns into the
old state. The states of both entities are not affected.

In the scenario shown in Fig. 11.15 a delayed signal offer appears first, and later
still a delayed message check which even has the same reference as the floating
corpse offer! The entity pi refuses the signal accept as it is not in the state of changing
the state (this was already seen in Fig. 11.14), while the entity pr discards the signal
check with improper reference; both entities remain in the old state. The floating
corpses produced no harm.

11.4 Robust Procedures for Distributed State Change

As was explained in Sect. 4.4, a protocol specification proceeds from the specification of the channel that interconnects protocol entities. That means that a protocol
that is declared as logically correct is guaranteed to operate correctly if the channel
properties are really such as assumed by the specification. It can however happen
that the channel properties change for some reason; in such a case the protocol may
not be able to correctly provide the service it should provide. However, the protocol
that is deemed robust may not be lost in such an unforeseen situation; it may not
enter a deadlock or livelock but must rather inform the users about the problem and
enter some well-defined state.

Different channel properties were assumed in Sects. 11.1–11.3; however, the
channel was assumed to exist and be usable. In the real world it may well happen that
a channel becomes unusable or even ceases to exist; a possible reason for this might
be a communication device or channel defect.

In Sects. 11.2 and 11.3 it was told how protocols solve the problem of message
losses using timers. In practice, however, the number of consecutive timer expirations is small. If however the channel for some reason ceased to be usable, timer
expirations would repeat with no limits, the protocol would enter a livelock and
could not provide the service. To prevent such situations, protocol entities usually
limit the number of consecutive timer expirations; they count them and, if the limit is
reached, they give up, return to the old state and inform the user about the problem
with an appropriate primitive (e.g. error.ind).

Counting consecutive timer expirations and reacting of a protocol entity is easy to specify and implement. A reader can add this himself/herself to the specifications of Sects. 11.2 and 11.3. They must however be warned that the timer expirations counter must be reset (set to 0) whenever a new state change is initiated, thence whenever a user requests it with the primitive req. This is necessary because the state change may be carried out many times during the lifetime of a system, but the timer expirations counting in one procedure may not continue from the previous one.

11.5   Connection Management

As was already discussed at the beginning of this chapter, the most frequent kind, or at least the best known case, of a distributed state is a connection or a session. The role of a connection is to organise the transfer of user information. Normally, protocol entities first set up the connection upon the request of a user; thereafter user messages are transferred through the connection from one user to another (a connection may also be regarded as a virtual channel that interconnects two users directly). When the user information transfer is no more needed, the protocol entities release the connection, again upon the request of one of the users. Hence the basic procedures of connection management are *connection setup* (also referred to as *connection establishment*) and *connection release* (sometimes also called *connection teardown*).

A connection setup and release are distributed state changes; procedures described in Sects. 11.1–11.3 are therefore used to this end.

In the real world it may well happen that a connection that has already been set up ceases to be usable, due to various reasons, such as a defect. It is therefore recommendable that a protocol which is to be robust discovers such inconveniences as soon as possible and informs the user about the problem.

When a connection-oriented information transfer is actively running it is not at all difficult to discover the termination of the connection usability; in this mode of operation the transfer is always bidirectional, as the information receiver must acknowledge the received information. If acknowledgments are not coming for some time, the connection is probably no more usable.

A connection may however sometimes be *idle* which means that it is prepared to transfer information at any time, but actually the information is not transferred. In such a case a protocol entity may periodically send to its peer special *heartbeat packets*. Some heartbeat packets can of course be lost in the channel; the peer entity will therefore inform its user about the problem only after it has not received several heartbeat packets in line. The peer entity uses a special timer, referred to as *keepalive timer*, to detect the absence of heartbeat packets.

The specification of a protocol that is testing an idle connection usability is shown in Figs. 11.16, 11.17, and 11.18.

Figure 11.16 shows the system structure and signal declarations. The constant d determines the period of sending heartbeat packets (which is given here the value

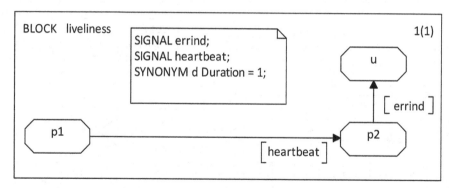

Fig. 11.16 Testing connection usability

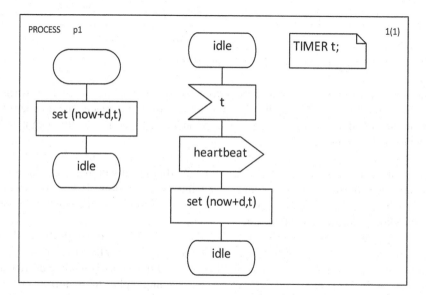

Fig. 11.17 Testing connection usability, sending heartbeat packets

1 for simplicity). This value must be known by both protocol entities; in practice, this value can be determined at the time of connection setup, or it can be fixed by the standard. In our specification this value is declared and defined as a global constant, seen by both processes.

In Fig. 11.17 heartbeat packets are seen being sent periodically with period d by the protocol entity p1.

Figure 11.18 specifies the receiving of heartbeat packets. The timer expiration time is slightly greater than a multiple of the period of sending heartbeat packets, $T_{to} = k \cdot d$, where k is the slightly increased multiple (3.1 in our example specification), and d is the period of sending heartbeat packets; this means that the keepalive timer will expire if the process p2 does not receive three heartbeat

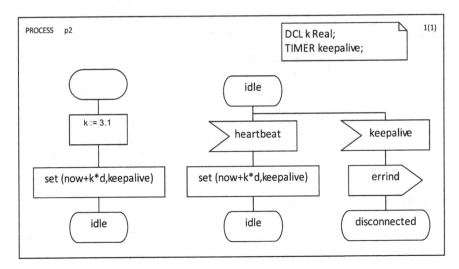

Fig. 11.18 Testing connection usability, receiving heartbeat packets

packets successively. Of course, any time the process p2 receives a heartbeat packet it must stop and immediately set again the timer (the reader should keep in mind that the SDL construct set, if applied to an active timer, means stopping it and immediately setting it again).

One must be aware that the constants k and d must be given appropriate values. If the period d is too small, the network will be overburdened with heartbeat packets; if too large, the reaction to a connection failure will be slow. If k is too small, longer bursts of errors might produce false alarms; if too large, the reaction to the connection failure will again be slow.

The procedure shown in Figs. 11.16, 11.17, and 11.18 can of course be used to test the connection availability in each direction independently. The connection can also be tested in both directions simultaneously if one entity is sending heartbeat packets to the other one, while the latter one is sending replies to the former one.

11.6   Collision Management

Two protocol entities may happen to propose a state change simultaneously or almost simultaneously; the term »almost simultaneously« means that one of the entities sends an offer to the other one soon after the other one has sent its own offer to it, but before it receives the offer of the other one. Such a situation is called a *collision of proposals*. A protocol that manages state changes must of course foresee also the possibility of collisions and specify the behaviour of protocol entities in such cases, too.

It is not possible to give some specific advice about how to act in case a collision occurs. The procedure is simple if both entities propose the same new state; in this case they just acknowledge to each other the state transition. If, however, the two entities propose new states that are incompatible or even contradictory, the protocol will probably specify a new state that will be most acceptable for both parties.

A connection is a distributed state with which both involved parties must agree. In case one of the peers requests the connection setup while the other one requests the connection release, the protocol will probably require the connection to be released. A careful reader might think that the previous statement is a nonsense: a connection setup is requested when there is no connection, while the connection release is requested when a connection is set up! A case where such unusual requests can really be proposed is as follows: a connection is set up, but then one peer requests the connection release, while the other one requests the connection reset (release and then immediately setup of a new connection).

11.7   Negotiations

Often various parameters are associated with a connection or a session; such parameters determine qualitative and quantitative properties of the connection/session and the mode of communication. Some examples of such parameters include the transfer rate, the maximum message length and the compression method.

These parameter values may either be fixed by the standard which specifies the protocol, or they are determined by protocol entities during the process of the connection/session setup. In this latter case there are two possibilities. One of them is that the values are determined by one of protocol entities (either the one that initiated the setup or the one that acts as a master). The other possibility is that parameters are determined with *parameter negotiations* between the peers.

The negotiation procedure is usually quite simple. One of the peers proposes to the other one a set of values or a value range that are acceptable to it. The other entity chooses one of these values or a subset from the proposed set which is/are acceptable to it, too, and lets its peer know about it. If the first entity receives a set of acceptable values from its peer, it must decide on the value to be used and communicate its decision to the peer.

The procedure for parameter values determination can also be more complicated. An example of such a complex procedure is the determination of a secret key that is to enable the secure communication in the connection; security procedures that are to allow for confidentiality, integrity and authenticity of communications may include the cooperation of a third party, such as a certificate agency, to help to set up a secure connection and determine the keys for the authentication and ciphering.

Chapter 12
Providing Reliable Message Transfer

Abstract This is the longest chapter of the book. The problems that can occur in a communication channel, such as errors, losses, message reordering and floating corpses, are presented first. Then it is explained what the term reliable message transfer means. Next the principle of channel coding and decoding is described. The difference between the error correction by receiver and the automatic repeat request protocol is discussed, and the advantages and disadvantages of both are listed. Their usability for the transfer of different kinds of information is discussed. Both the basic and the additional mechanisms of the automatic repeat request protocols are described. Sliding window protocols are then thoroughly described. Six variants of them are presented and formally specified (generalised, stop-and-wait, go-back-N, selective-repeat, selective-reject and multiple-timers protocols). Their respective advantages and disadvantages are listed. The performance simulation results for a characteristic communication scenario are shown and commented for five of these protocols. The description and specification of the generalised sliding window protocol is a special feature of this book, as they cannot be found in other books on communication protocols. The specification of the selective-repeat protocol with multiple timers and individual acknowledgments is also a peculiarity which cannot be found elsewhere. There are 52 figures, many of them formally specifying the protocols or displaying characteristic communication scenarios.

12.1 What Is a Reliable Message Transfer

In Sect. 4.5 the problems that can be encountered in a communication channel interconnecting protocol entities were discussed. It is the task of some protocols to solve these problems and hide them from their users; in this way a protocol can interconnect its users with a virtual channel with the properties which are better than the properties of the channel interconnecting protocol entities.

Errors and losses of messages, reorder of messages, duplicates and floating corpses were enlisted as channel deficiencies in Sect. 4.5. A protocol provides a *reliable message transfer* if the following conditions are fulfilled:

D. Hercog, *Communication Protocols*, https://doi.org/10.1007/978-3-030-50405-2_12

- Any user message is received by the receiving user exactly once (there are no losses and no duplicates).
- Any user message is received by the receiving user without errors.
- User messages are received by the receiving user in the same order as they were sent by the transmitting user.

Hence the term reliable transfer refers to the communication service and the message transfer between users. If protocol entities are interconnected with an unreliable channel (with the above listed deficiencies), it is the duty of a protocol to provide for the reliable transfer between users. In this chapter the protocols will be discussed that provide a reliable transfer between users, although the channel between protocol entities is not reliable.

12.2   Error Detection and Correction

As far as the protocol entities are concerned, a user message may contain any sequence of bits; a receiver of course cannot know which message it is going to receive from the transmitter (if it knew it, a communication system would not be needed!). Hence, a receiver can receive any sequence of bits, not knowing if the message is corrupted or not. If a protocol is to verify the correctness of transferred messages, a transmitting protocol entity must add additional bits, referred to as *redundant bits*, to the message to be transmitted; redundant bits help a receiving protocol entity to detect corrupted messages.

12.2.1   Channel Coding and Decoding

Let the length of an original message (an unprotected message) be m bits. This message can contain any sequence of bits (this is not quite true, as the protocol control information, conceived according to the rules of the protocol, is added to a user message, but our consideration will be correct anyway); hence, there can in principle be 2^m different unprotected messages. A transmitting protocol entity adds r redundant bits to the unprotected message according to a specified rule and transmits such a protected message into the channel. The process of adding redundant bits according to a specified rule is referred to as *channel coding*; this process is determined by the channel coding method. A protected message (hence an unprotected message together with the added redundant bits) is referred to as *codeword*. Although the length of a codeword is $(m + r)$ bits, a transmitting protocol entity may transmit only 2^m different codewords, as exactly one codeword can be assigned to any unprotected message because of the strictly prescribed channel coding rule.

If any bit or any combination of bits can be corrupted in the channel during the message transfer, the receiving protocol entity can receive any combination of

$(m + r)$ bits, hence anyone among $2^{m + r}$ different codewords. The codewords which a transmitter may transmit into a channel according to the channel coding rule are called valid codewords, while those which may not be transmitted by a transmitter but can be received by a receiver because of errors are called invalid codewords.

A protocol which allows for the detection of messages that were erroneously transferred through the channel determines the channel coding method as well as the location within a protocol data unit where the redundant bits are to be placed; these rules must of course be known by both the transmitter and the receiver. When a receiving protocol entity receives a protocol data unit it can verify if the received message meets the channel coding rule, hence if it has received a valid codeword. This procedure is referred to as *channel decoding*. The channel decoding allows for the *error detection*, i.e. it allows corrupted messages to be detected. If a receiver has received an invalid codeword, it knows that the message was corrupted in the channel, because a transmitter is allowed to transmit only valid codewords. However, if a receiver has received a valid codeword, it assumes that the message was not corrupted during the transfer. In the previous phrase the word »assumes« was used rather than »knows«! It may well happen that such a combination of transferred bits was changed (corrupted) during the message transfer that a valid codeword is received, but other than the codeword which was sent by the transmitter! Unfortunately, such an error cannot be detected by means of the channel decoding. Hence, it is evident that not all errors can be found with the process of channel (de)coding.

If a receiver discovers it has received an invalid codeword, it can search for a valid codeword that most resembles the received one (i.e. it searches for a valid codeword that differs from the received one in the least number of bits). In this way a receiver can correct a received message alone, without the interaction with the transmitter (this kind of error correction is based on the assumption that a lower number of bit errors is more probable than a higher number of bit errors). This mode of error correction is usually referred to as *forward error correction* and indicated with the acronym *FEC*. The author considers this term rather improper as an error cannot be corrected before it even happens! We will therefore prefer to call it *error correction by receiver*. Of course, the error correction by receiver also has its limitations. It may well happen that such a combination of bits has been corrupted during a message transfer that the received invalid codeword is more similar to a valid codeword other than the transmitted one. Hence the error correction by receiver method also cannot correct all errors.

There also exists another possibility for error correction, often referred to as *backward error correction (BEC)*. If this method is used, a receiving protocol entity only detects corrupted messages and explicitly or implicitly informs the transmitter about messages in error; the transmitter then retransmits those messages that have been corrupted. This method of error correction is also called *automatic repeat request (ARQ)*; protocols using this method are dubbed *automatic repeat request protocols* or *ARQ protocols*.

One must however be aware that only corrupted messages can be detected and corrected, not corrupted single bits.

We have just stated that there is no protocol capable of correcting all errors. Hence, a virtual channel interconnecting the users of an error correcting protocol still has some packet error rate which is referred to as *residual packet error rate*. Of course, the residual packet error rate is expected to be much lower than the packet error rate of the channel interconnecting protocol entities; were this not true, the error correcting protocol would be senseless. Mathematically the *residual bit error rate* can also be calculated from the residual packet error rate, using the formula (4.1); however, the notion of residual bit error rate has no firm physical background, as packets are always corrected rather than single bits.

All methods of channel coding require the addition of redundant bits to protocol data units; because redundant bits form a part of a message overhead and must be transferred through a channel just like everything else, the channel coding lowers the efficiency of a communication system as seen by the users. The efficiency is of course the lower, the more redundant bits are added to protocol data units. On one side it is therefore desirable to add as few redundant bits as possible, while on the other side a low ratio of the redundant error rate over the error rate in the channel is desired. Unfortunately, both desires are contradictory, as a lower ratio of residual error rate over error rate often requires more redundant bits, hence more overhead; in general one can say that a channel coding method which adds less overhead and at the same time offers a lower residual error rate is better. The efficiency of a coding method of course heavily depends on the coding method itself. It is however important to be aware that a coding method that allows for the error correction by receiver in general requires more overhead than a method which allows only for the error detection.

In past decades many channel coding methods were developed, some of them suitable for the error detection only and the others also for the error correction by receiver. Methods differ in their efficiency as well as in their consumption of processing and memory resources. Only general principles of channel (de)coding and error detection and correction by receiver are discussed in this book; a reader wanting to know more about this topic can find the treatment of many specific methods in many books and papers discussing this.

12.2.2   Minimum Hamming Distance

We have already mentioned that different codewords differ in one or more bits. If all codewords have the same length, the difference between them can be described with the metrics called *Hamming distance*. A Hamming distance between two codewords is defined as the number of bit pairs, laying at the same position in both codewords, which have different values. For example, the Hamming distance between codewords 110101 and 100101 is 1, between codewords 110101 and 100001 is 2, and between codewords 110101 and 100011 is 3. Now we can say that a receiver which has received an invalid codeword corrects it so that it finds a valid codeword with the least Hamming distance to the received one; if there are more valid

Fig. 12.1 Role of minimum hamming distance in assessment of capability for error detection and correction

codewords with the equal least Hamming distance to the received codeword, the receiver cannot decide what to do and therefore cannot correct the received message.

If a specific channel coding method is considered, the Hamming distances between different pairs of valid codewords can be different. The least possible Hamming distance between any two valid codewords of a channel coding method is called *minimum Hamming distance*. The minimum Hamming distance of a coding method will be denoted as d_{Hm}.

It is easy to understand that the capability of a channel coding method to detect and correct errors is the stronger, the larger is the minimum Hamming distance of the code, hence the more valid codewords differ between them. The minimum Hamming distance can therefore be used to assess the capability of a channel coding method to detect or correct errors. What has just been told (and has been based on our intuition) is going to be further explained in the continuation of this section.

In Fig. 12.1 two valid codewords, L and R, are shown, the Hamming distance between them being the minimum Hamming distance; in reality, the space of codewords is not one-dimensional, as shown in Fig. 12.1, as only the least favourable case is shown in this figure. Black circles between the two valid codewords represent invalid codewords, numbered in our figure. If bit values are changed one by one in the valid codeword L, d_{Hm} steps are needed to achieve the valid codeword R in the least favourable case. In these intermediate steps $d_{Hm} - 1$ invalid codewords were got, numbered from 1 to $d_{Hm} - 1$ in the figure. Hence, if a single bit is changed in the valid codeword L, the invalid codeword 1 is achieved; if two bits are modified, the invalid codeword 2 is achieved, etc.

Now let us assume we have received an invalid codeword containing c corrupted bits and want to correct the error in the received message. (If a receiver receives a corrupted message it can of course not know how many bits are erroneous; here we only want to find out the maximum value of c which still allows the error correction.) As the invalid codewords that are closer to L will be corrected to L and those which are more similar to R will be changed to R, it can easily be seen that the error correction is possible only if the relation (12.1) holds,

$$2 \cdot c \le d_{Hm} - 1. \tag{12.1}$$

Hence, c is the maximum number of bits which may be corrupted in a codeword so that this codeword can be corrected by receiver if a channel coding with minimum Hamming distance d_{Hm} is used. If more than c bits, as restricted in Eq. (12.1), were corrupted during the message transfer, the receiver would either erroneously »correct« the message or could not decide what to do, hence it would not correct the error.

If, however, one only wants to detect an error in a transferred message, it is clear, considering Fig. 12.1, that an error is detectable only if at most d bits are corrupted in the message where the condition (12.2) must be valid,

$$d \leq d_{Hm} - 1. \tag{12.2}$$

If more than d bits, as restricted by Eq. (12.2), were corrupted, the receiver could possibly receive a valid codeword, but not the same one as was sent by the transmitter.

The correction and detection of errors in received codewords can also be combined. Let us assume we have decided to correct only those codewords where at most c bits are corrupted where the value of c must fulfil the condition (12.1). In such a case an error can be detected only if at most $c + d$ bits were corrupted in the message, where the relation (12.3) must be fulfilled,

$$2 \cdot c + d \leq d_{Hm} - 1. \tag{12.3}$$

The validity of the condition (12.3) can be verified by partitioning all invalid codewords in Fig. 12.1 into three regions: c invalid codewords on the left side are corrected into L, c invalid codewords on the right side are wrongly corrected into R, while d invalid codewords in between are not corrected, they can only be detected.

A reader can verify the validity of Eqs. (12.1)–(12.3) by drawing Fig. 12.1 for different specific values of minimum Hamming distance d_{Hm}.

12.2.3   Error Correction

After the discussion given in previous sections one might ask which error correction method is better: the correction by receiver or the error detection by receiver with the automatic repeat request. There is no general answer to this question; the answer depends on the kind of information to be transferred as well as on the channel properties.

The error correction by receiver has the following advantages.

- The error correction is executed solely by the receiver, so it does not require the interaction between the transmitter and the receiver; the correction is therefore fast and requires approximately the same processing time for all transferred messages, were they corrupted or not, regardless of the need to correct them. The error correction by receiver therefore does not introduce variable delays.
- If bit error rates are high, the error correction by receiver is more efficient than the correction with automatic repeat request because corrupted messages are not discarded but are rather corrected.
- The error correction by receiver can be used even if protocol entities are interconnected with a unidirectional (simplex) channel.

The deficiencies of the error correction by receiver are the following.

- A receiver can only correct those corrupted messages which were received; the messages that were lost in a channel and were therefore not received can of course not be corrected; hence this kind of error correction cannot provide for a reliable message transfer.
- If bit error rates are low, the error correction by receiver is less efficient than the automatic repeat request method due to a larger number of redundant bits and consequently overhead that is required.

The error correction using the automatic repeat request method has the following advantages.

- The automatic repeat request protocols are capable of detecting and correcting not only corrupted but also lost messages, as well as those that arrive to the receiver in incorrect order; automatic repeat request protocols can therefore provide for a reliable transfer of messages.
- If bit error rates are low, the automatic repeat request protocols are more efficient than the error correction by receiver methods due to a relatively small number of added redundant bits.

Automatic repeat request protocols have the following deficiencies.

- The messages that have to be retransmitted several times may have much larger delays than those which are correctly received in the first trial. Hence the automatic repeat request protocols may introduce considerable variable delays into the message transfer.
- If bit error rates are high, many messages are corrupted; they must be discarded and retransmitted by the transmitter; due to this the efficiency of automatic repeat request protocols can be quite low if the error rate is high.
- The automatic repeat request protocols are especially inefficient when channel delays are large; in such cases transmitters become aware of errors very late, so they are also late to react to losses.
- An automatic repeat request protocol can only be used if a duplex or a half-duplex channel is available.

Various kinds of user information can be transferred through a communication network. In this text we will essentially distinguish between two different information classes: data and streams.

Data are the type of information that does not depend on time; therefore they do not need to be transferred in real time. When data are transferred, the delay variation does not present a problem (if it is not excessively large, of course); however, a reliable transfer of data is required (hence transfer of user messages without errors, losses, duplicates and without message reorder). An automatic repeat request protocol must therefore be used to transfer data to assure a reliable transfer. An ARQ protocol must be used in at least one layer of the protocol stack, preferably in the layer where the problems are encountered or in a higher one.

A *stream* is the kind of information represented by a sequence of information elements which are arranged in time according to some order, hence there is always some timing relation among the information elements (e.g. information elements can be generated and transferred with equal time intervals between them). A stream must be transferred in real time in order to retain the original timing relations; in other words, the elements of a stream must be transferred with equal delays, or almost equal delays (if the delay variation is not too large, special techniques to equalise delays are available, but the delay variation must be limited). A stream often contains some amount of redundancy; because of this property errors and losses are tolerated to some degree. Automatic repeat request protocols are not appropriate for the transfer of streams, because they introduce strongly variable delays. If needed, errors can be corrected by receivers, while losses are only detected and lost messages substituted by messages that are generated locally at the receiving side using interpolation algorithms.

In Sect. 12.2.1 we explained that all the errors can never be detected and corrected; after the error correction has been executed there is always some residual error rate remaining. In this section, however, we told that automatic repeat request protocols must be used when transferring data, in order to assure a reliable data transfer; unfortunately, ARQ protocols are much less efficient than the error correction by receiver if error rates are high. In communication systems where the error rates are very high (such as mobile radio networks) *hybrid automatic repeat request* methods are therefore usually used for the error correction; these methods are also known as *HARQ* protocols. If HARQ protocols are used, errors are first corrected by receiver to achieve residual error rates that are much lower than the error rates in the lower layer channel. Then the losses and the residual errors still present in messages are detected in the higher layer and corrected with an automatic repeat request method.

12.2.4   Block Coding and Convolutional Coding

The *block coding* is most usually used for data transfer; if the block coding is used, redundant bits r are added to a block of information bits m which can include both the user and the control information (see Fig. 12.2a). The added redundant bits r depend on the information bits of the block m. A protocol data unit hence consists of information bits m and redundant bits r; all of them are transmitted into the channel. This message (possibly modified due to errors) is then decoded by the receiver.

As was already told in Sect. 12.2.2 the strength of a channel code for detecting and correcting corrupted messages depends on the minimum Hamming distance of the code which in turn depends on the channel coding method as well as on the ratio of the number of redundant over the number of information bits. This means that in a case when too many bits are corrupted in a message (if the error rate is high such a case is quite likely) the receiver is incapable to detect or correct the errors. Furthermore, we told in Sect. 4.5 that quite often the probability of bit errors in a channel is

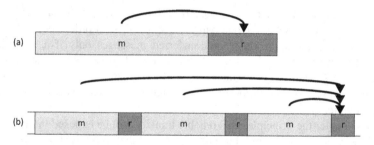

Fig. 12.2 Block (**a**) and convolutional (**b**) coding

not constant, errors can occur in bursts. Block codes are therefore especially vulnerable if errors occur in bursts, as a single burst of errors may well corrupt a large number of bits in a single message.

The vulnerability of channel coding to error bursts can be significantly mitigated by spreading the interdependence of information and redundant bits over a longer sequence of binary values. This is done if a *convolutional coding* is used. The convolutional coding is especially suitable for channel coding of streams (very long, theoretically infinitely long sequences of values). From time to time, usually periodically, groups of redundant bits are inserted into the sequence of information bits where a group of redundant bits depends on several precedent groups of information bits. The procedure is illustrated in Fig. 12.2b. The convolutional coding is most suitable for channel coding of streams and correcting errors in them by receivers. Besides this, it also has another deficiency: it can introduce a significant delay into the transferred stream; this delay is the greater, the more is the interdependence of information and redundant bits spread over the stream, as a receiver can correct information bits only after it has received also the redundant bits which depend on them.

12.3   Automatic Repeat Request Protocols

We have already told that automatic repeat request protocols (also referred to as ARQ protocols) not only allow for error correction in transferred messages but can also provide for a reliable transfer of user messages; hence they assure a message transfer without errors, losses and duplicates and also preserve the order of transferred messages. Due to all these capabilities they are indispensable when data are transferred. They are mostly used in data-link and transport layers of protocol stacks (according to the OSI reference model), but can also be employed in other layers.

The essential feature of automatic repeat request protocols is that the receiving protocol entity detects corrupted and lost messages and informs the transmitting protocol entity about the success or failure of transfer; the transmitter retransmits the messages that have been corrupted or lost. In most ARQ protocols the receiver discards corrupted messages, so there is no difference between corrupted and lost messages. Furthermore, the receiver must reveal duplicates and discard them, it also must pass user messages to the receiving user in the correct order.

One can talk about the order of transferred messages only if a sequence of messages is transferred. It is also possible to reveal losses only in such a case. This means that the data transfer must be somehow organised which is only possible if the data transfer is connection oriented. If the data transfer is connectionless, a reliable transfer cannot be assured.

In this chapter the principles of a reliable transfer will be exposed in the simplest possible way, separated from other procedures and methods which are also carried out by (possibly the same) protocols. Two assumptions will be supposed.

Firstly, a connection will be assumed to already be set up; hence only the transfer of user information in the data transfer phase will be described. One must however be aware that a connection can be in different states during its lifetime and that various values can be associated to it which determine it in more detail; the state of a connection and the parameters that determine it more precisely must be initialised at the beginning of the existence of a connection, hence at the time the connection is set up. In this chapter we will of course have to explain the initial conditions of a connection when a data transfer phase will be described.

Secondly, it will be assumed in the majority of this chapter that the user information is transferred only in one direction, e.g. from the user ut towards the user ur (see Fig. 4.16). The protocol entity that transmits information protocol messages will be called transmitting protocol entity or simply transmitter for short, while the protocol entity which receives information protocol data units will be referred to as receiving protocol entity or receiver for short; both users (ut and ur in Fig. 4.16) will also be called a transmitting and a receiving user, respectively. In reality the user information is often transferred in both directions simultaneously (which means that the protocol entities act both as transmitters and receivers at the same time); however, because the user information transfer in one direction is mostly independent from the transfer in the other direction, our discussion of automatic repeat request protocols will nevertheless be general enough. Not earlier than in Sect. 12.15 we will expose some peculiarities of the bidirectional user information transfer one must be aware of.

12.3.1 Basic Mechanisms

A logically correct automatic repeat request protocol must employ the following mechanisms.

A transmitting protocol entity must use a transmit *memory buffer* where those messages are stored which have already been transmitted but the transmitter does not yet know if they have also been received by the receiver. Such messages are referred to as *outstanding messages*.

A receiving protocol entity also must use a receive memory buffer where those messages are stored which have already been received but not yet passed on to the receiving user.

A transmitting protocol entity must channel encode protocol data units before transmitting them so that a receiver can detect errors in them and discard corrupted messages.

A transmitting protocol entity must number transmitted information protocol data units and include their *sequence numbers* into their respective protocol control information; the sequence numbers of received messages allow a receiver to discover lost messages and duplicates and also to put the received messages into the correct order. The sequence numbers of protocol data units are only a part of the protocol control information and do not concern users.

A receiving protocol entity sends *acknowledgments* for those information protocol data units it has successfully received and stored into its receive buffer; such acknowledgments are sometimes called *positive acknowledgments*. An acknowledgment must unambiguously specify which information protocol data units are to be acknowledged. Of course, acknowledgments can also be corrupted in the channel, so they must be channel coded as well. Acknowledgments can either be contained in special control protocol data units, or they can be included into information protocol data units that are sent by the sender of acknowledgments (this can of course be done only when the user information is transferred in both directions). Including acknowledgments into information protocol data units is referred to as *piggybacking*. With piggybacking the overhead that must be transferred through a channel can be somehow reduced.

A transmitting protocol entity must use one or more timers. An active timer (i.e. a timer which is running) guards one or more unacknowledged messages. If the transmitter receives the awaited acknowledgment, it stops the timer; if it does not receive the awaited acknowledgment, the timer expires and the transmitter retransmits one or more protocol data units. If the timer of the transmitting protocol entity expires, the transmitter is said to have timed out.

12.3.2   Acknowledgments

In general, a transmitter may transmit several information protocol data units before it receives acknowledgments of them. An *acknowledgment* must therefore precisely specify the reception of which of them it acknowledges. For this reason any acknowledgment contains one or two values specifying the sequence number or the range of sequence numbers of the information protocol data units to be acknowledged.

There exist three kinds of acknowledgments; they are going to be described in the following three paragraphs. Acknowledgment protocol data units will be symbolically denoted with the acronym ACK.[1]

The simplest kind of acknowledgment is an *individual acknowledgment*. Its abstract syntax is ACK(x) where x indicates the sequence number of the information message to be acknowledged. Hence an individual acknowledgment acknowledges the reception of a single, precisely determined information protocol data unit.

In practice, a *cumulative acknowledgment* has most usually been used up to now. A cumulative acknowledgment ACK(x) informs the transmitter that the receiver expects to receive the information protocol data unit with the sequential number x next and so acknowledges all the information messages with the sequential numbers up to and including the number $x - 1$.

A *block acknowledgment* has the abstract syntax ACK(x,y) and acknowledges the reception of the information protocol data units with the sequential numbers from x to y, both inclusive.

12.3.3 Timers

One of the basic and necessary conditions to assure the logical correctness of the operation of an automatic repeat request protocol is the use of a *timer* to guard any outstanding message; a timer expiration tells a transmitter that either an information protocol data unit or the acknowledgment of it has been lost. If more than one outstanding messages can be stored in the transmit buffer of a transmitter, each of them can be guarded with a separate timer, or a single timer can be used to guard all of them.

As a timer expiration indicates a loss of an information or a control protocol data unit, the purpose of the timer is to trigger the retransmission of the supposedly lost message; a timer with such a purpose is therefore referred to as *retransmit timer*. Although both transmitters and receivers may employ many timers with different purposes in practice, the retransmit timer will be mostly thought of in this chapter.

12.3.4 Additional Mechanisms

The employment of all the mechanisms that were enlisted in Sect. 12.3.1 suffices to assure the correct operation of an automatic repeat request protocol. However, the efficiency of a protocol may be further increased if some additional mechanisms are used, although they are not indispensable for a logically correct operation. The two

---

[1]ACK is also one of ASCII control characters which can be used for the same purpose as is described in this Chapter.

kinds of additional mechanisms which will be explained in the next two paragraphs are most usually used.

A *negative acknowledgment*, usually also referred to as *reject*, is a control protocol data unit which can be used by a receiver to explicitly inform the transmitter that a specific information protocol message has been corrupted or lost. Most usually a receiver sends a reject after it has received an information message with a sequence number that is higher than the one it expects (which may indicate the loss of an intermediate message). In principle, a receiver could send a reject immediately after it has received a corrupted information protocol data unit and discarded it, but only in the case if multiplexing is not used in the channel. If multiplexing is used, the protocol control information of the corrupted message contains the indicator of the connection with which it is associated; if the message is corrupted, the connection indicator may also be corrupted, so the receiver cannot be sure with which connection the message is associated and hence cannot know to which transmitter it should send a reject.

We told already that one of the basic problems of information transfer is a limited knowledge about the distributed state of a communication system. An *enquiry* is a control protocol data unit used by one protocol entity to ask the other one in which state it actually is.

12.3.5 Specification of Automatic Repeat Request Protocols

In this chapter several automatic repeat request protocols will be described in more detail and also formally specified. The communication system that will be specified will be very similar to the one shown in Fig. 4.16. The system shown in Fig. 4.16 and the systems to be specified in this chapter will essentially differ in the channel interconnecting the two protocol entities: while the channel in Fig. 4.16 was unidirectional (simplex), the channels of all systems specified in this chapter will be bidirectional (duplex or half duplex), allowing data transfer in both directions (information messages in one direction and control messages in the other one). Taking the limitation explained in Sect. 12.3 into account, information protocol data units ipdu will be transferred only from the protocol entity modelled in the specifications by the process trans to the protocol entity modelled by the process recv, while acknowledgments ack will only be transferred in the opposite direction. Information protocol data units will transport both user information of the type Datatype (this data type will not be defined in this book) and sequence numbers of the type Seqnum (a set of all nonnegative integer numbers). On the other hand, acknowledgments will transport only a sequence number (only protocols using cumulative or individual acknowledgments will be specified). As was told already, the protocol entity which transmits information messages and receives acknowledgments will be called transmitting protocol entity or transmitter for short, while the other one, receiving information messages and transmitting acknowledgments, will be referred to as receiving protocol entity or receiver.

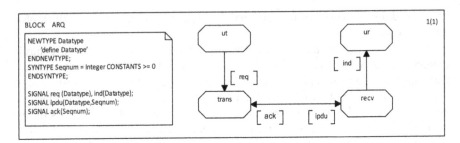

Fig. 12.3 Specification of communication system for data transfer according to ARQ protocols

We will formally specify only those protocols that employ solely basic mechanisms. Only protocol data units (signals) ipdu and ack will therefore be used in specifications. In Sect. 12.13, however, negative acknowledgments will also be used.

The specification of a communication system in the SDL language is shown in Fig. 12.3. Once more we emphasise that the data type Datatype is not specified in this figure (its specification is only informally indicated), while the data type Seqnum is specified as the set of all nonnegative integers. Processes ut and ur model the transmitting and the receiving user, respectively. The signals req and ind model the primitives data.request and data.indication, respectively, both of them bring a user message of the type Datatype with them. The signal ipdu models an information protocol data unit, bringing a user message of the type Datatype and a sequence number of the type Seqnum with it. The signal ack models a control protocol data unit (acknowledgment) that brings a sequence number of the type Seqnum with it.

The formal specifications of processes trans and recv for automatic repeat request protocols that will be given in the following sections will all refer to the specification of a communication system given in Fig. 12.3.

12.4   Stop-and-Wait Protocol

The simplest variant of an ARQ protocol is the *stop-and-wait protocol*. The most important advantage of this protocol is its remarkable simplicity which also means a simple implementation and consequently a low consumption of processor and memory resources in protocol entities. However, its deficiency is potentially low efficiency as the usage of the resources of a transfer channel is concerned. The stop-and-wait protocol is most heavily used in higher layers of protocol stacks, partly because of its simplicity, but also because the multiplexing is usually used in higher layers and the relative efficiency (see Sect. 6.4) of this protocol is not worse than the relative efficiency of many protocols that have a better protocol efficiency but are much more complex.

12.4.1 Basic Operation

In this section the stop-and-wait protocol using only the basic mechanisms, hence those that are indispensable to assure a logically correct operation, will be described. We will also assume that a single timer and cumulative acknowledgments are used. In reality, using more than one timer with the stop-and-wait protocol would be useless. There is also no essential difference between a cumulative and an individual acknowledgment with this protocol, except that the sequence number in a cumulative acknowledgment is by one greater than the sequence number of the last acknowledged message; block acknowledgments have no sense in this protocol. A transmitter and a receiver have a space to store a single message each in their transmit and receive buffer, respectively.

After a transmitter has received from the transmitting user a request for a user data transfer, it constructs an information protocol data unit containing the user message and the sequence number of the information message. A protocol may predetermine with which sequence number the counting of messages must begin after a connection setup (often, but not necessarily, this number is simply 0), or both protocol entities can determine the initial sequence number at the connection setup time. In any case, a receiver must know which the initial sequence number is, hence which is the first number it expects to receive. The protocol control information also contains other data besides the sequence number, such as the marking of the beginning of a message, the message type, the channel coding redundant bits and other information; for the purpose of our discussion in this chapter the sequence number is only relevant and will therefore be the only item to be contained in the control part of a message. The transmitter stores this message into the transmit buffer, transmits it into the channel, starts the retransmit timer and waits to receive an acknowledgment. The transmitted message must be kept in memory until the acknowledgment is received, as the transmitter cannot know if it will have to retransmit it after a timeout, were the acknowledgment not received. Only after it has received the acknowledgment it can delete it from the memory and write a new message into it. While the transmitter is waiting for the acknowledgment it may not transmit the next message as it has no room to store it (only one message can fit into the transmit buffer). After the transmitter has received the acknowledgment, it stops the timer and can run the same procedure to send the next information protocol message with the next sequence number, of course, when it receives the next request from the user or if that request is already present in the input waiting queue. If, however, the acknowledgment is not received, the timer expires and the transmitter retransmits the same information protocol data unit as previously (with the same sequence number of course), sets the timer again and waits for the acknowledgment.

A very important parameter of the protocol is the *timer expiration time* (hence the time difference between setting the timer and its expiration) which will be denoted as T_r here (as it is about the retransmit timer). The active timer guards an outstanding information protocol data unit and a timeout indicates that something went wrong either with the transmitted message or with its acknowledgment (one of them was

Fig. 12.4 Round trip time
in physical channel

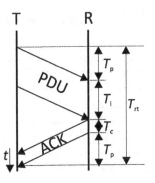

lost). Hence the timer may not expire before the transmitter receives the acknowledgment if neither the message nor the acknowledgment has been lost. The timer
expiration time must therefore be greater than the *round trip time* T_{rt},

$$T_r > T_{rt};$$
(12.4)

a round trip time is the time difference between the moment when the transmitter
transmits a message, and the moment when it receives its acknowledgment; it equals

$$T_{rt} = T_{di} + T_{dc},$$
(12.5)

where T_{di} and T_{dc} are the delays of the information protocol data unit and the control
protocol data unit (acknowledgment), respectively; here the time necessary for the
decision and processing in both protocol entities was neglected.

If both protocol entities are interconnected directly with a nonmultiplexed physical channel (the channel is used by a single connection), the round trip time and
consequently also the timer expiration time can be more precisely determined. As
can be seen from Fig. 12.4, the round trip time is in this case equal to the value given
in Eq. (12.6)

$$T_{rt} = T_i + T_c + 2 \cdot T_p = \frac{L_i}{R} + \frac{L_c}{R} + 2 \cdot \frac{D}{v},$$
(12.6)

where T_i, T_c, T_p, L_i, L_c, R, D and v are the information protocol data unit transmit
time, the control protocol data unit (acknowledgment) transmit time, the propagation
time of the electromagnetic signal in the physical channel, the information protocol
message length, the control protocol message (acknowledgment) length, the nominal
transfer rate of the physical channel, the physical channel length and the propagation
speed of the electromagnetic wave through the physical channel, respectively.
Equation (12.6) is in accordance with Eq. (12.5) if Eq. (6.18) is used for the delays
of an information protocol message and an acknowledgment. Again, the protocol
message processing times in both protocol entities were neglected.

If protocol entities are interconnected with a virtual channel, the round trip time cannot be calculated as it varies randomly; it can only be assessed statistically, based on the network operation in the past, or it can be measured.

The operation of a receiver is still simpler than the operation of a transmitter. After a connection has been set up, the receiver expects the information protocol data unit with the initial sequence number. Whenever it receives an expected information protocol message it passes the user information contained in it to the receiving user and then waits to receive the next information protocol message. If, however, it receives an unexpected message, it recognises it as a duplicate and discards it. In any case (even if it has received an unexpected message) it sends to the transmitter the acknowledgement specifying which sequence number it is currently expecting to receive. Namely, if a duplicate has been received, this means that the acknowledgment has been lost, the transmitter has timed out, has retransmitted the information message and is now waiting for the acknowledgment. If the receiver discovers an error in a received message, using the channel decoding procedure, it discards such a message; there is therefore no difference between a corrupted and a lost message.

12.4.2   Specification of Basic Operation of Stop-and-Wait Protocol

In this section the basic stop-and-wait protocol will be formally specified in the SDL language. The communication system was already specified in Sect. 12.3.5 in Fig. 12.3. Here the specifications of both protocol entities will be shown.

In Fig. 12.5 the specification of the transmitting protocol entity is shown.

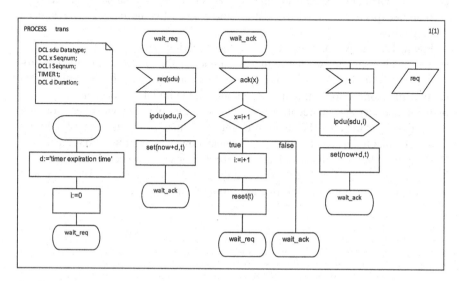

Fig. 12.5 Specification of transmitting protocol entity for stop-and-wait protocol

The variable i is used for counting information messages; during the process initialisation it is assigned the value 0, and thereafter it is increased by one after the reception of any acknowledgment; in a real communication system the variable i must be initialised every time a new connection is set up. A user message that the protocol entity gets from its user is stored into the variable sdu. Hence the variable sdu represents the transmit buffer. (We have said recently that the transmitter stores a protocol data unit into the transmit buffer, hence both the user and the control information; in this sense the transmit buffer here consists of the variables sdu and i together. It is the matter of protocol implementation whether the user and control information are stored together or separately in the transmit buffer; this is also the matter of the trade-off between the usage of memory, the amount of processing and the complexity of the protocol entity implementation—if they are stored separately, a protocol message must be reconstructed for each retransmission again.) The value brought by an acknowledgment is only temporarily stored into the variable x. The protocol entity employs the timer t. The timer expiration time depends on the channel properties and the length of protocol messages while the channel properties themselves depend on the mode of transmission and on the load imposed on the network. The timer expiration time will be discussed with more detail in Sect. 12.12; here it is only informally indicated.

The specification of the transmitter shown in Fig. 12.5 follows the description of the transmitter operation explained in Sect. 12.4.1. Two interesting and important details must also be emphasised here. If the transmitter receives a new request for the user data transfer from its user while waiting for the acknowledgement in the state wait_ack, it cannot immediately process this request, as was already told in Sect. 12.4.1; however, this request may not be discarded and lost, so the transmitter saves it in its input queue. If the transmitter receives an unexpected acknowledgment ($x \neq i + 1$) when in the state wait_ack it discards it. In the normal protocol operation such an unexpected acknowledgment cannot appear; it could however appear as a floating corpse.

Figure 12.6 shows the specification of the receiving protocol entity.

The receiver memorises the sequence number of the expected information protocol data unit in the variable i; the value of this variable must also be initialised at the time of a connection setup; it is important that the initial sequence numbers used by the transmitter and the receiver are equal. The variables sdu and x serve only as the temporary storage for the values brought by the message ipdu. Let us emphasise once more that the receiver always sends an acknowledgment after it has received a signal ipdu, whether it has received an expected or an unexpected message ipdu; however, only in the former case it forwards the user message to the receiving user and increases the counter of information protocol data units i.

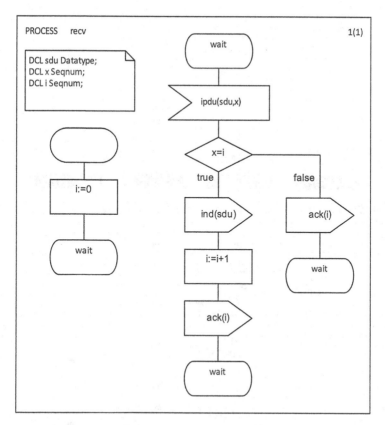

Fig. 12.6 Specification of receiving protocol entity for stop-and-wait protocol

12.4.3   A Few Characteristic Scenarios

In this section a few characteristic communication scenarios according to the stop-and-wait protocol, as specified in the previous section, will be shown. In all three cases two user messages, namely the characters 'a' and 'b', will be transferred from the user ut to the user ur. The sequence of the primitives ind('a'), ind('b') at the receiving side, corresponding to the sequence of primitives req('a'), req('b') at the transmitting side, clearly shows that the service was successfully implemented. As was already explained, the transmitting protocol entity transmits the second information protocol data unit only after it has received the acknowledgment of the first one, although it received both requests at the very beginning of the scenario. After the transmitter has transmitted the first message, it enters the state wait_ack; immediately then it receives still the next request, but saves it in the input queue, due to the save symbol used in the state wait_ack, and can process it only after it has returned into the state wait_req. A reader is encouraged to study the communication scenarios along with the protocol specification shown in the previous section.

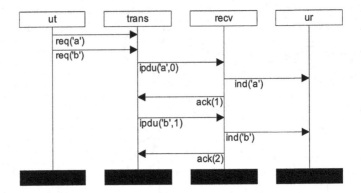

Fig. 12.7 Data transfer according to stop-and-wait protocol, no losses

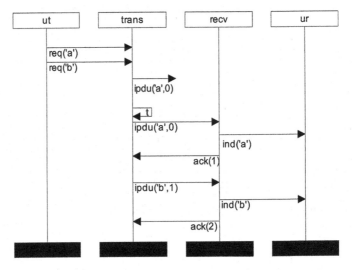

Fig. 12.8 Data transfer according to stop-and-wait protocol, information protocol message lost

The scenarios are shown as MSC diagrams which were generated with the simulator of the specification shown in the previous section.

In Fig. 12.7 the simplest case is shown where no message was lost.

In the scenario shown in Fig. 12.8 the information protocol data unit bringing the user message 'a' is lost. The transmitter times out and retransmits the message.

Figure 12.9 shows the scenario where the acknowledgment is lost. Also in this case the transmitter times out and retransmits the information protocol data unit; however, the receiver recognises the retransmitted message as a duplicate, because it has the same sequence number as the previously received message. The receiver discards the duplicate and does not pass it to the user, but sends an acknowledgment to the transmitter.

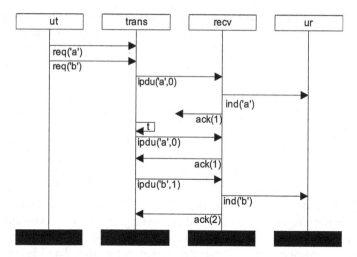

Fig. 12.9 Data transfer according to stop-and-wait protocol, acknowledgment lost

12.4.4 Efficiency

The primary purpose of automatic repeat request protocols is to provide a reliable transport of messages through a communication system in which some messages may be corrupted, lost or reordered. Hence the use of an ARQ protocol in a case where the two protocol entities are interconnected with a lossless channel has no sense. However, the protocol efficiency can most easily be calculated for the lossless case. If the channel is lossy, the efficiency is of course lower. This means that the protocol efficiency of a lossless system is the upper limit of the protocol efficiency of a lossy system; the protocol efficiency of a lossless system can also be used as an approximation of the protocol efficiency of a system with low losses. The protocol efficiency of a lossless system can therefore also be sometimes used to compare the efficiencies of some protocols providing a same service.

In this section the efficiency of the stop-and-wait protocol in the lossless case will be determined. The efficiency of this and other ARQ protocols in lossy systems will be treated in one of the following sections.

In Sect. 6.2 the efficiency of a protocol was defined as the maximum possible throughput, which means the throughput in the case where the transmitting protocol entity transmits a new information protocol data unit whenever it is allowed by the protocol to do so; in other words, this means that the waiting queue between the transmitting user and the transmitting protocol entity is never empty, there is always at least one request for a user data transfer in it.

If the channel is lossless and there is at least one request in the input queue at any time, the communication scenario, as shown in Fig. 12.4, repeats periodically with the period equal to the round trip time T_{rt}; during this period only the time needed to transmit an information protocol message T_i is profitably used. Because the traffic intensity can also be expressed as the ratio of the profitably used time over the total time, as was shown in Eq. (6.6), the protocol efficiency can be written as

$$\eta = \frac{T_i}{T_{rt}};$$
(12.7)

in the case where protocol entities are interconnected with a physical channel, this can be written as

$$\eta = \frac{T_i}{T_i + T_c + 2 \cdot T_p},$$
(12.8)

where T_i, T_c and T_p are the information protocol data unit transmit time, the acknowledgment transmit time and the electromagnetic wave propagation time, respectively, while T_{rt} is the round trip time. As can easily be seen in Eqs. (12.7) and (12.8), the efficiency of the stop-and-wait protocol is lower than 1 in any case. It can reach a value almost equal to 1 if the round trip time is only a little bit longer than the time needed to transmit an information protocol data unit; however, if the round trip time is much longer than the information message transmit time, the efficiency is low. If a physical channel is used, the following consideration is also possible. As acknowledgments are often much shorter than information messages (while an information message contains both user information and overhead, an acknowledgment contains only overhead), the time needed to transmit an acknowledgment can be neglected in Eq. (12.8) and one can write

$$\eta = \frac{1}{1 + \frac{2 \cdot T_p}{T_i}}.$$
(12.9)

If a channel is long and fast (i.e. it is physically long and its nominal transfer rate is high), hence the propagation time is long and the transmit time is short, the efficiency is low; if a channel is short and slow (i.e. its physical length is small and its nominal transfer rate it low), the protocol efficiency is high. A long and fast channel is frequently dubbed *long thick pipe* or *long fat pipe*, while a short and slow channel is often referred to as *short thin pipe*.

Regarding the fact that the efficiency in case of a long and fast channel is essentially different from the efficiency in case of a short and slow channel, a very important parameter of a channel with a strong impact on the protocol efficiency is the *bandwidth-delay product*. The bandwidth-delay product will be denoted here as RD; it is defined as the product of the nominal transmission rate and the round trip time, as can be seen in Eq. (12.10):

$$RD = R \cdot T_{rt}.$$
(12.10)

If Eq. (6.4) is inserted into Eq. (12.7), the stop-and-wait protocol efficiency can also be written in the form

$$\eta = \frac{L_i}{RD}.\qquad(12.11)$$

The efficiency which is experienced by users is of course still lower than the protocol efficiency by the overhead factor, defined in Eq. (6.10).

If the multiplexing is used in the channel, the relative protocol efficiency, as defined in Eq. (6.14) in Sect. 6.4, is more relevant than the protocol efficiency. Because the offered load and the throughput are equal in case of a lossless channel, the relative efficiency of the stop-and-wait protocol is 1 in such the case; hence, in systems employing multiplexed channels the stop-and-wait protocol is not worse than many more sophisticated protocols we will explain in the forthcoming sections; however, it is much simpler.

12.5   Continuous Protocols

Equations (12.7) and (12.8) show that the stop-and-wait protocol uses the transfer resources potentially very inefficiently, especially in the case of a long and fast channel. The reason for this inefficiency can easily be seen in Fig. 12.4: after a transmitter has transmitted an information protocol data unit it waits for the acknowledgment; during this time it does nothing and so loses the precious time as a communication resource; this is the more evident, the longer is the round trip time relative to the information message transmit time.

The efficiency problem can be solved so that a transmitter is allowed to transmit several information protocol data units one after another so that a new message is transmitted before the previous one is acknowledged. A receiver must of course send acknowledgments for correctly received messages; however, the acknowledgments are received at the transmitting side with a delay, so there may be several outstanding information protocol data units in the transmit memory buffer at the same time. Hence the transmit buffer must be large enough to store several messages concurrently. The transmitter may remove an information protocol message from the transmit buffer only after it has received the acknowledgment of that message.

An automatic repeat request protocol which allows the transmitter to transmit several information messages before they have been acknowledged is called *continuous protocol*.

In Fig. 12.10 the timing diagram of a transfer of a sequence of seven information protocol data units from the transmitter T to the receiver R and cumulative acknowledgments in the opposite direction according to the continuous protocol is shown. Transmitted information protocol messages are labelled with the syntax IPDU(x), where x indicates the sequence number of a message, while the acknowledgments are labelled with the syntax ACK(y), where y indicates the sequence number the

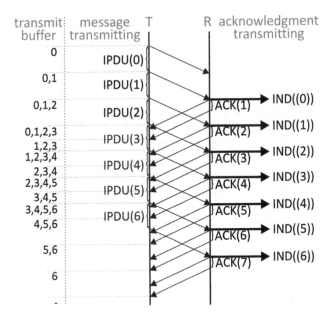

Fig. 12.10 Transfer of messages with continuous protocol

receiver is expecting to receive next. The goal of information messages is of course to bring also user messages with them; however, as we are mainly interested in the transfer control here, a simplified syntax is used, without quoting user information. The sequence numbers of the already transmitted, but not yet acknowledged messages in the transmit buffer are shown on the left edge of the figure. The primitives of the indication type which are used by the receiver to forward user messages to the receiving user are shown on the right edge of the figure. The syntax $IND((z))$ means that the user message received with the protocol message $IPDU(z)$ is being forwarded; the sequence number itself (z in this case) is a part of the protocol control information and is therefore not forwarded to the user. Primitives of the request type are not shown in the figure; they are assumed to be generated frequently enough, so a request primitive is present in the input queue whenever the transmitting protocol entity can process it (hence the input queue is assumed to never be empty).

In this section we have argued the point that a continuous protocol is more efficient than the stop-and-wait protocol. This may or may not be true. For sure this is true if the only communication resources that are of interest are the time and the transfer rate, which is evident if one considers the formulas (6.6), (6.8), (6.9) and (12.7); in Fig. 12.10 one can see that the efficiency of a continuous protocol can even be 1 (100%), of course, if the channel is lossless. However, an appropriate price must be payed for that! The transmitting protocol entity of a continuous protocol uses more memory for the storage of outstanding messages as the stop-and-wait protocol; furthermore, the memory management is more complex with a continuous protocol, so protocol entities also use more processing resources than is the case with the stop-and-wait protocol.

Besides a higher usage of memory and processing resources, a continuous protocol poses another requirement to protocol designers. While the protocol entities of the stop-and-wait protocol can be connected with a half-duplex channel, it is easy to see, both from the discussion in this section and from Fig. 12.4 and Fig. 12.10, that a continuous protocol requires a full duplex channel.

Up to now, only the transmit buffer of continuous protocols was considered. Of course, the receiver of a continuous protocol also can have a receive buffer capable to memorise more than one message. This is especially useful if information protocol messages do not reach the receiver in the correct order, so the receiver must reorder them before passing their user contents to the receiving user.

12.6 Sliding Window Protocols

When discussing continuous protocols we did not consider a basic limitation which must be taken into account when designing a physically realisable system that must be capable to operate in the real world. The amount of memory a system can use is always limited. A continuous protocol must therefore restrict the memory that can be used to implement the transmit buffer; the number of already transmitted but not yet acknowledged (outstanding) messages that can concurrently be stored in the buffer must therefore be restricted; the amount of memory needed to implement the receive buffer must also be restricted, and consequently also the number of messages in the receive buffer.

The number of messages that pass through the transmit and the receive buffer of the transmitter and the receiver, respectively, during the lifetime of a connection, is not limited. However, the number of messages that can be stored in the one or the other buffer at the same time must be limited; a protocol must therefore manage both buffers so that the memories they are implemented with can be reused. The transfer of protocol data units between both protocol entities must be adjusted to the mode of transmit and receive buffer management.

A continuous protocol which uses a transmit and a receive buffer of a limited size is called *sliding window protocol*.

12.6.1 Transmit Window and Receive Window

The *transmit window* is a succession of sequence numbers of information protocol data units which may at some moment be stored in the transmit buffer of a transmitting protocol entity. The sequence number that is stored at the beginning of the transmit window is the lowest unacknowledged sequence number. The *transmit window width* is the largest allowed number of outstanding messages that

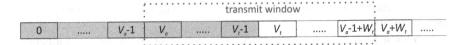

Fig. 12.11 Transmit window

may be stored in the transmit buffer at the same time; in other words, the transmit window width is the largest number of consecutive information protocol data units a transmitter may transmit, while having not received their acknowledgments. In this book the transmit window width will be denoted with the symbol W_t. The transmit window can be viewed as a symbolic image of the transmit buffer where the sequence numbers of messages are placed instead of the messages they denote. If in some moment the number of outstanding messages in the transmit buffer equals the transmit window width, the transmit window is said to be full; in such a case the transmitter may not transmit new information messages, it only may retransmit messages which are already in the window.

In Fig. 12.11 the position of a transmit window on the number line is shown for the moment when the transmitter has already transmitted V_t information protocol data units (with the sequence numbers from 0 to $V_t - 1$, inclusive); the sequence numbers of the already transmitted messages are shown on the grey background in the figure. The reception of V_a information protocol data units (from 0 to $V_a - 1$, inclusive) has already been acknowledged. Hence the transmit window begins with the sequence number V_a and ends with the sequence number $V_a - 1 + W_t$, hence it contains W_t sequence numbers.

With respect to the fact that the sequence number of the outstanding information message with the lowest number is always positioned at the beginning of the transmit window, the transmit window is shifted towards higher sequence numbers whenever the transmitter receives a new acknowledgment of the message(s) placed at the beginning of the transmit window. Hence, if the transmitter receives a cumulative acknowledgment ACK(a), acknowledging the reception of all information messages up to the sequence number $a - 1$ inclusive, the transmit window is shifted so that it begins with the sequence number $V_a = a$. The window may be shifted only in the direction of higher sequence numbers; if the transmitter receives an acknowledgment with the sequence number $a \le V_a$, the window is not shifted as the received acknowledgment is not a new one.

If the transmitter wants to transmit a new information protocol data unit, it transmits the message with the sequence number V_t, but may do so only if the condition

$$V_t < V_a + W_t \tag{12.12}$$

is fulfilled; if the condition (12.12) is not fulfilled, it means that the transmit window is full and the transmitter is not allowed to transmit new information messages, as it has no room to store them.

Fig. 12.12 Transfer of messages according to sliding window protocol with transmit window width 3

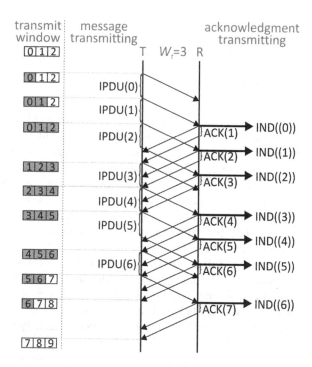

We told already that the operation of a protocol entity must begin in a well-defined state; this must of course also be true for the position of the transmit window. If the protocol determines that the counting of sequence numbers must begin with the initial sequence number 0 (which was also assumed in Fig. 12.11), the values of variables V_a and V_t are both initialised to 0 during the protocol entity initialisation.

Similarly as Fig. 12.10, Fig. 12.12 also shows the timing diagram of the transfer of a sequence of seven information protocol data units from the transmitter T to the receiver R and the corresponding cumulative acknowledgments in the opposite direction; however, in Fig. 12.12 the transmit window width $W_t = 3$ is assumed. The same syntax is used to label protocol messages and primitives as in Fig. 12.10. On the left edge of the figure the contents of the transmit window are shown each time the window is moved or its contents are modified. The sequence numbers of outstanding messages are shown on the grey background. In the upper left corner of the figure the initialisation of the transmit window is shown. One can see in Fig. 12.12 that the transmitter is not allowed to continue the transmission of information messages after it has finished the transmission of the messages IPDU (2) and IPDU(5), respectively; in both cases the window is full and the condition (12.12) is not fulfilled, so the transmitter has to wait for a new acknowledgment when the window is shifted towards higher sequence numbers and the condition (12.12) is again fulfilled for the next sequence number.

The *receive window* is a succession of sequence numbers of the information protocol data units which may at some moment be stored in the receive buffer of a

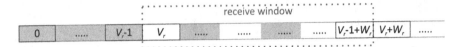

Fig. 12.13 Receive window

receiving protocol entity. The sequence number that is stored at the beginning of the receive window is the lowest sequence number which has not yet been received; this means that the receiver cannot pass user messages contained in information protocol data units from that sequence number on to the receiving user. The *receive window width* is the largest allowed number of received messages that may be stored in the receive buffer at the same time. In this book the receive window width will be denoted with the symbol W_r. The receive window can also be viewed as a symbolic image of the receive buffer as it contains the sequence numbers of received messages instead of the received messages themselves.

Figure 12.13 illustrates the position of the receive window on the number line in the moment when the receiver has already received V_r information protocol data units from the sequence number 0 to the sequence number $V_r - 1$ inclusive; hence the receive window begins with the sequence number V_r and ends with the sequence number $V_r - 1 + W_r$. The sequence numbers of already received messages are shown on the grey background. If the receiver does not receive messages in the same order as they were transmitted, there are sequence numbers that have already been received and those that have not yet been received intermixed in the receive window; in any case, however, the message with the sequence number V_r has not yet been received. If the receiver receives an information protocol data unit with the sequence number s, hence IPDU(s), it stores it into its receive buffer only if the condition

$$V_r - 1 < s < V_r + W_r \tag{12.13}$$

holds; if the condition (12.13) is not valid, the message must be discarded, as there is no room in the receive buffer to store it. In the case $s < V_r$, a message has been received which had previously already been received and also acknowledged.

If the receiver receives the information protocol data unit with the sequence number equal to V_r, it passes the user information contained therein to the receiving user and shifts the receive window by one sequence number towards higher numbers (it increases the value of V_r by 1). If the sequence number V_r is immediately followed by other sequence numbers that have also already been received, the receive window can be shifted by more numbers, so that there is again a not yet received number at the beginning of the window. The receive window also may be shifted only towards higher sequence numbers.

At the time of a new connection setup the initial position of the receive window must also be determined, and hence the value V_r must also be initialised. At the time of a connection setup, the variables V_t and V_a of the transmitter and V_r of the receiver must all have the same value (the first sequence number to be transmitted by the transmitter must of course be equal to the first sequence number expected by the receiver). Hence, if the protocol requires the sequence numbering to begin with the value 0, the receiver also must wait for the sequence number 0 after a connection has been set up (this was also assumed in Fig. 12.13); the variables V_t and V_a of the transmitter and V_r of the receiver must therefore all be initialised to 0 in such the case.

In this section we have shown how the transmit and receive windows slide along the sequence number line; this is the reason why the continuous protocols with limited sizes of transmit and receive buffers are called *sliding window protocols*.

12.6.2   Counting Protocol Data Units

The transmit and the receive protocol entities count information protocol data units. Sequence numbers are processed, e.g. using Eqs. (12.12) and (12.13), they are stored in memory, they are also transferred through a network as a part of the protocol control information.

In the world of digital systems only a limited number of different values can always be used; the maximum number of different values is 2^n if n is the length of the word containing a binary-encoded value. This means that an ARQ protocol, too, can use only a limited number of sequence numbers $M = 2^m$, where m is the number of bits used to record a sequence number. This number depends on the length of the word used by computers on which protocols are implemented, and, most importantly, on the number of bits used to record a sequence number within a protocol data unit, hence in the transfer syntax of protocol data units; one must be aware that a sequence number is a part of the protocol control information within a protocol data unit, hence a part of a message overhead, and must therefore be transferred through a channel; in this way it impacts the efficiency of data transfer between the users of a system.

As we have already told, the number of information protocol data units that are transferred between protocol entities within the frame of a connection is not limited; however, the number of different sequence numbers is limited, due to the restriction mentioned in the previous paragraph. Hence the procedure called *counting modulo M* must be used to count sequence numbers. Counting modulo M is a cyclic counting with the period M: after M values have been counted the counting begins from the beginning again: 0, 1, 2, ..., $M - 2$, $M - 1$, 0, 1, ... To allow for a simpler implementation in computers the modulus which is a power of 2 is most usually used: if m bits are used to store a sequence number, the counting modulo $M = 2^m$ is used.

The purpose of a sequence number is that it unambiguously identifies an information protocol data unit within the whole communication system (consisting of the transmitter, the channel and the receiver). That means that any protocol entity must at any time know without any doubt which protocol data unit a sequence number identifies. The least favourable position of both windows, as far as sequence numbers are concerned, is shown in Fig. 12.14; the transmit and the receive window

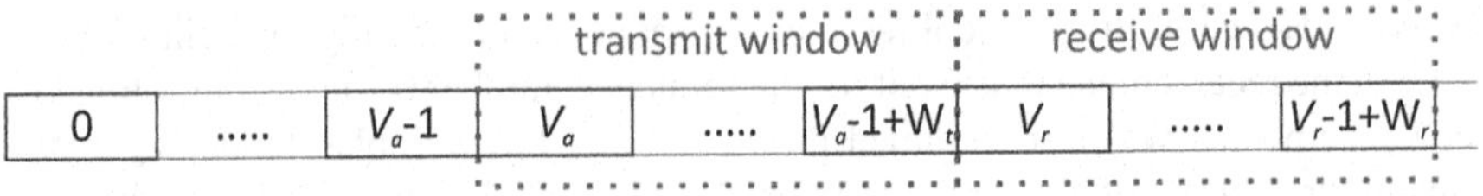

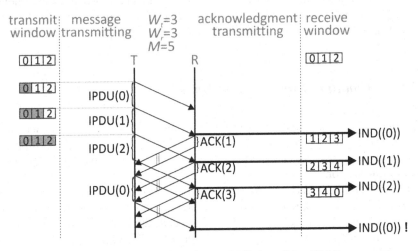

Fig. 12.14 Most unfavourable position of transmit and receive window

Fig. 12.15 Example of incorrect operation due to unfulfilled condition (12.14)

are positioned on the number line side by side, without overlapping. Such a situation can occur if the transmitter transmits a whole window of information protocol data units with sequence numbers from V_a to inclusive $V_a - 1 + W_t$ (hence its window becomes full); the receiver successfully receives all these messages and acknowledges them, but, unfortunately, all the acknowledgments are lost. Because the transmitter does not receive any acknowledgments it may not shift its window, because maybe it will have to retransmit them; it also may not transmit new messages because its window is full. The receiver which has received all those messages, up to and including the one with the sequence number $V_a - 1 + W_t$, has moved its window and is prepared to receive new massages with sequence numbers from $V_r = V_a + W_t$ to $V_a - 1 + W_t + W_r$, inclusive. Evidently, both protocol entities must distinguish all these messages; hence they must distinguish $W_t + W_r$ messages.

From what has just been told it follows that both protocol entities must distinguish $W_t + W_r$ information protocol data units; hence sequence numbers must be counted modulo M, where

$$M \geq W_t + W_r. \qquad (12.14)$$

In Fig. 12.15 an example of incorrect operation of the sliding window protocol is presented, showing that the protocol is not logically correct if the condition (12.14) is not fulfilled. The transmit and the receive window widths are $W_t = W_r = 3$; hence, the counting should be done modulo 6 or more to fulfil the condition (12.14). However, we use counting modulo 5 in our example, which is not correct according to the condition (12.14)! The transmitter transmits three messages and thus fills its window; the receiver receives all of them, sends acknowledgments and shifts its window, so that the window contains the sequence numbers 3, 4 and 0 after the reception of IPDU(2) (if counting modulo 5 is used, 4 is followed by 0!). When the transmitter times out it assumes that the message IPDU(0) has been lost and retransmits it. The receiver receives the retransmitted message; as the sequence number 0 is then inside its window, the receiver forwards the user message contained therein to the user, not being aware that it has received the same message for the second time. Thus, the user receives a duplicate which means a logical error in protocol operation (the reception of the duplicate by the user is emphasised by an exclamation mark in the figure). If in this example counting were done modulo 6, the receive window would contain sequence numbers 3, 4 and 5 after the receiver has received the message IPDU(2), so the receiver would recognise the retransmitted IPDU(0) as a duplicate (number 0 would not be in its window) and would simply discard it—the operation would be correct in this case.

For the sake of a more simple presentation of protocols an infinite counting modulus will be assumed in the examples and specifications in this book; in real-world implementations, however, the modulus is limited and the condition (12.14) must be taken into account. In practice, all calculations and evaluations concerning sequence numbers must be done modulo M.

Unfortunately, even if the condition (12.14) is fulfilled, two different messages with the same sequence number may happen to be present in the system at the same time if one of them is a floating corpse. This means that a message was delayed for an abnormally long time, the transmitter timed out and retransmitted the supposedly lost message; the receiver received the retransmitted message and acknowledged it, so both protocol entities moved their windows and began reusing sequence numbers. Then the old message appeared as a floating corpse when its sequence number was already reused for another message. Such a floating corpse can of course produce a confusion and incorrect operation of the protocol, so many protocols try to prevent such situations if there is a possibility of floating corpses in the communication channel.

Two countermeasures are available to prevent the harmful effects of floating corpses. Firstly, the maximum lifetime during which messages may stay in the network (often referred to as *time to live*) must be limited. If a packet stays in the system for a longer time than is its predetermined time to live, the network may simply discard it. The time during which a packet stays in the network can be measured by means of *timestamps* which are added to the protocol control information at transmit time; unfortunately, this method does not work well if the clocks of different network elements are not well synchronised and the time mismatch of different clocks is comparable with or even greater than the transfer delays of packets. A simpler method is *hop counting* where a hop is a packet transfer step between two neighbouring network elements; hence a hop count is the number of hops made by a packet in the network. When the maximum lifetime of packets in the network is known or assessed, one must care that sequence numbers cannot be reused during the lifetime of packets, neither within the frame of a single connection nor in successive connections. If sequence numbers in successive connections are not to be reused within the duration of the packet lifetime, the initialisations of sequence numbers in different connections cannot be independent.

12.6.3   Acknowledgments and Timers with Continuous Protocols

In Sect. 12.3.2 acknowledgments were discussed; three kinds of acknowledgments were enlisted: individual, cumulative and block acknowledgments. While block acknowledgments have no sense and there is no essential difference between individual and cumulative acknowledgments with a stop-and-wait protocol, all three kinds of acknowledgments can be used with sliding window protocols. In the past, cumulative acknowledgments were by far most used, as they allow for a simple protocol implementation; furthermore, if cumulative acknowledgments are used, it is not necessary to acknowledge the reception of every information protocol data unit separately, as a single acknowledgment can be used to acknowledge several successive information messages at the same time, thus imposing a lower load upon the communication network.

In Sect. 12.3.3 the use of timers was discussed; we told that a transmitting protocol entity can use one or more timers. If a single timer is used, it must be active whenever the transmitter has at least one outstanding information protocol data unit in its transmit buffer. If several timers are used, a separate timer is activated for each information message transmitted; this means that the number of timers that are needed must be at least equal to the transmit window width, as this is also the maximum number of already transmitted but not yet acknowledged (i.e. outstanding) messages.

Not every combination of acknowledgments and timers has sense.

With the stop-and-wait protocol the use of a single timer only has sense; the acknowledgments are essentially individual, though they can also be considered cumulative.

The use of cumulative acknowledgments and a single timer go very well together along with a continuous protocol. Whenever the transmitter transmits an information protocol data unit it starts the timer if it is not already active; consequently, the timer is active whenever at least one outstanding information message is in the transmit buffer. If the transmitter receives a new acknowledgment allowing it to shift the transmit window towards higher sequence numbers, it stops the timer and immediately restarts it if there is still at least one outstanding message in the transmit buffer. The reason for stopping the timer after having received a new acknowledgment is that this means that the message which has been at the beginning of the transmit buffer up to now was successfully received (this may also be true for more messages that follow it, depending on the cumulative acknowledgment received). However, if the timer expires, this means that the message at the beginning of the transmit buffer was probably lost and must be retransmitted (indeed, it is also possible that the acknowledgment was lost).

The use of more timers only has sense if individual acknowledgments are also used, as each of the individual timers guards an individual outstanding information protocol data unit.

12.6.4 Specification of Basic Sliding Window Protocol with Cumulative Acknowledgments and Single Timer, Lossless Channel Case

In this section the sliding window protocol using cumulative acknowledgments and a single timer will be described and also formally specified in the SDL language. The channel will be assumed to be lossless which means that the timer will never expire; in this section we will therefore not tell how the transmitting protocol entity reacts to a timer expiration. Although the timer will be declared in the formal specification and it will also be managed as specified by the protocol, the declaration and the use of the timer will be redundant in this section as a timer is really not needed if the channel is lossless. Of course, such a specification seems to be senseless, indeed, as not only a timer but also an automatic repeat request protocol is not needed if there are no losses in the channel. We must, however, immediately tell that there are several variants of sliding window protocols which differ primarily in how the transmitter reacts to a timer expiration, which means the loss of one or more messages. These variants will be discussed in the sections that follow where the specification given in this section will be supplemented with the reaction of the transmitter to a timer expiration. According to the author's opinion this kind of discussion will be the easiest way to understand this topic which indeed may not be most easy.

In this section we will specify the functionalities of the transmitting and the receiving protocol entities of the system already specified in Fig. 12.3. Hence the specification given in this section will be a supplement to the specification shown in Fig. 12.3. In the sections that follow this specification will be further supplemented by specifying the reaction of the transmitter to a timeout.

In Figs. 12.16–12.18 the specification of a transmitting protocol entity is shown.

In Fig. 12.16 the declarations of the variables and the timer as well as the initialisation of the transmitting protocol entity are shown. The transmit window width is informally »nondefined« in this figure; in real world this value may either be specified by the protocol, or it may be determined at the time of a connection setup (one must be aware that the window width impacts the consumption of memory, but also, as will soon be explained, the protocol efficiency). As will be explained in the next section, the transmitting protocol entity must also know the receive window

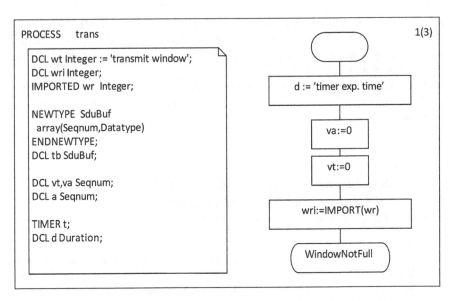

Fig. 12.16 Specification of transmitting protocol entity (declarations and initialisation)

width of the receiver; in our specification this value is therefore »imported« from the receiving protocol entity process; to this end, the SDL constructs IMPORTED and IMPORT are used in the process trans, and the SDL constructs EXPORTED and EXPORT are employed in the process recv. In real life, however, the receiver communicates this value to the transmitter at the connection setup time, if it is not already predefined by the standard. Because of the restriction imposed by the rules of the SDL language the receive window width is named wri in the process trans, while in the process recv its name is wr. The transmit buffer is the array tb and has an infinite length (as specified in the definition of the data type Seqnum in Fig. 12.3, an index of this array can be any nonnegative integer); in practice, however, this buffer has a finite size, as sequence numbers are counted and processed using the modulo arithmetic. The value of the variable va indicates the beginning of the transmit window, while the value of the variable vt determines the sequence number of the information protocol data unit to be next transmitted; before a data transfer can begin both these values must be assigned the same value, 0 in our case (once more, we must emphasise that normally the values of these two variables are initialised at a connection setup time, as several connections can be set up during the lifetime of a protocol entity, and each connection must begin with a well-defined initial state—in our specification, however, these values are initialised during the initialisation of the process). The variable a only serves as a temporary store for received acknowledgments. In our specification we admit various channels to interconnect both protocol entities; the timer expiration time is therefore informally »nondefined«. In reality the timer expiration time depends on various circumstances; more about this topic will be told in Sect. 12.12.

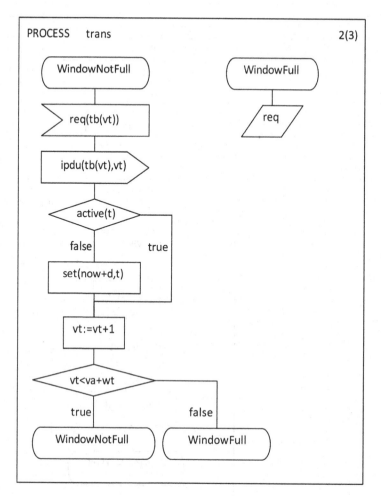

Fig. 12.17 Specification of transmitting protocol entity (transmitting information protocol messages)

In Fig. 12.17 the transmitting of information protocol data units is specified. Here the difference between the stop-and-wait protocol and the sliding window protocol must be emphasised. While a stop-and-wait protocol transmitter may transmit a new information message only in the state wait_req if it receives a request, but after having transmitted it only waits for the acknowledgement and may not transmit new messages (it is in the state wait_ack), a sliding window protocol transmitter may transmit only when its window is not full (it is in the state WindowNotFull in Fig. 12.17), but may not transmit new messages when its window is full (it is in the state WindowFull in Fig. 12.17). Therefore the transmitter must check after each transmission whether its window is full or not, according to the condition (12.12). If it receives from its user a request for a data transfer when in the state WindowNotFull

Fig. 12.18 Specification of
transmitting protocol entity
(receiving
acknowledgments)

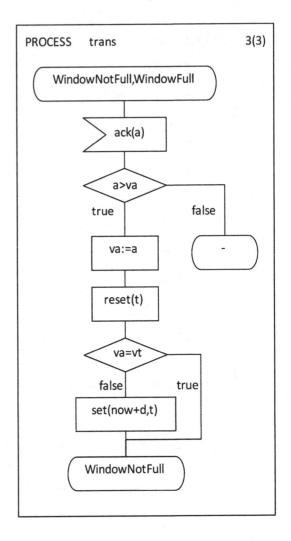

Fig. 12.18 Specification of transmitting protocol entity (receiving acknowledgments)

it stores the user message into the appropriate location in the transmit buffer, transmits the information message ipdu (which contains the user message and the sequence number), starts the timer (if it is not already running, as was explained in Sect. 12.6.3), and increments the variable vt (so the next information message will have a higher sequence number). If, however, it receives a request when in the state WindowFull, it saves the request in the input queue to be processed later.

A transmitter clearly cannot transmit more than one message into the physical channel at the same time; it only can begin the transmission of a message after the transmission of the previous message has been terminated. In our model of the communication system the protocol entity of the lower layer is assumed to provide for the transmission of a message and its transfer through the physical channel; in our specification the lower-layer protocol entity is included in the process of the channel which is not explicitly shown in Fig. 12.3, but was explicitly modelled by the author to allow him to simulate the system and thus verify the correctness of the specification. Hence the process trans (which models the transmitting protocol entity) transmits a message by simply adding it to the input queue of the channel; in this way it can »transmit« several messages one immediately after another. This mode of transmission is usual in the higher layers of a protocol stack. If this mode is used, a transmitter, after it has transmitted a message, cannot control the message any more, it cannot cancel or disrupt its transmission. Using this technique, the specification of the transmitting protocol entity was considerably simplified, as only those mechanisms were included into the specification which are executed by the protocol entity of the specified layer, but not the transfer through the physical channel. However, one must also bear in mind that in a real system the timer expiration time must be adjusted to the mode of a message transfer through the channel. In our specification the definition of the timer expiration time is only informally indicated.

Figure 12.18 specifies how the transmitter receives acknowledgments; it does this independently from information message transmissions. The transmitter can therefore receive an acknowledgment regardless of its state (WindowFull or WindowNotFull); the activities of the transmitter do not depend on whether the acknowledgment has been received in the state WindowNotFull or WindowFull. The transmitter always verifies if it has received a new acknowledgment ($a > va$) and, if this is true, appropriately shifts the transmit window, stops the timer and restarts it if there are still outstanding messages in the transmit buffer ($vt > a$). After the window has been moved towards higher sequence numbers it is of course not full, were it full or not before the reception of the acknowledgment. If the received acknowledgment is not a new one ($a \leq va$), the transmitter discards the acknowledgment and stays in the same state.

Similarly as in the case of the stop-and-wait protocol, the operation of the receiver is simpler than the operation of the transmitter also in the case of the sliding window protocol. The specification of the functionality of the receiving protocol entity is shown in Figs. 12.19 and 12.20.

In Fig. 12.19 the necessary declarations are shown and the initialisation of the receiving protocol entity is specified. As the receive window width wr is concerned, the same comment can be given as for the case of the transmitting protocol entity. The array rb represents the receive buffer (in this specification it also has an infinite length, as was already the case in the specification of the transmit buffer). The array rcvd is associated to the receive buffer rb; the array rcvd indicates which information protocol data units have already been received and stored into the receive buffer (this is necessary because the receiver may also receive information messages out of order); during the initialisation all the elements of the array rcvd must be assigned the initial value false. The value of the variable vr indicates the beginning of the receive window; its initial value is 0 in our case (in any case it must be equal to the initial value of the variable vr in the transmitter). The variables sdu and x only serve as the

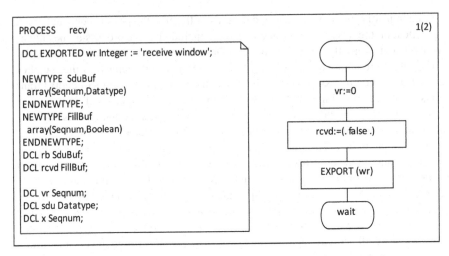

Fig. 12.19 Specification of receiving protocol entity (declarations and initialisation)

temporary storage for the user message and the sequence number, respectively, received with the information protocol data unit ipdu.

Let us once more emphasise here that the SDL constructs IMPORTED and IM PORT in the process trans as well as EXPORTED and EXPORT in the process recv only formally serve to inform the transmitter about the value of the receive window width, which in the real life is either predetermined by the standard or is determined at the time of a connection setup.

In Fig. 12.20 the receiving of information messages and transmitting of acknowledgments is specified. The receiver may only be in one state which is named wait in our specification. If it receives an information protocol data unit with a sequence number which is currently inside the receive window (vr $\le$ x $<$ vr + wr), the user message is stored into the appropriate location in the receive buffer and this fact is also recorded in the array rcvd; then all the user messages which are already successively present (without holes) at the beginning of the receive buffer (the array rcvd helps to find them) are forwarded to the user with the indication primitive, and the receive window is shifted towards the higher sequence numbers so that the sequence number at the new beginning of the window has not yet been received (the receiver looks for the first element of the array rcvd having the value false). Has it received a sequence number inside or outside the receive window, the receiver sends to the transmitter the acknowledgment telling which sequence number it is expecting to receive next (of course, this is the sequence number at the beginning of the receive window).

Fig. 12.20 Specification of receiving protocol entity (receiving information messages and transmitting acknowledgments)

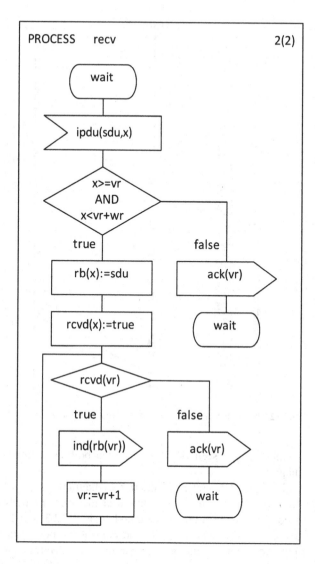

12.6.5   *Characteristic Scenario*

If a continuous protocol is used, several protocol data units can concurrently be present in the physical channel, being transferred in both directions simultaneously; a timing diagram is therefore more appropriate to illustrate a communication scenario than a time-sequence diagram or an MSC diagram. A scenario can of course also be illustrated as an MSC diagram if the channel process is also shown, in addition to the processes modelling both protocol entities. In this section an example

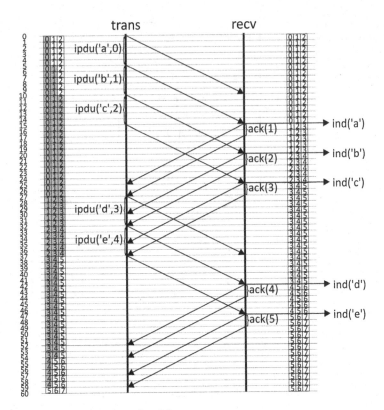

Fig. 12.21 Timing diagram of transfer of five user messages

communication scenario will be shown both as a timing diagram and as an MSC diagram.

In Figs. 12.21 and 12.22 the timing diagram and the MSC diagram, respectively, are shown for the transfer of five user messages (namely characters from 'a' to 'e') according to the protocol specified in Fig. 12.3 and Fig. 12.16–12.20, with the transmit times of information and control protocol data units equal 5 and 2 time units, respectively, the propagation delay of electromagnetic signal in the physical channel equal 10 time units, and the transmit and receive window widths both equal 3 (as a matter of fact, the receive window width is not at all important if messages are transferred through a lossless channel which also preserves the order of messages). The timing diagram shown in Fig. 12.21 was drawn by hand while the MSC diagram shown in Fig. 12.22 is the output result of the simulation of this scenario. In the model specified in Fig. 12.3 and Figs. 12.16–12.20 the transmission of protocol data units and the transfer of them through the physical channel were not explicitly shown, as both these procedures were modelled within the process specifying the channel which also was not shown in this book (the SDL language is meant and suitable for the specification of protocol entities rather than physical channels, and besides this the operation of protocol entities is of primary interest in this book).

Fig. 12.22 MSC diagram
of transfer of five user
messages

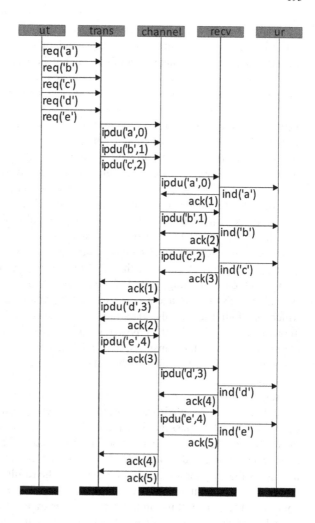

Furthermore, in both diagrams all the requests for user data transfer are assumed to
be present in the input waiting queue of the transmitting protocol entity before the
actual transfer starts; this fact is explicitly shown only in Fig. 12.22. In both figures
the forwarding of user messages to the receiving user using indication primitives is
also shown. While the timing diagram is drawn on the correct timing scale (time
units are shown as horizontal lines labelled with values on the left edge), the MSC
diagram shows only the sequence of events. Events correspond to the beginnings of
transmissions and the ends of receptions of protocol messages into and out of the
channel, respectively, in the timing diagram in Fig. 12.21, and the handovers of
protocol messages into the channel and out of it in the MSC diagram in Fig. 12.22.
One can easily see that both diagrams are equivalent, as the contents are concerned.

In the timing diagram shown in Fig. 12.21 the sliding of the transmit and the
receive window is also shown on the left and on the right side, respectively. The

outstanding sequence numbers (already transmitted but not yet acknowledged) in the transmit window are drawn on grey background. In the receive window of Fig. 12.21 the already received sequence numbers cannot be seen; as soon as an expected information protocol data unit is received by the receiver (its sequence number is at the beginning of the receive window), the user content of this message is forwarded to the user, the acknowledgment is sent and the window is shifted. Hence, as long as the receiver is receiving information messages in the correct order (in the same order in which they have been transmitted), the receive window has no essential role, the receiver needs the room only to store a single message. In the sections that follow, however, those messages that have already been received by the receiver but have not yet been forwarded to the user will be shown on grey background.

Protocol data units can be transmitted into the channel in two different ways. In the first mode, which is normally used in the protocol layers that are higher than the data-link layer and which also was assumed in our specifications in the SDL language, a protocol entity transmits a message simply by adding it into the input queue of the protocol entity of the adjacent lower layer of the protocol stack; by doing this the transmitting entity loses the ability to cancel or interrupt the transmission, as soon as the message was passed into the waiting queue. In the second mode, which is often used in the data-link layer, especially if the data-link layer is implemented jointly with the physical layer, the protocol entity can keep the control of the message until the transmission is finished; consequently, the entity can, after it has already decided to transmit the message, »change its mind« due to changed conditions, and cancel or interrupt the transmission. In timing diagrams of this book messages will be shown as if they are directly transmitted into the channel (not through a queue), but their transmissions will never be interrupted (once an entity begins to transmit a message, it will continue to transmit it, until the transmission is finished). The author's goal in choosing this mode of presentation was to make diagrams as simple and as understandable as possible, and, at the same time, produce the results that are not too much different from the results which are obtained when messages are transmitted into waiting queues of lower layers. In Fig. 12.21 the control of the transmitting protocol entity over transmitted messages can be seen from the fact that the sequence number 1 appears on the grey background only when the message with this sequence number begins to be actually transmitted.

Although in the timing diagram of Fig. 12.21 the events of the reception of the information message, the forwarding of the user message to the user and the transmission of the acknowledgment appear to have happened at exactly the same time (in this figure the time was quantised with time units) while in the MSC diagram of Fig. 12.22 they do not appear at the same time, it is clear that in the real world two events cannot happen exactly at the same time, especially if the protocol entity has a single processor. What is really important is that the transmissions of the signals ind and ack follow the reception of the signal ipdu, while the order of the transmissions of the signals ind and ack is not relevant; in Fig. 12.22 these two events are shown in the same order as were specified in the specification of Fig. 12.20. We should mention here once more that the times needed to process messages and make decisions were neglected which is clearly evident in the timing diagram of Fig. 12.21 .

12.6.6 Efficiency and Transmit Window Design

In Sect. 12.4.4 we explained the efficiency of the stop-and-wait protocol for the case of a lossless channel in Eqs. (12.7) and (12.8). In this section the efficiency of the sliding window protocol in the case of a lossless channel will be shown. Comparing the scenarios in Figs. 12.4 and 12.12, one can easily see that the efficiency of the protocol shown in Fig. 12.12 is three times as high as the efficiency of the protocol shown in Fig. 12.4. This statement can be generalised to say that the efficiency of the sliding window protocol is W_t-times as high than the efficiency of the stop-and-wait protocol, where W_t is the transmit window width, if there are no losses in the channel. However, one must be aware that the efficiency cannot be higher than 1; if the transmit window width is sufficiently large, the transmitter may be transmitting all the time, which means that the efficiency equals 1. Hence the efficiency of the sliding window protocol in the lossless case equals

$$\left\{ \begin{array}{l} W_t \cdot T_i \leq T_{rt} \rightarrow \eta = \dfrac{W_t \cdot T_i}{T_{rt}} \\[2mm] W_t \cdot T_i \geq T_{rt} \rightarrow \eta = 1 \end{array} \right\}, \tag{12.15}$$

or, if the protocol entities are interconnected with a physical channel,

$$\left\{ \begin{array}{l} W_t \cdot T_i \leq T_i + T_c + 2 \cdot T_p \rightarrow \eta = \dfrac{W_t \cdot T_i}{T_i + T_c + 2 \cdot T_p} \\[2mm] W_t \cdot T_i \geq T_i + T_c + 2 \cdot T_p \rightarrow \eta = 1 \end{array} \right\}. \tag{12.16}$$

If there are no losses in the channel, the efficiency can therefore be 1 or 100% if

$$W_t \geq \frac{T_{rt}}{T_i} \tag{12.17}$$

holds, or in the case of a physical channel,

$$W_t \geq \frac{T_i + T_c + 2 \cdot T_p}{T_i}. \tag{12.18}$$

The efficiency of the sliding window protocol in the lossless case, as given in Eq. (12.15), can also be expressed with the bandwidth-delay product which was defined in Eq. (12.10), so we can write

$$\left\{ \begin{array}{l} W_t \cdot L_i \leq RD \rightarrow \eta = \dfrac{W_t \cdot L_i}{RD} \\ \\ \quad\quad W_t \cdot L_i \geq RD \rightarrow \eta = 1 \end{array} \right\}. \tag{12.19}$$

We also can remark here that the expression $W_t \cdot L_i$ means the transmit window width in bits (instead of the number of messages).

The minimum transmit window width which allows the protocol efficiency to equal 1 if there are no losses can also be expressed with the bandwidth-delay product:

$$W_t \geq \frac{RD}{L_i}. \tag{12.20}$$

Textually, the condition (12.20) can be explained with the words that the transmit window width in bits must at least be equal to the bandwidth-delay product of the channel.

The results obtained from Eqs. (12.17), (12.18) and (12.20) must of course be rounded to the nearest upper integer value to gain the efficiency 1, as the transmit window width is a natural (nonnegative integer) number.

As we have emphasised several times already, the use of automatic repeat request protocols has no sense if there are no losses in the channel; this means that the equations for the protocol efficiency in lossless conditions are useless in practice, except as approximations for the cases where the losses are very low. However, these equations can be used to compare various sliding window protocols; if one protocol is more efficient in a lossless environment, it can also be expected to be more efficient in a lossy environment.

In a system with a lossy channel the efficiency 1 can of course never be obtained, even if the condition (12.17), (12.18) or (12.20) is fulfilled. In spite of this, this condition is usually used to design the transmit window width. If this condition is fulfilled, the protocol efficiency is almost equal to 1 in the very low loss conditions; if losses are higher, even transmit window widths that are substantially larger than those required by the above equations do not provide higher efficiencies. On the other hand, one must be aware that the product $W_t \cdot L_i$ also means the amount of memory that is needed to implement the transmit buffer.

To some extent, the protocol efficiency can therefore be improved by enlarging the transmit window width, but the memory consumption is increased at the same time; this is very important to be aware of if small, not very powerful computers are used to implement the protocol, especially if several protocol entities setting up several concurrent connections are implemented in a single computer. This problem is especially acute in systems with large bandwidth-delay products which require large window widths.

Similarly as was stated for the stop-and-wait protocol in Sect. 12.4.4, the relative efficiency of the sliding window protocol also equals 1 if there are no losses. That means that sliding window protocols do not necessarily have advantages over the much simpler stop-and-wait protocol if a multiplexed channel is used, except, of course, if in a multiplexed system only a small number of multiplexed communication processes are concurrently active.

12.6.7 *Sliding Window Protocols with Lossy Channels*

In Sect. 12.3.1 the basic mechanisms were enlisted that any automatic repeat request protocol must use to be logically correct. Two of these mechanisms were only mentioned, but not much was told about the use of them; these two mechanisms are the receive buffer and the retransmit timer expiration. While we discussed already in the previous section the importance of the size of the transmit buffer and its symbolic image—the transmit window, nothing has yet been told about the importance of the size of the receive buffer and the receive window. In the sliding window protocol specification in Sect. 12.6.4 the timer was used, but the reaction of the transmitter to a timer expiration was not specified; with respect to the fact that a lossless channel was assumed there, there was nothing wrong with that, as the timer with a properly dimensioned timer expiration time never expires if there are no losses.

Several kinds of sliding window protocols exist; they differ in the receive window width, the kind of acknowledgments used, the number of retransmit timers and the way the transmitter reacts to a timer expiration. If there are no losses in the channel, all these protocols behave like the protocol we specified in Sect. 12.6.4. Hence, different sliding window protocols differ in their reactions to message losses. In the sections that follow some of these protocols will be discussed. The specifications of all protocols that use cumulative acknowledgments and a single retransmit timer will be given as supplements to the specification given in Sect. 12.6.4 in Fig. 12.16–12.20.

12.7 Generalised Sliding Window Protocol with Cumulative Acknowledgments and Single Timer

In this section the *generalised sliding window protocol* will be presented; this protocol employs the transmit and the receive window with arbitrary widths W_t and W_r, respectively, cumulative acknowledgments and a single retransmit timer. Although both windows can in principle have arbitrary widths, only the cases where $W_r \leq W_t$ holds have sense; if the receive window width is larger than the transmit window width, a part of the receive buffer stays unused all the time which is wasteful from the viewpoint of the usage of memory resources; however, the protocol is logically correct also if the receive window is larger than the transmit window.

Table 12.1 Sliding window protocols

Protocol	W_t	W_r	M_{min}
Generalised	$W_t \geq 1$	$1 \leq W_r \leq W_t$	$W_t + W_r$
Stop-and-wait	1	1	2
Go-back-N	$W_t > 1$	1	$W_t + 1$
Selective-repeat	$W_t > 1$	W_t	$2 \cdot W_t$

The protocol is named generalised because it is easy to see that the stop-and-wait protocol (Sects. 12.4 and 12.8), the go-back-N protocol (Sect. 12.9) and the selective-repeat protocol (Sect. 12.10) are only special cases of the generalised protocol. All these protocols essentially differ only in their respective transmit and receive window widths. The differences between these protocols are shown in Table 12.1; in this table the minimum modulus for information message counting is shown which, according to Eq. (12.14), guarantees a logically correct operation.

At this place, let us mention that the discussion of the generalised sliding window protocol is a peculiarity of this book. Up to now, all the books treating communication protocols and communication systems have discussed only the stop-and-wait, go-back-N and selective-repeat protocols, declaring all of them to be special cases of the sliding window protocol; none of these texts, however, offered a general specification valid for all of them. Only the stop-and-wait, go-back-N and selective-repeat protocols also have found their place in the communication practice. Yet the paper in which the generalised protocol was defined and specified also showed that the generalised protocol can be at least as efficient as the others, while using less memory resources.

If there are no losses in the communication system, the generalised sliding window protocol functions exactly as we have already described and specified in previous sections. In this section we therefore still have to explain what happens if a protocol data unit is lost in the channel and consequently the transmitting protocol entity times out. After the timer has expired the transmitter of course cannot know whether an information message or an acknowledgment has been lost; in any case it retransmits the oldest outstanding (not yet acknowledged) information protocol data unit, hence the message with the sequence number located at the beginning of the transmit window (V_a). Because, however, the receive window width in the case of the generalised protocol may be smaller than the transmit window width, it may well have happened that the transmitter has already transmitted some information protocol data units with the sequence numbers which are in the transmit window, but not in the receive window, so the receiver discarded them even if it received them without errors. These messages, too, must be retransmitted after the timer has expired. If the assumption that the information message from the beginning of the transmit window has been lost is valid, then the same sequence number is located at the beginnings of both transmit and receive windows, as the receiver may not shift its window until it receives the expected message. Let us notice once more here that the transmitter must be acquainted with both window widths.

The specification of the sliding window protocol which was shown in Figs. 12.16–12.20 must therefore be supplemented with the specification shown in

Fig. 12.23 Generalised
sliding window protocol
(reaction of transmitter to
timer expiration)

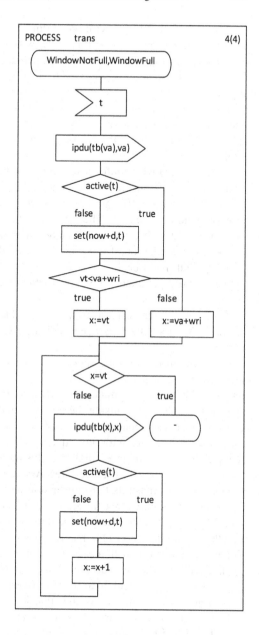

Fig. 12.23, specifying the reaction of the generalised protocol transmitter to a timer expiration. In this figure one can see that the transmitter retransmits the information protocol data units ipdu(va) and ipdu(x), va + wri$\leq$x < vt (of course, it is also possible that no x fulfils this condition!).

Let us summarise: the generalised sliding window protocol is specified in Figs. 12.3, 12.16–12.20 and 12.23.

The operation of the generalised sliding window protocol in case when there are losses in the channel will be illustrated with two scenarios. Both of them will be shown in the form of a timing diagram where the contents of the transmit and receive windows will also be shown, as well as the timer expirations. In the transmit window the outstanding messages will be shown on grey background, while in the receive window the already received but not yet forwarded to the user messages will be shown on grey background. The running of the timer will be indicated with a dotted line; a cross will indicate the stopping of the timer, and an arrow will show the timer expiration.

In both scenarios the information message transmit time will be 5 time units, the acknowledgment transmit time 2 time units, and the time of signal propagation through the channel 10 time units. The transmit window width will be 3 and the receive window width 2. The protocol entities will be assumed to be directly interconnected with a physical channel. The timer expiration time will be 1 time unit longer than the round trip time, as required by Eq. (12.6). The requests for the transfer of all five user messages (characters 'a' – 'e') will be assumed to already be present in the input queue of the transmitter at the beginning of the scenario (at time 0).

In Fig. 12.24 the scenario is shown in which an information protocol data unit was lost; the loss of the message is indicated with two vertical lines drawn across the message.

After the information message ipdu('a',0) has been lost the receiver successfully receives the message ipdu('b',1) and stores it into its receive buffer; however, when it receives the message ipdu('c',2) it discards it as it has no room to store it (the sequence number 2 is outside the receive window). Because cumulative acknowledgments are used the receiver sends the acknowledgment ack(0) after having received the messages with the sequence numbers 1 and 2, as it is still waiting the message with the sequence number 0 (this number is at the beginning of its receive window). As the transmitter hasn't received any new acknowledgment it times out at time 28, so it retransmits the information messages with sequence numbers 0 and 2, because it assumes that the message 0 has been lost and that the receiver has received and stored the message 1 and discarded the message 2. One may not forget that the transmitter knows the receive window width. The loss of a single information message (the one with the sequence number 0) is the most probable scenario for the transmitter. When the receiver at last receives the message ipdu('a',0) it forwards the user messages 'a' and 'b' to the user (the latter message has already been in the buffer) and informs the transmitter that it is now expecting the message number 2. From here on the scenario does not differ from the lossless scenario.

In Fig. 12.25 the scenario is shown where an acknowledgment has been lost.

In this scenario the receiver receives the information protocol data units with the sequential numbers 0, 1 and 2, passes the user messages contained therein to the receiving user, sends the acknowledgments and shifts the receive window accordingly. Unfortunately, the acknowledgment ack(1) is lost in the channel, so the transmitter times out before it receives the acknowledgment ack(2) and retransmits the information messages numbered 0 and 2. It can transmit the information protocol

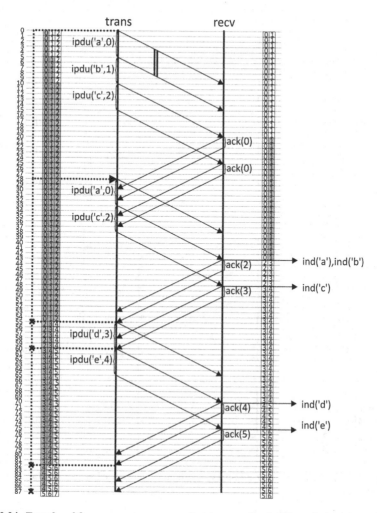

Fig. 12.24 Transfer of five user messages according to generalised sliding window protocol with information message lost

messages numbered 3 and 4 only after it has finished the transmission of the message 2 (meanwhile it has received the acknowledgment ack(3) and moved the transmit window). In Fig. 12.25 the timer seems to expire at the same time 65 as the transmitter receives the acknowledgment ack(4) (the timer expiration time is 28 time units). In the physical world two different events do never occur at exactly the same time; if they, however, did, the transmitter could not process both of them and react to both of them at the same time; in our figure the time is quantised with time units. From the time 65 on there are therefore two possible scenarios. If the transmitter receives and processes the acknowledgment ack(4) before the timer expires, it stops the timer and waits for the acknowledgment ack(5). If, however, it times out before it receives and processes the acknowledgment ack(4), the

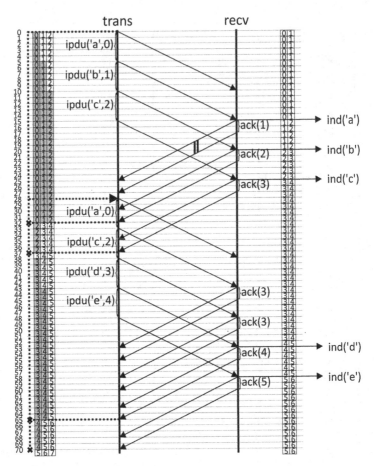

Fig. 12.25 Transfer of five user messages according to generalised sliding window protocol if acknowledgment was lost

transmitter retransmits the information message with the sequence number 3, but this one is discarded by the receiver, as number 3 is no more in the receive window when this message is received. Which one of the above scenarios actually happens in a real system depends on the system implementation. What is important is that the protocol is logically correct in both cases (in both cases the receiver receives all user messages in the correct order and without duplicates), only in the second scenario it is slightly less efficient. In a real system such ambiguities are best avoided if possible. In our case there would be no such ambiguity if the timer expiration time would be longer than the round trip time by at least 5 time units (information message transmit time); in such the case the transmitter would receive the acknowledgment ack(2) early enough, so it wouldn't time out (but would still time out if more than one acknowledgment were lost in line, which is less probable). In Fig. 12.25 the acknowledgment ack(4) was assumed to arrive in time to prevent the timeout.

12.8   Alternating Bit Protocol

In Sect. 12.4 the stop-and-wait protocol was presented and explained. The specification of this protocol can also be obtained by putting both window widths equal to 1, hence $W_t = W_r = 1$, in the specification of the generalised sliding window protocol in Figs. 12.16–12.20 and 12.23; in fact, this was already told in Table 12.1. The reasoning is very simple: if the transmit window width equals 1, the transmitter may transmit only one information protocol data unit to fill its window, so it must wait to receive the acknowledgment (allowing it to move the window) before transmitting the next one.

According to Eq. (12.14) information protocol data units may be counted modulo 2 if the stop-and-wait protocol is used; sequence numbers can therefore be written with a single bit whose value alternates (0, 1, 0, . . .) from one message to the next one. The stop-and-wait protocol is therefore usually also referred to as *alternating-bit protocol*.

The alternating bit protocol is probably the protocol that has most frequently been theoretically treated, because it also is the simplest one.

12.9   Go-Back-N Protocol with Cumulative Acknowledgments and Single Timer

The *go-back-N protocol* is a relatively simple protocol which employs cumulative acknowledgments and a single timer. This protocol usually uses some additional mechanisms that somehow improve its efficiency and is often used in the data-link layer of a protocol stack. Although the efficiency of this protocol may be considerably better than the efficiency of the stop-and-wait protocol and somewhat worse than the efficiency of the selective-repeat protocol (the latter protocol is going to be explained in the next section), the relative efficiency of the go-back-N protocol is quite bad; this protocol is therefore not very suitable for the environments where the multiplexing is used (e.g. in the transport layer).

The go-back-N protocol is a special case of the generalised sliding window protocol with the receive widow width equal 1.

If there are no losses in the channel all sliding window protocols using cumulative acknowledgments and a single timer, including the generalised and the go-back-N protocols, operate in exactly the same way. The differences between protocols can only be noticed when a protocol data unit is lost and consequently the transmitter times out. As was told already, the window width of the go-back-N receiver equals 1. That means that the receiver has the room in its receive buffer to store only one message already received but not yet forwarded to the user. Consequently, the receiver must receive information protocol data units in the correct order. If the receiver receives an unexpected message, it must discard it. If an information message is lost in the channel during the transfer, the receiver must therefore discard

Fig. 12.26 Go-back-N
Protocol (reaction of
transmitter after timeout)

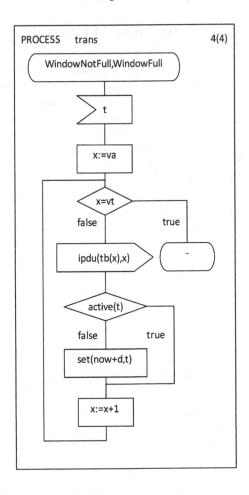

Fig. 12.26 Go-back-N Protocol (reaction of transmitter after timeout)

all the messages that follow the lost one, even if they have been received without error. Because of this the transmitter, after it has timed out, must retransmit the message which it has at the beginning of its window and also all the following messages which it has already transmitted. At a first glance, the retransmission of the messages that have not been lost seems to waste the resources of the channel, thus decreasing the protocol efficiency and even more the protocol relative efficiency; however, this is the price which must be payed for the lower demand for memory and the lower complexity of the protocol.

Because the go-back-N protocol is a special case of the generalised protocol, the specification of the go-back-N protocol can simply be obtained by putting the value of the receive window width to 1 (wr = 1) in the specification of the generalised protocol (Figs. 12.16–12.20 and 12.23). However, due to this fact, the specification of both transmitting and receiving protocol entities can be simplified so that the reaction of the transmitter to a timeout is specified by Fig. 12.26, while the specification of the stop-and-wait receiver (shown in Fig. 12.6) is used also as the specification of the go-back-N receiver (both protocols have the receive window width equal to 1). Of course, the receiver does not need to inform the transmitter about its receive window width, as it is implicitly known to be 1 because the protocol is the go-back-N.

We can summarise: the go-back-N protocol is specified in Figs. 12.3, 12.16–12.18, 12.26 and 12.6; however, in Fig. 12.16 the declaration of the receive window width wr can be omitted as it is not needed in the go-back-N protocol specification; furthermore, the constructs IMPORTED and IMPORT in the transmitter specification can also be omitted.

In Fig. 12.26 it can easily be seen that the transmitter, after it has timed out, retransmits all the information protocol data units it currently has in its transmit buffer (hence the one at the beginning of the transmit window and all that follow it).

In Fig. 12.26 one also can see that the transmitter, after it has timed out, somehow goes back to the beginning of the transmit window to transmit and count the information protocol data units; in the theoretical analysis of the protocol the authors assume that the transmitter goes back by N sequence numbers, hence the name of the protocol »go-back-N«.

In Figs. 12.27 and 12.28 similar scenarios of the transfer of five user messages are shown as in Figs. 12.24 and 12.25, only the go-back-N protocol is used this time; hence the receive window width equals 1, while all the other parameters are the same as in Figs. 12.24 and 12.25; an information protocol data unit was lost in the first scenario and an acknowledgment was lost in the second scenario. A similar comment could be given here as was added to Figs. 12.24 and 12.25. The purpose of these scenarios is to allow a reader to compare the operations of the generalised and the go-back-N protocols.

To allow a reader to more easily understand the scenario of Fig. 12.28, a short comment will nevertheless be added. The peculiarity of this scenario is that an acknowledgment was lost, rather than an information protocol data unit. After the transmitter has timed out and has begun to retransmit the messages present in its transmit buffer, the acknowledgment ack(2) is received by the transmitter and the transmitter moves its transmit window; it finishes with the transmitting of ipdu('a',0) first, but then continues with the transmission of ipdu('c',2), because the sequence number 1 is no more in its transmit window, and the information message associated with it is also not in the transmit buffer. From time 37 to time 38 there is no outstanding message in the output buffer, so the timer is not active during this time period. Again, it was assumed that at time 65 the acknowledgment ack (4) was received and processed before the timer could expire.

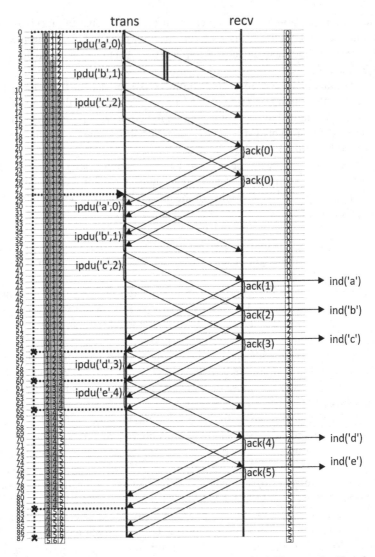

Fig. 12.27 Transfer of five user messages according to go-back-N protocol with information protocol data unit lost

12.10 Selective-Repeat Protocol with Cumulative Acknowledgments and Single Timer

The *selective-repeat protocol* is the protocol that is usually considered the most efficient (as far as the usage of the channel resources is concerned) among the classical sliding window protocols using cumulative acknowledgments and a single retransmit timer; however, it requires the largest amount of memory to implement

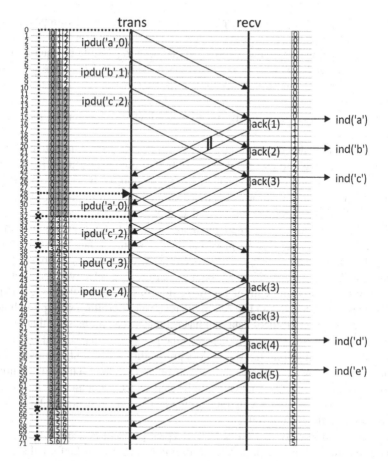

Fig. 12.28 Transfer of five user messages according to go-back-N protocol with acknowledgment lost

the transmit and the receive buffer. Although its efficiency is only slightly higher than the efficiency of the go-back-N protocol if the order of messages is preserved in the channel, it can be much more efficient than the go-back-N protocol if the order of messages is not preserved in the channel; while the go-back-N receiver discards all unexpected messages, the selective-repeat receiver stores them in its buffer until it receives the missing messages, and then passes them to the receiving user in the correct order. The selective-repeat protocol is therefore often used in the transport layer of networks that employ the connectionless mode of transfer in the network layer. A characteristic example of such use is the protocol TCP which is used for data transfer in the transport layer of the Internet. One must also be aware that the relative efficiency of the selective-repeat protocol is much better than the relative efficiency of the go-back-N protocol, which is an additional reason to prefer the use of the selective-repeat protocol in higher layers of a protocol stack where the multiplexing is usually used.

While we have pretended in the previous paragraph the selective-repeat protocol to be the most efficient among the classical sliding window protocols, the generalised sliding window protocol was shown in literature to possibly be equally or even more efficient, while using less memory resources; unfortunately, the generalised protocol has not yet been widely used up to now, so it cannot be considered classical.

The selective-repeat protocol can be viewed as a special case of the generalised sliding window protocol with the receive window width equal to the transmit window width.

As we have told already, all sliding window protocols using cumulative acknowledgments and a single timer operate in exactly the same way if there are no losses and no message reorder in the channel. In this section we must therefore explain what does the transmitting protocol entity do after a timeout indicating the loss of a message (either an information protocol data unit or an acknowledge).

The transmitter, after it has timed out, assumes the oldest outstanding information protocol data unit (hence the one placed at the beginning of the transmit buffer and having the sequence number V_a) to have been lost, which in fact is the most probable scenario; as both window widths are equal in case of the selective-repeat protocol, the transmitter also assumes all the following information protocol messages to have been successfully received and stored in the receive buffer; this assumption holds if only one information protocol data unit has been lost. The probability that only one message has been lost is indeed higher than the probability that several consecutive messages have been lost (at least if the loss probability is constant and losses do not occur in bursts). After a timeout the transmitter therefore retransmits only the information protocol data unit which is located at the beginning of the transmit window; then it can continue with transmitting new information protocol messages if the transmit window is not full.

We have told already that the selective-repeat protocol is a special case of the generalised sliding window protocol. According to Table 12.1 the formal specification of the selective-repeat protocol can be obtained by using the same value for both window widths (wr = wt) in the specification of the generalised protocol (Figs. 12.16–12.20 and 12.23). Due to this fact, the specification of the transmitter's reaction to a timeout can be simplified so that Fig. 12.29 is substituted for Fig. 12.23, while both window widths have the same value.

To summarise, the specification of the selective-repeat protocol is given in Figs. 12.3, 12.16–12.20 and 12.29; in the specification of the transmitting protocol entity in Fig. 12.16 the declaration of the receive window width wr can be omitted, one also can omit the constructs EXPORTED and EXPORT in the specification of the receiver and IMPORTED and IMPORT in the specification of the transmitter.

The selective-repeat protocol is so named because the transmitter retransmits only a selected information protocol data unit, which most probably was lost, after it has timed out.

In Figs. 12.24 and 12.25 and Figs. 12.27–12.28 the transfer of five user messages was shown, using the generalised protocol and the go-back-N protocol, respectively, if an information message or an acknowledgment had been lost. Similar scenarios for

Fig. 12.29 Selective-repeat
protocol (reaction of
transmitter to timeout)

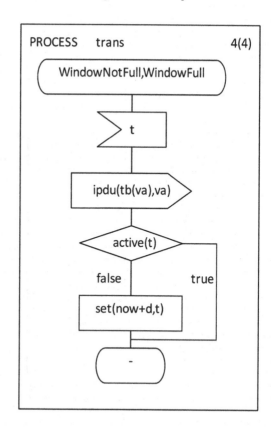

the selective-repeat protocol can be seen in Figs. 12.30 and 12.31. A reader who has understood all the previous scenarios does not need any further explanations.

12.11 Selective-Repeat Protocol with Individual Acknowledgments and Multiple Timers

If a sliding window protocol employs cumulative acknowledgments and a single retransmit timer, the transmitting protocol entity has a limited information at its disposal about what is going on in the system when losses occur. The consequence of this limited knowledge of the system state is not only the assumption that an information message has been lost rather than an acknowledgment, but also the assumption that only one information message has been lost. When the loss rate is high such an assumption is often wrong; it may also be wrong when the loss rate is not high, especially if losses occur in bursts. In spite of such false assumptions the protocol using cumulative acknowledgments and a single timer is logically correct, but can be less efficient, especially if the loss rate is high or if losses occur in bursts. The consequences of such wrong assumptions will be shown for the case of the

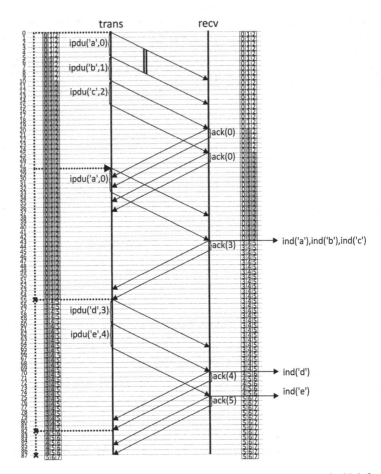

Fig. 12.30 Transfer of five user messages according to selective-repeat protocol with information protocol data unit lost

selective-repeat protocol using cumulative acknowledgments and a single timer if two consecutive information protocol data units are lost. Such a scenario is shown in Fig. 12.32 using the same data as were used in Figs. 12.24, 12.25, 12.27, 12.28, 12.30 and 12.31.

As can be seen in Fig. 12.32 the transmitter times out as it has received no acknowledgments. According to the rules of the selective-repeat protocol it then retransmits only the message ipdu('a',0) because it is not aware that the message ipdu ('b',1) has been lost, too. When the transmitter finally receives the acknowledgment ack(1) it stops and immediately restarts the timer; therefore it can detect the loss of the message number 1 only almost three round trip times after this loss occurred.

The transmitter can have much more complete information about the communication system state at its disposal if it guards each outstanding information message with a separate timer. The *multiple-timers selective-repeat protocol* is therefore

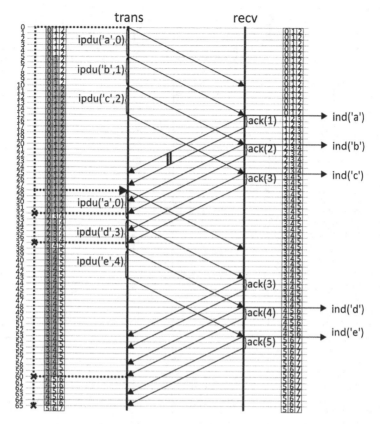

Fig. 12.31 Transfer of five user messages according to selective-repeat protocol with acknowledgment lost

much more efficient than the usual selective-repeat protocol if the loss rate is high, which is especially evident if the round trip times are long and the timer expiration times are consequently also long. If the multiple-timers selective-repeat protocol is used, the transmitter uses a special timer for each outstanding information message and the receiver sends individual acknowledgments. When the transmitter transmits an information protocol data unit it starts the timer associated with the sequence number of that message; it stops the timer when it receives the acknowledgment of the message which is guarded by that timer. If, however, a timer expires, the message guarded by it is retransmitted. Because at most W outstanding information protocol data units can be stored in the transmit buffer (where W is of course the window width) W timers are needed for the operation of the transmitter. This is not a problem in modern protocol implementations because timers are usually implemented simply with counting time units of the system clock of the computer on which the protocol entity is implemented.

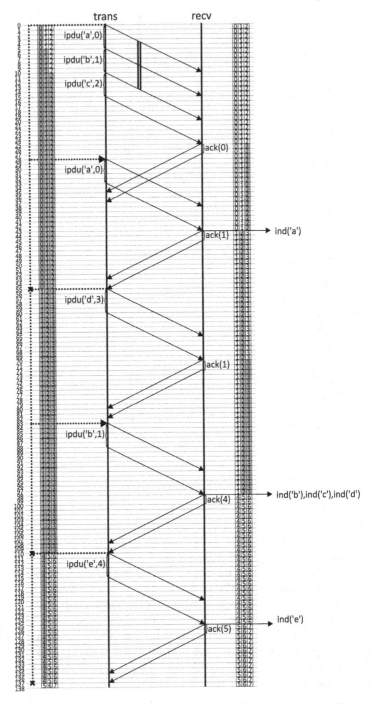

Fig. 12.32 Transfer of five user messages according to selective-repeat protocol with two consecutive information protocol data units lost

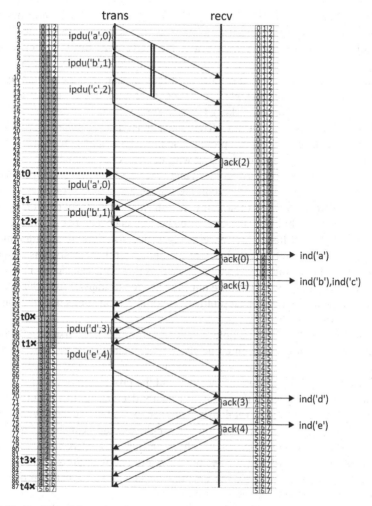

Fig. 12.33 Transfer of five user messages according to multiple-timers selective-repeat protocol with two consecutive information protocol data units lost

If the multiple-timers selective-repeat protocol is compared with the stop-and-wait protocol, one can see that the operation of the multiple-timers selective-repeat protocol resembles the concurrent operation of W stop-and-wait protocols.

In Fig. 12.33 a similar scenario is shown as in Fig. 12.32, only the selective-repeat protocol with individual acknowledgments and multiple timers is used here. Because several timers are used this time, only timer expirations are indicated with arrows and timer deactivations are indicated with crosses for better clarity; both timer expirations and timer deactivations are marked with labels tx, where x indicates the sequence number with which the timer is associated. Of course, a timer is started whenever the protocol data unit with the sequence number associated to it is transmitted. Inspecting Fig. 12.33, one can easily see that the multiple-timers selective-repeat transmitter is more busy with the management of the transmit buffer and the transmit window; because acknowledgments are individual and are not necessarily received in the correct order, the transmitter must memorise which messages in its transmit window are already acknowledged and which are not. As the consequence of this, the transmitter in Fig. 12.33 moves the beginning of its window from the sequence number 1 to the sequence number 3 at time 60, after it has received the acknowledgment ack(1), because the information message number 2 is already acknowledged at that time. The reader must also be aware that individual acknowledgments are used in this protocol, so the acknowledgment ack(x) acknowledges the information protocol data unit with the sequence number x.

Comparing the scenarios of Fig. 12.32 and Fig. 12.33 one can quickly notice that, in case of the multiple-timers protocol, the transmitter senses soon after the round trip time has elapsed the loss of not only the information protocol data unit with the sequence number 0 but also the one with the sequence number 1. Hence, it can quickly react to both losses, not excessively losing the time as a communication resource; due to this fact the efficiency of the multiple-timers protocol is better than the efficiency of the common selective-repeat protocol. This can clearly be seen also in Figs. 12.32 and 12.33, as 125 time units are needed to transfer five user messages with the classical selective-repeat protocol and only 75 time units with the multiple-timers protocol.

A better efficiency of the multiple-timers protocol must of course be payed for. A multiple-timers protocol transmitter needs W timers, and the operation is also more complex, requiring more memory and more processing time. However, with respect to the fact that both the amount of memory and the processing power of computers are steadily increasing while their price is steadily decreasing, this is not a problem nowadays. It is true that several messages can be acknowledged with a single cumulative acknowledgment while only a single message can be acknowledged with an individual acknowledgment which means that cumulative acknowledgments may impose a lower burden upon the communication network; this fact, however, is not so important if acknowledgments are piggybacked into information protocol data units being transferred in the opposite direction.

The only difference between the operations of both receivers is that the multiple-timers protocol receiver transmits individual acknowledgments while the classical selective-repeat protocol receiver sends cumulative acknowledgments. The difference between the operations of both transmitters is more significant. A multiple-timers transmitter must memorise which information messages in its transmit window have already been acknowledged and which have not; with respect to this fact it inspects its transmit window after each reception of an acknowledgment and determines if and how much to shift its transmit window.

In Figs. 12.34–12.39 the formal specification of both protocol entities of the selective-repeat protocol with multiple timers and individual acknowledgments is shown; the communication system was already specified in Fig. 12.3.

In Figs. 12.34–12.37 the specification of the transmitting protocol entity is shown. In the text of this section only those features of the specification will be

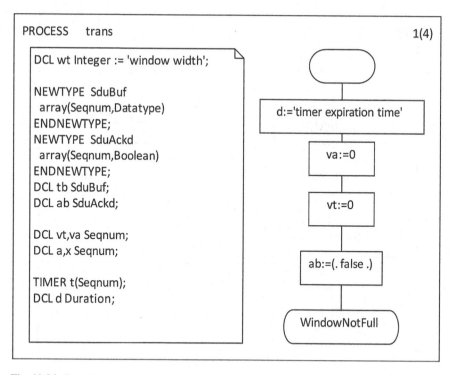

Fig. 12.34 Specification of transmitting protocol entity (declarations and initialisation)

mentioned which have not yet been seen in the specification of the generalised sliding window protocol.

In Fig. 12.34 the declaration of variables and timers as well as the initialisation of the protocol entity are shown. The array ab of the data type SduAckd must be mentioned here; in this array the transmitter marks the acknowledgments it has already received. The standard SDL language allows an array of timers to be declared and used, just like any other array. In our specification of multiple-timers protocol an array of timers is declared as t(Seqnum); although an infinitely long array of timers is declared here, a finite length is used in real-life implementations. Theoretically, W timers suffice where W is the window width, but an implementation with M timers (where M is the modulus) might be easier to be carried out. The specification in this book is more similar to the implementation with M timers.

Whenever the transmitter transmits an information protocol data unit it starts the timer which is associated with the sequence number of the transmitted message; this can be seen in Fig. 12.35.

In Fig. 12.36 the receiving of acknowledgments is shown. Because the acknowledgments are individual and do not necessarily come in the correct order, the transmitter verifies if and how much it must move the transmit window whenever it receives an acknowledgment, using the array ab for this purpose.

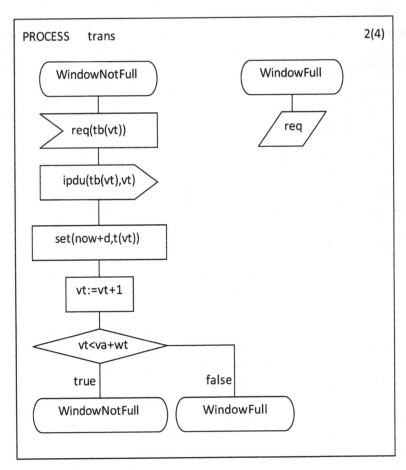

Fig. 12.35 Specification of transmitting protocol entity (transmitting information protocol data units)

In Fig. 12.37 the reaction of the transmitter to a timeout is specified. Of course, that particular message must be retransmitted which is associated with the timer that has expired.

In Figs. 12.38 and 12.39 the receiving protocol entity is specified. Figure 12.38 is very similar to Fig. 12.19, only the receive window width is not »exported«. One must, however, be aware that both the transmit and the receive window widths have to be equal, as this is the selective-repeat protocol!

The sending of acknowledgments is shown in Fig. 12.39; it is different from that specified in Fig. 12.20, as individual acknowledgments are being sent in Fig. 12.39.

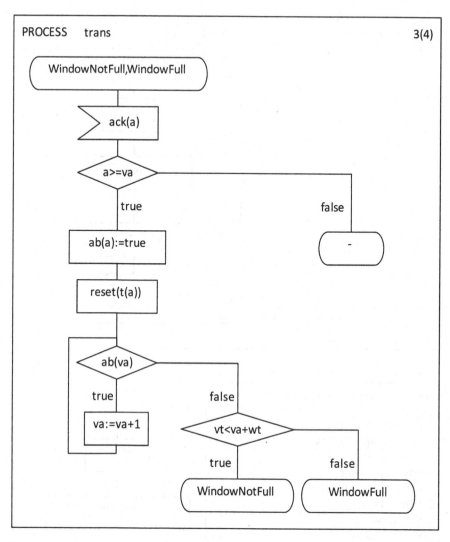

Fig. 12.36 Specification of transmitting protocol entity (receiving acknowledgments)

12.12   Retransmission Timer Expiration Time

In Sect. 12.4 the rule that is used to determine the *timer expiration time* was given in Eq. (12.4): the timer expiration time must be somehow longer than the round trip time. Although this equation was written in Sect. 12.4.1 where the stop-and-wait protocol was discussed, this equation can be used to determine the timer expiration time of all automatic repeat request protocols, the stop-and-wait protocol as well as all the continuous protocols. The expression »somehow longer« is not very precise.

Fig. 12.37 Specification of
transmitting protocol entity
(reaction to timeout)

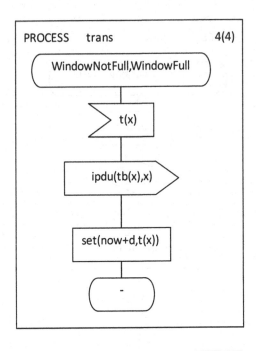

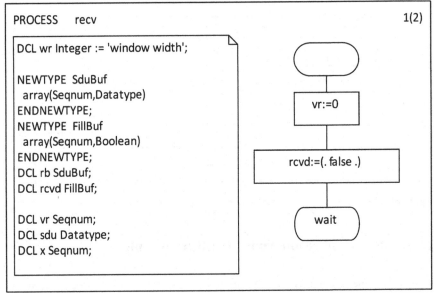

Fig. 12.38 Specification of receiving protocol entity (declarations and initialisation)

Fig. 12.39 Specification of
receiving protocol entity
(receiving information
protocol data units)

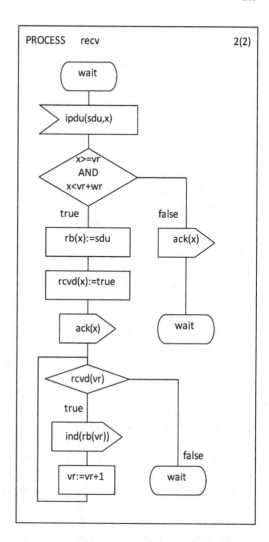

In technical sciences such imprecisions are usually better avoided. However, one must be aware that the use of Eq. (12.4) is not critical, although important, when designing a protocol. An ARQ protocol is logically correct even if Eq. (12.4) is not fulfilled or it is »too much« fulfilled (this means that the timer expiration time is much longer than the round trip time)!

If the timer expiration time is shorter than the round trip time, the transmitter may sometimes or even often unnecessarily time out and retransmit a message. If the loss rate in the channel is low, many messages are unnecessarily retransmitted which has a harmful effect on the efficiency; however, because the protocol can detect and discard duplicate messages, the logical correctness is preserved in spite of this. On the other side, if the loss rate is high, the protocol efficiency can even be improved if the timer expiration time is short; if the transmitter times out too early an information protocol data unit is transmitted several times which increases the probability that at least one message is received successfully. For this reason some protocols which expect very high loss rates in the channel (e.g. the protocols that are used in the wireless parts of mobile networks) transmit each message in several copies, not using a timer for this purpose.

If the timer expiration time is much longer than the round trip time, the transmitter detects losses quite late; consequently it also reacts to losses late and thus loses the precious time which again decreases the efficiency. The protocol is nevertheless logically correct in spite of such a timer setting.

How precisely (how tightly) Eq. (12.4) can be fulfilled depends on how precisely the round trip time is known. The less precisely the round trip time is known, the more safety margin one must keep in Eq. (12.4) for the timer expiration time.

In the simplest case the protocol entities are interconnected with a physical channel which is not multiplexed; hence the channel is used by a single communication process (a single connection). In this case the information transfer time through the channel can easily be previewed and computed; the information protocol data unit and the acknowledgment transmit times must also be added to it. Hence Eq. (12.6) can be used to calculate the round trip time. As the time of message processing in both protocol entities was neglected in this equation, this must also be added to the result of Eq. (12.6), or it must be taken into account within the safety margin, indicated by the relation » > « in Eq. (12.4). Even if the channel parameters are exactly known, the processing time cannot be known if the protocol is implemented on a computer in which multiple processes are concurrently running; the processing time must therefore somehow be assessed in this case. One must also consider the calculation of the information messages transmit time in Eq. (12.6); if all the information protocol data units do not have the same length (which is often the case), the maximum allowed message length (which is almost always defined by the protocol) must be taken into account if Eq. (12.4) is to be fulfilled.

When the possibility of a direct interconnection of protocol entities with a physical channel was mentioned in the previous paragraph, we had in mind that there are no waiting queue and no separate transmitter into the physical channel between the transmitting protocol entity and the physical channel; hence the transmitting protocol entity directly controls the message transmission into the channel. In such a case the protocol entity knows when the transmission of a message is finished and only then can start the transmitting of the next message, only then it also starts the retransmit timer. Equation (12.6) may be used only in such a case. Usually, however, a protocol entity transmits a protocol data unit simply by loading it into the waiting queue between itself and the protocol entity in the adjacent lower layer; this means that at most W information protocol data units can be transmitted one immediately after another (W is the transmit window width). In such a case the time $(W - 1) \cdot T_i$ can be elapsed in the most unfavourable case (where W is the transmit window width and T_i is the information message transmit time) before the transmitter in the lower layer actually begins the transmission. Because the higher layer transmitter starts its timer in the moment when it passes a message to the lower layer with a request primitive, the time a message spends in the queue must obviously be taken into account when determining the round trip time. There is still another difference between the cases mentioned in this paragraph: if a protocol entity controls the transmission into the channel, it can abort the transmitting if it finds out that the message will not be of any use to the receiver (e.g. if the go-back-N protocol is used, the information message that is currently being transmitted will not be useful if the timer expires, as it will have to be retransmitted in any case). If, however, a message is simply added into a waiting queue, the transmitter cannot control it any more and cannot cancel it; the message will be transmitted when it comes through the queue to the lower layer transmitter.

If a channel is used by several connections in the statistical or packet multiplexing mode, the common resources of the channel are shared by all connections. All data traffic must pass through a statistical or packet multiplexer at the input of the channel; there is a waiting queue at the input of the multiplexer where messages suffer randomly long delays due to the random nature of the communication traffic. Therefore the round trip time varies randomly, too, so it only can be assessed (e.g. with respect to the number of connections), or it can be measured, so that the necessary timer expiration time is based on the measurement results.

A network which employs the packet mode of data transfer contains a multitude of packet and/or statistical multiplexed channels as well as network elements (switches and routers) and waiting queues associated with them. The communication traffic of such a system is even more unpredictable than the traffic of a single statistical or packet multiplexed channel. Round trip times in such a system can only be statistically assessed based on the past traffic conditions, or they can be measured.

The round trip time measurement has been mentioned several times already in this section. Although the measurement of a delay between two different network elements is not quite easy due to the mismatch of the clocks of different network elements, the measurement of a round trip time is quite easy as the time difference between two events in a same network element are measured. There is, however, a different problem: the communication traffic is rapidly and heavily changing in a network based on the packet mode of information transfer. Consequently, the round trip time that was measured some time ago can already be obsolete in the moment we want to use it to determine the timer expiration time; such an obsolete value can have a detrimental effect which even can be just opposite to the goal we want to achieve. Such rapidly varying measured values are therefore smoothed with a digital low pass filter; in other words, one can say that the measured values are averaged during some time period. In this way the values of the round trip time are obtained which vary more slowly; unfortunately, if some quick changes in the network traffic happen to occur, the filtered values cannot tightly follow the actual traffic changes. A simple example of a digital filter for averaging measured values of the round trip time is the recursive Eq. (12.21), where M_i is the value measured in the i-th step, A_{i-1} and A_i are the assessments of the average value of the measured quantity in the $(i-1)$-th and the i-th step, respectively, and $0 \leq g \leq 1$ is the factor of filtering. If the value of the parameter g is small, the filtering is weak; if the parameter g has a big value, the filtering is strong. The choice of an appropriate value of the parameter g is therefore essential for a successful filtering. The initial value A_0 of the average value assessment must of course also be chosen; the index i is then running from 1 on.

$$A_i = g \cdot A_{i-1} + (1 - g) \cdot M_i. \tag{12.21}$$

In the formal specifications of ARQ protocols which are presented in this chapter the transmitting of protocol data units is specified as simply as possible; the duration of message transmissions (or the transmissions of the signals ipdu and ack) is not of interest here and is therefore not specified, as the process of message transmitting into the physical channel is the task of the physical layer; we also know that an input queue is implicitly associated with any SDL process. Hence protocol data units are transmitted by simply adding them to the waiting queue of the channel (in the specification of the communication system shown in Fig. 12.3 the process modelling a channel is not explicitly shown; however, it was explicitly specified in the variant of the specification which was simulated). We wanted all the specifications shown in this book to be as general as possible, hence usable with different varieties of channels, so the timer expiration time was always only symbolically shown.

12.13 Negative Acknowledgments

In Sect. 12.3.1 the basic mechanisms were explained that must necessarily be used by automatic repeat request protocols to assure their logical correctness. Among these mechanisms, the timer was indicated as the one that guards the already transmitted but not yet acknowledged (outstanding) messages. If a timer expires, the transmitting protocol entity assumes that an information protocol data unit has been lost and retransmits it. Let us emphasise once more the importance of the timer expiration time which should fulfil the condition (12.4). However, we told in Sect. 12.12 that the timer expiration time must be the more longer than required by condition (12.4), the less accurately the round trip time is known. The timer expiration time can therefore sometimes be very long due to a broad range of message lengths or due to a substantial delay variation because of a strong communication traffic variation. As the maximum message length and the longest possible message delay must be taken into account when using Eq. (12.4) in order to fulfil this condition, the timer expiration time is too long in most cases, which has a negative impact on the protocol efficiency. Sometimes there are also some other reasons to use a timer expiration time which is longer than the one required by Eq. (12.4); an example of such a reason was shown in Fig. 12.25 in Sect. 12.7 where an acknowledgment was lost. For all these reasons it is desirable that a receiving protocol entity explicitly informs the transmitting protocol entity about a message loss as soon as it becomes aware of it itself. A *negative acknowledgment*, also referred to as *reject*, serves this purpose. A reject message can be received by the transmitter much sooner than the transmitter's timer expires; consequently, the transmitter can react to a loss much sooner.

One must be aware that the use of a timer is mandatory in any case if the protocol is to be logically correct, even if negative acknowledgments are employed; it may well happen that a reject message is lost, or the receiver is even not able to detect an information message loss. However, the use of negative acknowledgments can improve the protocol efficiency, comparing it with the efficiency of the basic protocol without the use of negative acknowledgments.

At a first glance, one might consider appropriate that the receiver which has just received a corrupted message and discarded it immediately sends a reject to the transmitter. However, this is possible only if multiplexing is not used in the channel. If both protocol entities are interconnected with a multiplexed channel, there is no way for the receiver to know with which connection the corrupted message is associated, as the connection indicator is an integral part of the protocol control information within the message; with channel decoding the receiver can only detect that an error has occurred, but cannot know which part of the message has been corrupted.

Most usually a negative acknowledgment is sent immediately after the receiver has received an *out-of-sequence message*, hence a message with a sequence number that is higher than expected, which clearly means that a previous message has been lost; however, this method can only be used with continuous protocols in cases where the message order is preserved in the channel. If the order of messages is not preserved in the channel, the receiver may not immediately react to an out-of-sequence message, as the missing message might have been overtaken by another message.

Automatic repeat request protocols are usually designed so that most message losses are resolved with negative acknowledgments, as long timer expiration times are used. If a message loss cannot be resolved in this way, the timer must do its job to assure the logical correctness of the protocol.

Various protocols may employ different rules that determine when the receiver should send a negative acknowledgment. In this section a variant of the selective-repeat protocol using cumulative positive acknowledgments, a single timer and negative acknowledgments will be presented and formally specified. The channel will be assumed to preserve the order of transferred messages.

Whenever the receiving protocol entity receives an information protocol data unit with a sequence number which is within the receive window but not at its beginning (which means that the receiver has received an unexpected message), it transmits a reject rej(r) for any sequence number r at the beginning of the receive window which has not yet been received and which has not yet been already rejected; this means that the receiver rejects any missing information message only once. Whenever the receiver receives an information protocol data unit that it is expecting or one that is not within the receive window, its reaction is identical as in the case of the common selective-repeat protocol; hence it forwards the already received user messages at the beginning of the receive window to the receiving user and discards the messages which do not fit into the receive window. In any case, it sends the cumulative acknowledgment ack.

Whenever the transmitting protocol entity receives a negative acknowledgment rej(r), it retransmits the information protocol data unit with the sequence number r unless this message has already been acknowledged meanwhile ($r \geq V_a$). If, however, the transmitter times out, it retransmits the information message with the sequence number V_a, but only if the transmit window has not been shifted since the most recent start of the timer (if the window has been shifted since the last timer start, this means that the transmitter has already received the rejection of the message guarded by the timer, has retransmitted it and has also received its acknowledgment).

The specification of the selective-repeat protocol with negative acknowledgments to be presented in this section is an extension of the basic selective-repeat protocol which was specified in Sect. 12.10; hence these two specifications are very similar. In spite of this the complete specification of the transmitting and receiving protocol entities of the protocol with negative acknowledgments will be given here, in order to enable a reader to read and understand the specification more easily. The specification of the whole communication system is such as was shown in Fig. 12.3 with the addition of the signal declaration,

$$\text{SIGNAL rej(Seqnum)};$$

this signal models a negative acknowledgment that is sent by the receiving protocol entity recv to the transmitting protocol entity trans.

The transmitting protocol entity is specified in Figs. 12.40–12.44. A careful inspection of this specification reveals the declaration of the variable sva of data type Seqnum which is used by the transmitter to memorise the value of the variable

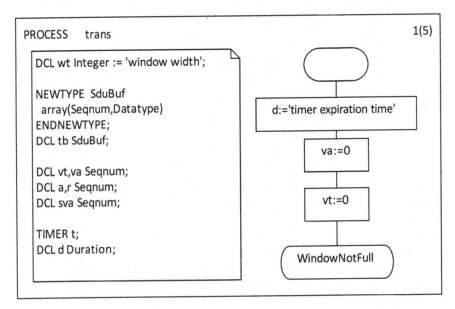

Fig. 12.40 Specification of transmitting protocol entity (declarations and initialisation)

Fig. 12.41 Specification of
transmitting protocol entity
(transmitting information
protocol data units)

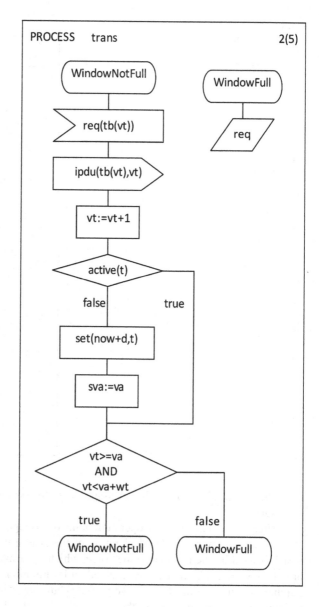

va in the moment when the timer was last started; this value must be refreshed
whenever the timer is started. The reaction of the transmitter to a negative acknowledgment rej is of course brand new in this specification. We must emphasise the
testing of the condition »sva = va« after a timeout. If this condition is not fulfilled,
this means that the transmit window has been shifted since the last starting of the
timer because the transmitter has received a new positive acknowledgment; in such a

Fig. 12.42 Specification of transmitting protocol entity (receiving acknowledgments)

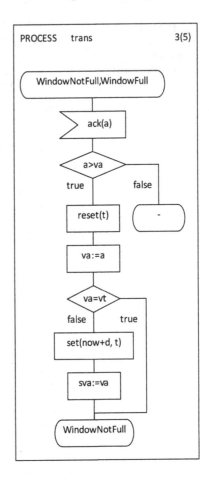

case the retransmission of the message with the sequence number va is not necessary.

In Figs. 12.45 and 12.46 the operation of the receiving protocol entity is specified. Here a reader may notice the declaration of the array rjtd of the data type FillBuf which is used by the receiver to memorise which information protocol data units it has already rejected. After it has received an information message with a sequence number which is within the receive window but not at its beginning, it transmits negative acknowledgments of those messages at the beginning of the receive window which have not yet been received and also have not yet been rejected. In this way the receiver may reject any missed message only once; if the reject or the retransmitted message is lost the transmitter times out.

At the beginning of this section we said that the use of negative acknowledgments improves the protocol efficiency especially if the message length and delay vary substantially which results in an inaccurate knowledge about the round trip time and consequently also an excessively long timer expiration time. However, a protocol

Fig. 12.43 Specification of
transmitting protocol entity
(receiving rejects and
retransmitting information
protocol data units)

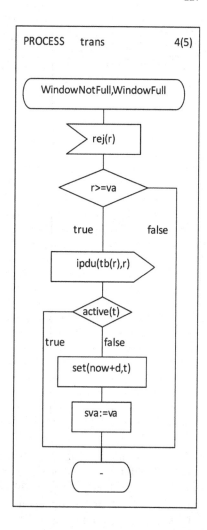

that employs negative acknowledgments can be more efficient than the basic protocol even if the round trip time is exactly known, so the condition (12.4) can be used
with only a tight safety margin (meaning that the timer expiration time is only
slightly longer than the round trip time). This will be demonstrated with an example
in this section and with simulation results in a later section.

In Figs. 12.47 and 12.48 the transfer of five user messages (characters 'a'–'e') is
shown, using the basic selective-repeat protocol and the selective-repeat protocol
with negative acknowledgments, respectively; in both cases the information message
transmit time is 5, the acknowledgment transmit time is 2 and the signal delay in the
channel is 10 time units. In both cases the second information protocol data unit is
lost. As the transmit time and the channel delay are known, the round trip time can be
calculated using Eq. (12.6). In the scenario shown in Fig. 12.47 the timer expiration
time is by one time unit longer than the round trip time, while in the scenario shown

Fig. 12.44 Specification of transmitting protocol entity (retransmitting information protocol data units after timeout)

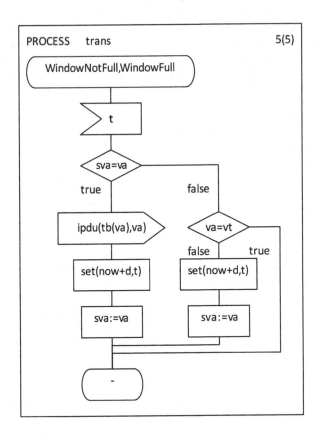

in Fig. 12.48 a very long timer expiration time is assumed, as the message loss is resolved with the negative acknowledgment in this case. Both scenarios are similar to the scenario shown in Fig. 12.30, only the second information message is lost in this case (if the first information message were lost, the difference in protocol efficiencies would not be seen).

In Fig. 12.47 the message transfer with the basic selective-repeat protocol is shown; this protocol was explained in Sect. 12.10. One can see that the transmitter detects the loss of the message ipdu('b',1) only almost two round trip times after it begun transmitting this message, because meanwhile it received the acknowledgment ack(1), stopped the timer and restarted it. Due to this fact it reacts lately to the message loss which causes a lower efficiency.

In Fig. 12.48 a similar scenario is shown, but using the selective-repeat protocol with negative acknowledgments which was described in this section and specified in Figs. 12.40–12.46. In this scenario a very long timer expiration time was chosen, so the timer never expires. In this case the receiver sends a negative acknowledgment rej to the transmitter as soon as it senses a message loss (receives an unexpected message); the transmitter receives the negative acknowledgment much before it would time out if the timer expiration time were the same as in the scenario of

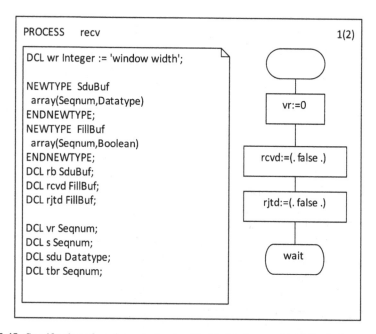

Fig. 12.45 Specification of receiving protocol entity (declarations and initialisation)

Fig. 12.47; therefore the transmitter reacts to the message loss much sooner. Even if a shorter timer expiration time, slightly longer than required by the condition (12.4), were used, the transmitter would time out and retransmit the message, the protocol would nevertheless be logically correct, as the receiver would detect and discard the duplicate.

Comparing the scenarios of Figs. 12.47 and 12.48 one can easily see that the protocol employing negative acknowledgments can be more efficient: while 97 time units are needed to transfer the five user messages with the basic protocol, only 79 time units are needed if the protocol enhanced with negative acknowledgments is used. Of course, the properties of a protocol may not be assessed according to a single communication scenario; however, the practice and the simulation results, as will be shown in a later section, prove the protocol with negative acknowledgments to be more efficient than the basic protocol.

12.14   Enquiry Messages

We told already in this book that one of the basic problems of communication systems is the fact that none of the communicating entities of a communication system precisely knows the state of other entities in the whole distributed system. An entity knows about the states of other entities in the system only the information they send to it with protocol messages; however, any message reaches its destination after

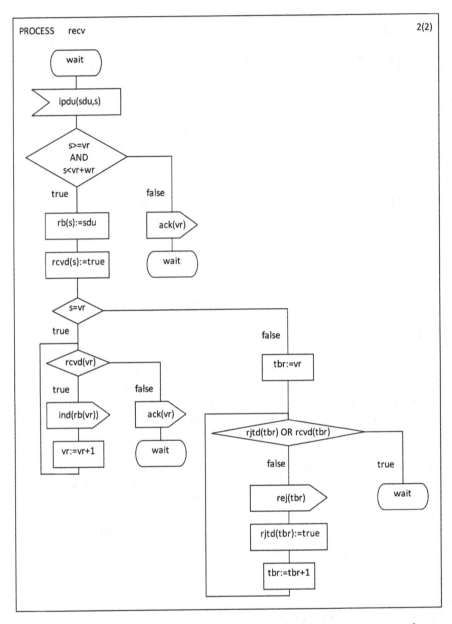

Fig. 12.46 Specification of receiving protocol entity (receiving information messages and transmitting acknowledgments)

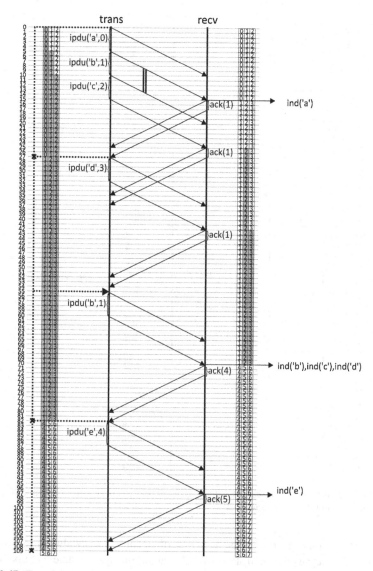

Fig. 12.47 Transfer of five messages with basic selective-repeat protocol (second information message lost)

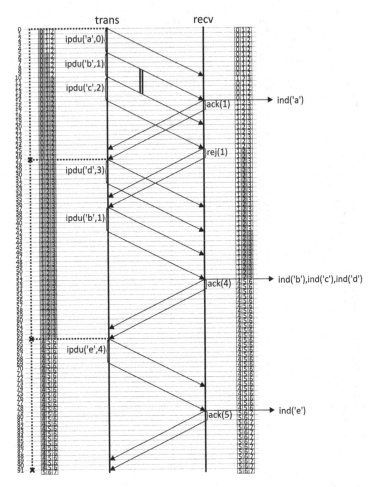

Fig. 12.48 Transfer of five messages with selective-repeat protocol with negative acknowledgments (second information message lost)

a delay. Due to this uncertainty the decisions of an entity are not always so good and successful as they would be if its knowledge about the system state were better.

If a protocol entity being in a certain state does not know which of the several possible scenarios has actually taken place in the system and in which state its peer entity is, it can send an *enquiry message* (or simply an *enquiry*) to its peer; the enquiry message is often symbolically indicated with the acronym ENQ.[2] The peer entity must respond to the enquiry; in this way the uncertainty can be resolved much

---

[2]By the way, ENQ is also one of the ASCII control characters which serves the same goal as described here.

quicker that if the reaction of an entity were based on an assumption about the most probable scenario.

Sometimes automatic repeat request protocols employ enquiry messages. We told already that in the case when the transmitter has timed out there is no way for it to know whether an information protocol data unit or an acknowledgment has been lost. We saw that the protocol is logically correct if the transmitter retransmits the information message in any case; however, because information messages may be much longer than control messages (the former contain both user and control information, while the latter contain only control information), much shorter time can be needed to transfer an enquiry and receive the response to it than to retransmit a possibly very long information message. If the receiver of an automatic repeat request protocol with cumulative acknowledgments receives an enquiry, it must immediately retransmit the acknowledgment it sent most recently which allows the transmitter to unambiguously find out what has actually happened. Because a much longer time may be needed to retransmit an information message and the corresponding acknowledgment than to transmit an enquiry and the corresponding acknowledgment, the protocol that employs enquiry messages may be more efficient than the basic protocol.

12.15   Bidirectional Information Transfer and Piggybacking

For the sake of a more simple treatment of protocols we have assumed up to now that the user information (and hence also the information protocol data units) is trans-ferred only in one direction, from the so-called transmitting protocol entity towards the so-called receiving protocol entity, while the acknowledgments are transferred only in the opposite direction (from the receiver towards the transmitter). In practice, however, information messages, and so also acknowledgments, can well be trans-ferred in both directions simultaneously. Although the transfers in both directions are mostly independent one from the other, some peculiarities of the bidirectional information transfer must also be taken into account.

If both protocol entities are interconnected with a half-duplex channel, the information can be transferred in both directions, but not at the same time; in any given moment the information can be transferred either in one direction or in another. If the information is to be transferred in both directions, both protocol entities must agree on when each one may transmit. Essentially, this is the problem of *medium access control* which will be treated more thoroughly in Chap. 14. As we already know, only the stop-and-wait protocol can be used with a half-duplex channel. One possibility to transfer information bidirectionally via a half-duplex channel is that one entity sends an information message, then the other protocol entity, after it has received it, sends the acknowledgment and its own information message, and so on. The other possibility is that one entity plays the role of a *master* which controls the channel and so may allow or disallow the other one (a *slave*) to transmit its own user information. Both entities also can use a special control packet usually referred to as *token* to control the direction of data transfer; only one protocol entity can possess the token in a given moment and may therefore transmit information; when it no more needs the token it passes it to the other entity and so allows it to transmit its own information.

If user information is transferred in both directions simultaneously, the technique referred to as *piggybacking* is often used which we already mentioned in Sect. 12.3.1; as was already explained there the piggybacking means that acknowledgments are simply added (inserted) to information protocol data units which are transferred in the opposite direction as are the information messages that are being acknowledged. One must however be aware that a protocol which employs piggybacking in spite of this needs explicit acknowledgments which do not contain any user information; a protocol entity transmits information protocol data units only when users request user messages to be transferred. If there are no user requests, the protocol entity sends no information messages; in spite of this it must acknowledge information protocol data units it receives. A receiver can use a special *acknowledge timer*. If it receives an information message which must be acknowledged when there is no user information to be transmitted within an information protocol data unit, but expects to receive it in a near future, it starts the acknowledge timer and waits for a user request. If it receives a user request, it stops the timer and sends an information message with a piggybacked acknowledgment; if the acknowledge timer expires, it sends an explicit acknowledgment. The purpose of waiting for user information is of course to decrease the traffic burden imposed by explicit acknowledgments on the communication network; exactly this is also the purpose of piggybacking in general. One must be aware that the length of an information message without acknowledgment and an explicit acknowledgement combined is longer than the length of an information message with an included acknowledgment, as any protocol message, either information or control, contains some additional control information, such as the indication of the message type, the indication of the message beginning/end, redundant bits of channel coding... If piggybacking is used, this additional control information of an information message and an acknowledgment are joined.

If user information is transferred in both directions and acknowledgments are piggybacked into information protocol data units, the retransmit timer expiration time of the transmitter must be adjusted to this because the transmit time of an information message (with an included acknowledgment) is in most cases longer than the transmit time of a simple acknowledgment. In Fig. 12.49a the least favourable scenario is shown for the case where both protocol entities are interconnected with a physical channel and an acknowledge timer is not used (the least favourable scenario is of course the scenario with the longest round trip time). In this scenario the entity B receives an information protocol data unit from the entity A just in the moment when it begins transmitting its own information message; in this moment it is of course too late to include the acknowledgment into the message being transmitted, so the acknowledgment can only be included into the next

Fig. 12.49 Round trip time
if piggybacking is used

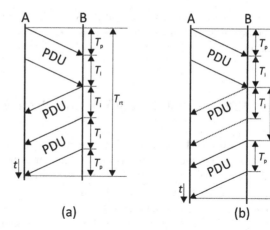

(a)   (b)

information message (if there were no next message an explicit acknowledgment
would be sent, which is in general shorter than an information message). As can be
seen from the figure the retransmit timer expiration time in the case of such a
scenario must be

$$T_r > 3 \cdot T_i + 2 \cdot T_p, \tag{12.22}$$

where T_i is the information message transmit time, T_p is the propagation time of a
signal in the channel and T_r is the retransmit timer expiration time. In Fig. 12.49b the
least favourable scenario is shown for a case similar to the one of Fig. 12.49a, but the
acknowledge timer is used. In this case the entity B starts the acknowledge timer
after it has received an information message; after the timer expires it sends either an
explicit acknowledgment or an information message with the piggybacked acknowledgment (this requires a longer time to transmit). Of course, also in this case the
entity B may happen to receive the message to be acknowledged just in the moment
when it begins transmitting an information protocol data unit; in such a case it will
send the acknowledgment only after it has finished transmitting the previous message, even if the acknowledge timer expiration time is shorter than the information
message transmit time. Hence the retransmit timer expiration time must be

$$T_r > 2 \cdot T_i + 2 \cdot T_p + \max(T_i, T_a), \tag{12.23}$$

where T_a is the acknowledge timer expiration time, the function max returns the
bigger value of both arguments, and the symbols T_r, T_i and T_p have the same
meaning as in Eq. (12.22).

The retransmit timer expiration time must be appropriately adjusted to the
bidirectional information transfer even if piggybacking is not used. In this case
only explicit acknowledgment messages are used to acknowledge information
messages; for this reason the last protocol message PDU in Figs. 12.49a, b must

be replaced with an acknowledgment ACK, while in Eqs. (12.22) and (12.23) one information message transmit time T_i must be substituted by the acknowledgment transmit time T_c.

12.16 Efficiency of Sliding Window Protocols

In Chapter 6 the performance of communication protocols was discussed. In this context the importance of the efficient usage of communication resources was emphasised. Three different performance measures were defined and more thoroughly discussed, based on the consumption of the transfer rate and time as communication resources. The efficiency of a protocol was discussed in Sect. 6.2 and defined in Eq. (6.9); the protocol efficiency represents the share of transfer rate and time, which a protocol is capable to use for the information transfer. If all the resources of a communication channel are at the disposal of a protocol, hence if multiplexing is not used, the protocol efficiency depends only on the protocol and channel properties. The relative efficiency of a protocol was discussed in Sect. 6.4 and defined in Eq. (6.14); the relative efficiency is primarily used to assess the burden a protocol imposes on communication resources (transfer rate and time) of a multiplexed channel. The relative efficiency does not depend on the burden other communication processes impose on the same multiplexed channel. The third important measure used to assess the performance of a protocol is the delay; delays were discussed in Sect. 6.5. Although the notion of the delay as a time difference between two events is very simple by itself, the delay of a message transfer can nevertheless be defined in different ways, depending on what the event of a message transmission and the event of a message reception mean for us.

In this chapter automatic repeat request protocols and, more specifically, sliding window protocols are treated. Several variants of these protocols were described. Their efficiencies were calculated for the specific case when a lossless channel is used; this efficiency is the upper limit of the protocol efficiency if a lossy channel is employed. Now we are going to say also something about the performance of sliding window protocols if protocol entities are interconnected with a lossy channel.

The analytic treatment of the performance of sliding window protocols in case of a lossy channel is difficult and requires a very good mastery of mathematics and complexity. It is based on the probability theory, as losses occur randomly. More or less simplified models of protocols and communication channels are therefore often treated. Usually the simulation of the protocol performance is used instead; although simplified models of communication channels are simulated as well, the development of simulation models is much simpler and much cheaper than the development of analytic models. In this book some results of the performance simulations of sliding window protocols using lossy channels will be shown and commented. Numerous scenarios were simulated by the author which differed in the protocol type, the protocol parameters and the parameters of the channel interconnecting protocol entities. Here, however, only some selected characteristic results will be presented with the aim to offer a reader some impression and feeling for the performance properties of sliding window protocols. Of course, the simulation results of some specific communication scenario may not be simply generalised; the aim of our discussion of the results that are shown here is, however, to emphasise some more general performance properties of sliding window protocols.

In the continuation of this section we will show the results of simulations that were run with the following communication channel and protocol parameters. The simulated communication system was such as is shown in Fig. 12.3 with both protocol entities interconnected with a physical channel. The nominal transfer rate of the channel was $R = 10$ Mb/s and the propagation delay of the channel was $T_p = 1$ ms. All information protocol data units had the same length $L_i = 105$ o, while the length of acknowledgments was $L_c = 5$ o. According to Eqs. (12.6) and (12.10) the bandwidth-delay product was therefore $RD = 20,880$ b. With all simulated continuous protocols the transmit window width was $W_t = 25$, which assures the protocol efficiency $\eta = 1$ in case of a lossless channel, according to Eq. (12.19). The retransmit timer expiration time was slightly longer than the calculated round trip time. The losses of the channel were modelled with bit errors that occur randomly, but with constant probability (constant bit error rate). In all scenarios the bit error rate ran from $ber = 10^{-7}$ to $ber = 10^{-3}$, which means the packet error rate from $per = 8.4 \cdot 10^{-5}$ to $per = 0.57$, according to Eq. (4.1) (the packet error rate for information protocol data units). The maximum throughput and consequently the protocol efficiency was found so that the simulated transmitter transmitted a new information protocol data unit whenever it was allowed to do so by the protocol. The delay of a message was measured from the moment when the transmitter began transmitting the message into the channel to the moment when the receiver passed the user message to the receiving user.

The following was assumed when simulating the communication system.

The simulation models of the stop-and-wait protocol (labelled ABP—alternating bit protocol in the diagrams), the go-back-N protocol (labelled GBN), the selective-repeat protocol (labelled SRP) and the multiple-timers selective-repeat protocol with individual acknowledgments (labelled MT) were derived from the formal specifications given in Sects. 12.4.2, 12.6.4, 12.9, 12.10 and 12.11, respectively; only the basic mechanisms were used that are mandatory for a logically correct operation of these protocols. The simulation model of the selective-repeat protocol with negative acknowledgments and enquiries (labelled SRQ in the diagrams) was derived from the specification given in Sect. 12.13 with the addition of enquiry control messages. In fact, these protocols are simpler than many realistic protocols which are commonly used in practice that often use more additional mechanisms, serving some other purposes.

In our simulated scenarios all user messages were equally long. In real communication systems the length of user messages varies; the message lengths and the length distributions depend not only on the protocol but also on the applications that use the communication system.

As was already told, the message losses were modelled with bit errors that occur with a constant probability. In real systems the bit error rates usually vary; often

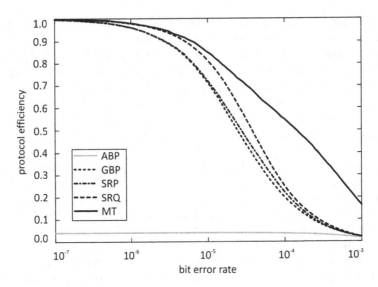

Fig. 12.50 Efficiency of sliding window protocols as function of bit error rate

errors occur in bursts. Furthermore, message losses quite often are not due to bit errors but to the lack of storage or processing resources in the communication system; such losses strongly depend on the traffic load upon the communication system.

The simulated channel was assumed to preserve the order of transferred messages. In higher layers of some protocol stacks this assumption is often not valid. If message reorder can occur in the channel, the ratio of the efficiencies of different protocols can be different.

From what has already been told one can see that it is not possible to conceive a universal model of a communication system that would be valid in all environments and all circumstances. The topic of this book are the general principles of information transfer on which communication protocols are based; such a treatment of communication systems as is used here, though quite general and simplified, is therefore adequate and in accordance with the goals of this work. If, however, one wanted to research the properties of a more specific communication system and wished to obtain more accurate simulation results, they should develop more accurate simulation models which would be adapted to the peculiarities of the studied system.

In Fig. 12.50 the dependence of the protocol efficiency on the bit error rate is shown for the stop-and-wait protocol (ABP), the go-back-N protocol (GBN), the selective-repeat protocol (SRP), the selective-repeat protocol with negative acknowledgments and enquiries (SRQ), and the multiple-timers protocol with individual acknowledgments (MT). As expected, the efficiency of all protocols decreases with the increasing bit error rate. One can see that the efficiency of the stop-and-wait protocol is by far the worst in this diagram which is expected as its

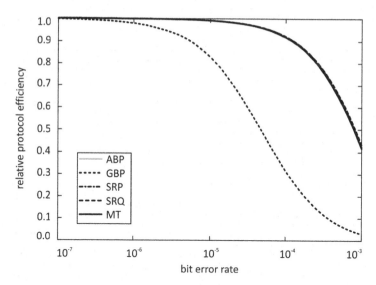

Fig. 12.51 Relative efficiency of sliding window protocols as function of bit error rate

transmit window width (1) is much smaller than the one which is needed to achieve the efficiency 1 (namely 25). The efficiency of all continuous protocols can be seen to be nearly 1 at very low bit error rates, as the transmit window width 25 was used. It is, however, necessary to notice that the stop-and-wait protocol yields much better results if the bandwidth-delay product is small, as can be seen from Eq. (12.11); in such a case the stop-and-wait protocol can even be more efficient as the go-back-N protocol if the loss rate is high, as the go-back-N protocol wastes much time retransmitting messages which have already been correctly received in such conditions. Among continuous protocols the go-back-N protocol yields the worst results; however, the basic selective-repeat protocol is not much better (especially not if the bandwidth-delay product is big). One can, however, see that the use of negative acknowledgments and enquiries can much improve the efficiency (though not necessarily!). In all scenarios that were simulated by the author the multiple-timers protocol with individual acknowledgments yielded the best results, as was already anticipated in Sect. 12.11.

In Fig. 12.51 the dependence of the relative efficiency on the bit error rate is shown for the same protocols as in Fig. 12.50. The relative efficiency decreases with the increase of the bit error rate, too. In this diagram the go-back-N protocol stands out, having much lower relative efficiency than all the others; this was expected, as the go-back-N transmitter usually retransmits messages which have already been successfully transferred but have been discarded by the receiver because they did not arrive in the correct order. All the other four protocols, including the stop-and-wait protocol, have almost the same relative efficiency.

In Fig. 12.52 the normalised message delay is shown for the same protocols as the function of the bit error rate. The actual delays were normalised with the minimum delay of a message transfer which was given in Eq. (6.18); the normalised delay in

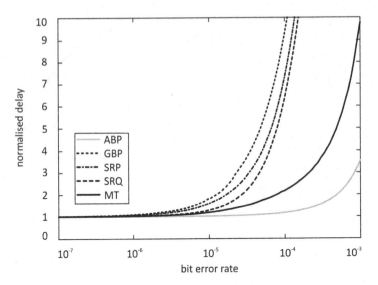

Fig. 12.52 Message delay of sliding window protocols as function of bit error rate

case of a lossless channel (bit error rate 0) is therefore 1 with all protocols, as any message is received by the receiver at the first trial if there are no losses. If the loss rate increases, the delay also increases, of course, as in average a message must be transmitted more times to be successfully received if the bit error rate is higher. In the scenario that is presented in this section the results of the diagrams shown in Figs. 12.50 and 12.52 agree in the sense that a protocol with a higher efficiency also yields shorter delays; the stop-and-wait protocol is an exception as it offers the shortest delays due to the fact that any information protocol data unit has all the resources of the channel at its disposal, until it is successfully received. Regarding other protocols, the selective-repeat protocol is better than the go-back-N protocol in this scenario, the selective-repeat protocol with negative acknowledgments and enquiries is still better, and the multiple-timers protocol is the best (excluding the stop-and-wait protocol). In all scenarios that were simulated by the author the shortest delay was achieved with the stop-and-wait protocol, the multiple-timers protocol was the second best, while the order of other protocols was not in all cases the same as in the scenario shown in Fig. 12.52.

In Figs. 12.50 and 12.52 one can see that the performance of the selective-repeat protocol with negative acknowledgments and enquiries can be better than the performance of the basic selective-repeat protocol, although all the messages were equally long and the timer expiration time was only slightly longer than the round trip time in this scenario. A protocol that employs negative acknowledgments and enquiries can be expected to be even more advantageous if a longer timer expiration time were used. This hypothesis was verified with simulations as well. The assumption was fulfilled in most cases, although not in all of them. In Table 12.2 the ratios of the efficiency η and delay Td of the selective-repeat protocol with negative

Table 12.2 Ratio of efficiency and delay at different timer expiration times with $ber = 10^{-4}$

K_t	$\eta_{srp\text{-}rej\text{-}enq}/\eta_{srp}$	$Td_{srp\text{-}rej\text{-}enq}/Td_{srp}$
1	1.2	0.98
2	1.6	0.58
4	2.4	0.37

acknowledgments and enquiries (*srp-rej-enq*) and the basic selective-repeat protocol (*srp*) at the bit error rate $ber = 10^{-4}$ are shown, where the timer expiration time was used that was K_t-times longer than the minimum timer expiration time, calculated with Eqs. (12.4) and (12.6).

Chapter 13
Flow and Congestion Control

Abstract First we explain what a congestion is, why and how congestions occur in communication networks, and why the control is needed. The difference between the flow control and the congestion control, as well as the difference between the congestion avoidance and the congestion recovery are described. Various control methods are discussed and classified with respect to the mode of control; their advantages and disadvantages are assessed. The role of feedback, either explicit or implicit, is emphasised. Then several more specific methods of flow/congestion control are discussed, such as the polling, the deferring of acknowledgments, the deferring of retransmissions (exponential backoff), the requests to stop transmitting, and the variable window width control method. The exponential backoff method and the variable window width control method are pointed out as the most important ones. Four figures are in this chapter.

13.1 Losses, Congestions and Control

As was already told in Chap. 6, two communication entities together with the channel interconnecting them constitute a system that looks quite simple (at least at the first glance). The transmitting protocol entity generates communication traffic into this system and the receiving protocol entity consumes this traffic from the system. This system uses communication resources (such as transmission rate, time, processing power, memory) for information transfer. If the operation of a system is based on the packet mode of information transfer (only such systems are treated in this book), the system assigns communication resources to individual communication processes or even to individual packets dynamically, according to their current needs. The transfer capacity of the system is limited with the limited amount of communication resources that are at disposition for information transfer; our definition of protocol efficiency that was given in Eq. (6.9) stems from this very limitation. A communication system that is logically correct works well as long as the information transfer required by the transmitter does not claim for more resources than are available in the system. If the offered load exceeds for a rather short time period the

D. Hercog, *Communication Protocols*, https://doi.org/10.1007/978-3-030-50405-2_13

amount that can normally be transferred with the available resources, the messages which cannot be carried in a timely manner are stored in the waiting queues which are always used by the network elements of a system based on the packed mode of transfer; if, however, such an overload lasts too long, the waiting queues fill up and messages are lost. The losses that are due to the lack of resources are the more severe the higher is the traffic intensity through the system. This kind of losses is characteristic for the communication systems which are based on the packet mode of information transfer.

A communication network is an expansion of a simple system which has just been described. In a network many network elements with waiting queues associated with them are operating concurrently. A network as a whole as well as all of its components also have a limited amount of communication resources at their disposal. If the communication traffic generated by traffic sources into the network claims for more resources than the network and its components have at their disposal and if such an overload lasts for too long, some waiting queues fill up and losses occur.

In a system based on packet mode of operation the probability of losses can be arbitrarily low if the system has an appropriate amount of communication resources at its disposition. However, this probability can be zero only if the amount of communication resources is so large that this amount is sufficient to carry any amount of traffic that can be generated by sources; unfortunately, such an amount of resources would be excessively costly and noneconomic as a substantial part of resources would be unused most of the time if this amount were available all the time. In practice, communication systems are therefore designed so that they have sufficient resources at their disposal most of time; however, sometimes they can happen to be insufficient which should occur with a probability that is low enough. The nature of the traffic in a communication system is random; therefore it may well happen in a system where the average traffic intensity is relatively low that a burst of traffic occurs which is high and long enough to fill up waiting queues and cause losses. Furthermore, losses may also be due to an uneven distribution of traffic intensity throughout the network, although the total traffic in the network is rather low. Of course, losses can occur in any network element, as all network elements, terminals as well as relay nodes, employ waiting queues.

In the communication networks used for data transfer automatic repeat request protocols are very usually employed; as we have already told, an ARQ transmitter detects losses and responds to them by retransmitting supposedly lost messages. The use of ARQ protocols in the networks with packet mode of data transfer forms a so-called *positive feedback*; this means that the offered load increases with the increase of the loss rate which in turn increases with the increase of the offered load (the higher is the loss rate, the more messages have to be retransmitted, thus increasing the offered load). In such systems the throughput increases with the offered load if the latter is low; however, the throughput increases the more and more slowly, the higher is the offered load. If the offered load increases more and more, the losses also increase more and more, until the loss rate becomes so high that the grand majority of messages which are transferred from a transmitter towards a

Fig. 13.1 Typical
throughput versus offered
load dependence of
potentially unstable systems

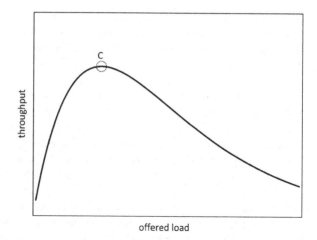

receiver are retransmitted messages and almost no messages are successfully received by the receiver; the throughput begins to decrease with the further increase of the offered load. The network is said to be congested, it is in the state of *congestion*. A congested network is unstable; in this state the throughput can only decrease until it drops to zero value—the network finds itself in a state called a *deadlock*. A network in a deadlock state cannot return into a stable state without the help of special recovery procedures.

Hence, all those communication networks in which the loss rate increases with the offered load and the offered load increases with the loss rate are potentially unstable as they include a positive feedback.

In Fig. 13.1 a typical dependence of the throughput on the offered load of a potentially unstable system is illustrated; the average values of both traffic intensities are plotted. The point of congestion where the throughput begins to decrease with further increase of the offered load is marked with the letter C. In the area left of the congestion point the system operation is stable, while it is unstable in the area right of the point C. However, one must be aware that the operation of a network can become unstable even if the average offered load is lower than the offered load of the congestion point, if a burst of a higher traffic intensity appears which lasts for a time period that is long enough (which can well happen with a higher or a lower probability, due to the random nature of the communication traffic); the probability for this to happen is the higher, the closer the average offered load is to the congestion point C. Due to this fact, the maximum throughput of the point C can never be achieved in practice; as soon as the offered load approaches the congestion point the network becomes unstable sooner or later and enters the congestion state.

A congestion in a network is of course undesirable, as the throughput drops while the delays increase in a congested network which means a much worse quality of service. A communication network and communication protocols must therefore be designed so that the offered load does not reach the values that are too close to the congestion point C; the closer is the average offered load to the congestion point the higher is the probability that a congestion occurs. The most important task of the *flow/congestion control* is therefore to try to keep the average offered load enough lower than the one in congestion point; this activity is referred to as *congestion avoidance*. What does the term »enough lower« mean in this context depends on how reliably one wants to keep a network in the stable state. If the working load comes closer to the point C, a higher throughput can be obtained, but with a higher risk of entering the congestion. The decision of a network designer which point of the offered load/throughput curve to consider as a normal network operation must therefore depend on the environment in which the network is to operate, as well as on the network performance and quality of service to be achieved.

As was told already, a congestion can occur in spite of the fact that the congestion avoidance activities have been used in order to retain the network in the state of stability. In the unwanted case that a congestion has occurred, the network must be somehow helped to return from the congestion state to the stable state of operation. This also is the task of the flow/congestion control methods, usually called *congestion recovery*.

It must be emphasised that the primary task of flow/congestion control is to prevent the occurrence of a congestion if possible; congestion recovery must only be used as an emergency procedure, should the congestion avoidance fail and a congestion occur in spite of prevention activities.

Although the title of this chapter is Flow and Congestion Control, there is however a difference between the flow control and the congestion control. In spite of this, both kinds of control are always treated together, as both of them employ similar methods.

Towards the end of this section, the goal of flow/congestion control can be summarised as follows: if longer overloads of the network occur, the flow/congestion control methods try to prevent the network to become congested; if a congestion occurs in spite of this, the flow/congestion methods return the network into the stable state of operation. If only short traffic overloads occur in a network, the waiting queues are used to resolve this situation.

13.2 Flow Control

The *flow control* is used to avoid congestions or to recover from them in a receiving protocol entity. Hence the flow control operates between two endpoints of a communication process. The flow control is relatively simple procedure, as the receiver (where a congestion can occur) and the transmitter (which can perform the control) communicate directly between them using some protocol; the flow control is the task of this protocol. If a protocol of the data-link layer of a protocol stack (according to the OSI reference model) is in question, the flow is controlled between two neighbouring network elements, hence the receiver can be a relay network element or a terminal. In case of a protocol of a higher layer (above the network layer) the receiver is a network terminal and the flow control is performed between two network terminals.

13.3 Congestion Control

In a communication network which uses the packet mode of information transfer, protocol entities are interconnected with virtual channels; a virtual channel consists of several network elements that operate in a lower layer of a protocol stack. In these elements, too, losses and congestions can occur. The *congestion control* tries to avoid these congestions or recover from them; hence the congestion control is used to prevent congestions in the network that interconnects both protocol entities. As the transmitting and the receiving protocol entities do not directly communicate with the network elements where congestions can occur (according to the rules of the protocol stack design), the congestion control can be more complicated than the flow control. The congestion control is usually used in higher layers of a protocol stack (normally above the network layer).

13.4 Modes of Control

The basic reason for a congestion to occur is the demand for communication resources that exceeds the amount of resources which are currently available in the network, in a part of it or in some of its elements. The most evident and the simplest solution to this problem is to simply discard some messages. If a waiting queue of a network element is already full, this action can of course not be avoided. Unfortunately, discarding messages is not a long-term measure if automatic repeat request protocols are employed in the network, as transmitters will detect losses sooner or later and begin retransmitting discarded messages which will in turn increase the offered load and consequently exacerbate the congestion which will make the situation still worse. It is, however, possible to discard some messages even before a waiting queue fills up, hoping that traffic sources will detect the losses and relieve the load upon the network, using the congestion control methods. Messages to be discarded can be chosen randomly; this procedure is therefore referred to as *random early drop—RED*.

A more long-term procedure is to limit the traffic intensity in the network and between the protocol entities and to adapt this limitation to the current state in the network. Either the amount of messages that can be in the network at the same time or the transmission rate can be limited and modified; of course, both these possibilities depend one on the other.

The flow/congestion control theory and practice are based on the general *control theory*. In control theory two kinds of control are distinguished: the *closed-loop control* is based on the *feedback* information, while the *open-loop control* uses no feedback. In case of the flow/congestion control the feedback provides the information about the state of the network and the possibilities of forthcoming or already existing congestions; this information is used by the network elements which are capable of executing the flow/congestion control. The network elements that are most capable to act are of course transmitters. The feedback information can be explicit or implicit.

If the open-loop control (without feedback) is used, a transmitter has no information about the state of that part of the network where a congestion could possibly occur, or even already has occurred. A transmitting protocol entity can therefore execute only those activities which are beneficial in any case, were there a congestion or not; an example such activity is restricting the offered load to a level that is considered appropriate to maintain a successful operation of the network and a good quality of service. Should, however, a congestion occur, the network cannot react to prevent it or recover from it.

If the closed-loop control with explicit feedback is used, an entity which approaches the state of congestion or has already its waiting queue filled up (e.g. a receiver) sends a report about its state or even an explicit instruction how to act to the entity that can control the flow or congestion (usually a transmitter). This explicit information poses an additional traffic load upon the network, so it is usually piggybacked into information protocol data units that are being sent towards traffic sources. The closed-loop control with explicit feedback is appropriate for the flow control where both protocol entities, the one which experiences a (threatening) congestion and the one that can act, communicate directly by interchanging protocol messages. The closed-loop control with explicit feedback is, however, less appropriate for congestion control. Congestions occur in a layer that is below the layer where controlling entities can act; as we already know, a direct communication between entities in different layers of a protocol stack is not allowed according to the rules of the protocol stack design.

Due to ever higher demands for a better quality of service in modern communication networks, the flow/congestion control is becoming more and more important nowadays; therefore a combination of the congestion control and the flow control with explicit feedback is sometimes used. A network element of a lower layer that senses an approaching congestion may report this situation to a protocol entity of the same layer in a terminal, which in turn reports this to the protocol entity in the higher layer of the same terminal, using an appropriate primitive. The control proper is then carried out by the protocol entities in the higher layer.

If the flow/congestion control with implicit feedback is used, the transmitting protocol entity which executes the control receives no explicit reports about the state of that part of the network where a congestion could possibly occur. It must therefore infer the situation in the network from the various signs it senses in its local environment. To this end, it can use the measurement of round trip times (which increase when a congestion approaches) or the sensing of message losses (in case of a congestion the loss rate increases substantially). When implicit feedback is used the information about the network state can be unreliable; the assumptions of a transmitting protocol entity about the actual reasons for the round trip time or the loss rate increase can therefore not be correct, and consequently its decisions on how to carry out the control can also be inappropriate. The control which is based on the implicit feedback is mostly appropriate to control congestions which occur in a layer that is lower than the layer where the control is executed.

Due to the random nature of the communication traffic in a network that is based on the packet mode of information transfer, the traffic conditions can vary strongly and rapidly in such a network. The information that a protocol entity has acquired implicitly can therefore already be obsolete when used; the use of such obsolete information can even be counterproductive. A lowpass filter is therefore sometimes used to average the information about the network state. Equation (13.1) can be used to this end

$$A_i = \alpha \cdot A_{i-1} + (1 - \alpha) \cdot M_i, \tag{13.1}$$

where M_i is the latest reported (measured) value, A_{i-1} and A_i are the previous and the new value of the running average, respectively, while the parameter $0 < \alpha < 1$ determines how much the reported values are to be smoothed: if the parameter α is small, the smoothing is weak, and if the parameter α has a big value, the smoothing is strong.

13.5   Polling

The *polling protocol* represents the simplest example of the flow control. Messages are exchanged between a *master* entity and a *slave* entity. The communication is controlled by the master which can poll the slave to transmit whenever it is itself prepared to receive messages.

13.6   Automatic Repeat Request Protocols and Flow/ Congestion Control

In Chap. 12 automatic repeat request protocols were discussed in detail from the point of view of the assurance of a reliable data transfer. In practice, ARQ protocols usually also execute the flow/congestion control to assure a better quality of service (which requires a fluent data transfer in addition to a reliable transfer); an important reason to incorporate the flow/congestion control into automatic repeat request protocols is that both reliability assurance and flow/congestion control can easily and elegantly be merged into a single ARQ protocol. This is due to the fact that some basic ARQ mechanisms, such as the acknowledgments, the retransmissions and the sliding window, basically intended to assure a reliable transfer, can also be used to control flow and congestions.

13.6.1 Deferring Acknowledgments

When we discussed automatic repeat request protocols we saw that the average rate with which a transmitter transmits new information protocol data units depends also on the rate with which it receives the acknowledgments, because the transmitter moves its window whenever it receives a new acknowledgment. Consequently, one might expect a straightforward method of executing the flow control by a receiver to defer sending acknowledgments when the receiving queue fills up, thus throttling message transmitting at the transmitting side. However, one must be careful! If acknowledgments are used for the flow control, they play two different roles simultaneously, namely acknowledging the message receptions and controlling the flow: the processes of assuring a reliable data transfer and controlling the flow thus become interdependent. The retransmit timer expiration time must be adjusted to the deferring of acknowledgments to prevent false timer expirations due to deferred acknowledgments; this could still deteriorate the situation. The flow control by deferring acknowledgments is unusual in that the receiver controls the data flow.

13.6.2 Deferring Retransmissions

In many networks losses are mainly due to the lack of communication resources and consequently full waiting queues. In such networks frequent timeouts indicate an approaching or even an already existing congestion somewhere in the network. In such situations a transmitter can defer retransmissions and thus relieve the burden upon the network. The usual procedure is that a transmitter increases by a certain multiplicative factor the retransmit timer expiration time or the postponement time of a message retransmission after each consecutive timeout (both actions have a similar effect). This procedure is referred to as *exponential backoff*. This algorithm is based on the assumption that the number of consecutive timeouts is the higher, the more serious is a congestion. However, after the transmitter has received a new acknowledgment, the timer expiration time must be reduced by a certain factor, to a previous value or even to the initial value, as the reception of a new acknowledgment means a relief or even the end of the congestion. Unfortunately, the exponential backoff procedure does not help or even deteriorates the situation if message losses are not the consequence of the lack of resources but are rather due to bit errors, as it unnecessarily increases the time after which the transmitter reacts to losses. The exponential backoff algorithm is an example of the closed-loop control with implicit feedback.

13.6.3   Request to Stop Transmitting

The flow control with *requests to stop transmitting* is a kind of the closed-loop control with explicit feedback and binary control of transmission rate (a transmitter may or may not transmit). The method can be used with any automatic repeat request protocol. Three control protocol data units are used which will be called here *acknowledgment*, *wait acknowledgment* and *enquiry*, and denoted with the acronyms *ack*, *wack* and *enq*, respectively. In various specific protocols these messages may have different names.

If a receiver estimates it can no more receive new information protocol data units because its waiting queue is (almost) full, it sends a control message wack to the transmitter; with this message it can acknowledge the messages it has already received, but at the same time forbids the transmitter to further transmit new information protocol data units until it receives a positive acknowledgment ack. When a transmitter receives a message wack it must cease transmitting new information messages; thereafter, it sends an enquiry enq to the receiver from time to time. While receiving answers wack to its enquiries it may not transmit new information messages; however, as soon as it receives an answer ack it may continue to send new information protocol data units. Whenever a receiver receives a control message enq it must respond either with the message wack (if it still cannot receive new messages) or with the message ack (if it is already prepared to receive new information protocol data units). Here we must still explain why the enquiry enq is really necessary. If the transmitter only waited to receive an ack message after it has received a wack and the ack were lost, the protocol would enter a deadlock: the transmitter would not be allowed to transmit and would just wait for an ack (which actually was sent but lost on the way), while the receiver would think it has allowed the transmitter to transmit and would wait for new information messages; hence, both the transmitter and the receiver would wait each other, while none of them could do anything and nothing would happen. Of course, a message ack as a reply to an enq can also be lost; however, in such a case the transmitter continues to send enquiries until it receives an ack successfully which happens sooner or later.

In Fig. 13.2 an example use of the control messages ack, wack and enq is shown. A large window width and a lossless channel are assumed; the transmitter T sends seven information protocol data units (with sequence numbers from 0 to 6) to the receiver R. After the transmitter has received the message wack it stops transmitting further information messages and starts a special timer; when this timer expires it transmits the control message enq and continues transmitting enquiries until it receives a positive acknowledgment from the receiver. After it has received an ack it may continue to transmit information messages. One can see that the receiver must send the wait acknowledgment message early enough because all messages are delayed when transferred through the channel. If the receiver in Fig. 13.2 sent a wack message only when its receive buffer was already full, some information protocol data units, already on their way towards the receiver, would be lost. The flow control with a request to stop transmitting is most suitable to be used with the

Fig. 13.2 Example of flow control with request to stop transmitting

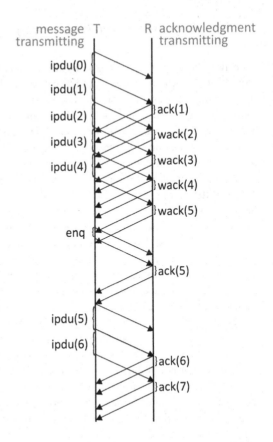

stop-and-wait protocol; with a continuous protocol it is better to use the variable window width control method which is going to be explained in the next section.

13.6.4   Variable Window Width Control

The request to stop transmitting flow control which was discussed in the previous section is a binary control: only two possibilities are available—either the transmitter may transmit or it may not transmit. In this section a more sophisticated method of flow/congestion control will be explained which allows the transmitter to transmit with more different average transmission rates which depend on the state of the network.

In Sect. 12.6 the sliding window protocols were discussed; in Eq. (12.15) the protocol efficiency and the maximum average transfer rate were shown to be proportional to the transmit window width, with the upper bound, of course, determined with Eq. (12.17). The dependence of the average rate of data transmission on the transmit window width is illustrated in Fig.13.3 where the transmitter

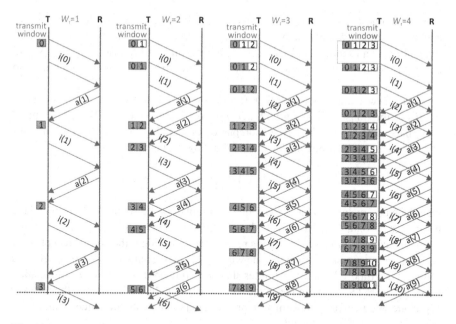

Fig. 13.3 Transfer of information protocol data units with four different transmit window widths

transmits information messages towards the receiver, the channel is lossless, and four different scenarios are shown with four different transmit window widths, namely 1, 2, 3 and 4. Transmit window widths that are larger than 4 have no sense in this case, as the transmit window width $W_t = 4$ already assures the maximum actual transfer rate and the protocol efficiency 1 (in the lossless case, of course). The abstract syntax i(s) is used for the information protocol data units and the abstract syntax a(s) is used for the acknowledgments where s is the sequence number of a message or an acknowledgment. At the left edge of each scenario the state of the transmit window is shown whenever it changes. The sequence numbers that have already been transmitted (or are just being transmitted) are shown on the grey background.

In Fig. 13.3, as well as in Eq. (12.15), one can see that the average transfer rate is proportional to the transmit window width; in the case of $W_t = 4$ this proportionality cannot be seen, as the actual transfer rate is bounded with the maximum transfer rate, which can also be seen in Eq. (12.15). In the same time period (which is delimited with the dashed line at the bottom of the figure) the receiver received 3, 6, 9 and 10 information messages in case of transmit window widths 1, 2, 3 and 4, respectively.

By modifying the transmit window width (only up to the upper limit, as defined in Eq. (12.17), of course) one can change the maximum average transfer rate and also the maximum number of messages that can be in the system at the same time. This is the basic principle of the *variable window width control*.

If the congestion control with implicit feedback is used, a transmitter measures or counts communication parameters (such as round trip times, lost packets, timeouts. . .) and modifies, if necessary, the transmit window width with respect to the measured/counted values; consequently, the average transmit rate is also modified.

If the flow control with explicit feedback is used, a receiver sends so-called *credits* to the transmitter along with acknowledgments; a credit tells the transmitter how many data the receiver is still prepared and capable to receive. In such cases cumulative acknowledgments have the following syntax:

$$ack\,(a,c),$$

where the parameter a contains the sequence number of the acknowledgment and the parameter c contains the credit. If the transmitter receives a cumulative acknowledgment ack (a,c), this means that the receiver acknowledges the correct reception of all information messages up to the one with the sequence number $a - 1$ inclusive and at the same time tells the transmitter that it is prepared to receive c information messages from the one with the sequence number a inclusive on. Hence the receiver is prepared to receive the information protocol data units with the sequence numbers from a to $a - 1 + c$, inclusive.

If the variable window width control is used, the functionalities of the reliable data transfer and the flow/congestion control are separated; acknowledgments are used to acknowledge the correct reception of information messages, while credits are employed to control the data flow. We can still add to this that the receiver determines the credit depending on the state of its waiting queue if the flow control with explicit feedback is used, while in case of the congestion control with implicit feedback the credit is determined by the transmitter itself depending on the state it can sense in its environment.

In Fig. 13.4 an example of the variable window width flow control with explicit feedback is shown. As was already told, the receiver (denoted as R in the Fig. 13.4) sends to the transmitter (labelled with the letter T) control protocol data units a(a,c) where a is the sequence number of a cumulative acknowledgment and c is the credit; the transmitter sends information protocol data units i(s) to the receiver where s is the sequence number of an information message. In this example, as can be seen in the figure, the transmit window width is 4 at the beginning of the scenario which allows the transmitter to transmit all the time, never filling up its window (in the absence of losses, of course). Unfortunately, this is too fast for the receiver, which is incapable to process all messages on time, so it requests the decrease of the transmit window width first to 3 and then even to 2; consequently, the transmitter reduces the average transmission rate by not transmitting from time to time (whenever its window is full). As soon as the receiver considers itself capable to receive more new messages and process them on time, it allows the transmitter to increase the window width to 3 by sending the message a(6,3) to it; as there is no need for the receiver to send a new acknowledgment, it only reports the credit change with the message a(6,3).

Fig. 13.4 Example variable
window width flow control

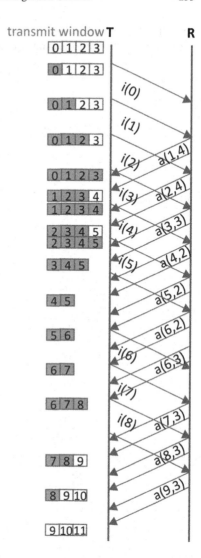

If a protocol (such as a transport layer protocol) operates between two terminals, interconnected with a network, congestions can occur in the receiving terminal as well as in the intermediate network; both flow control and congestion control are therefore necessary. The transmitter can modify its window width both with respect to the credit received from the receiver and with respect to the conditions it senses in its environment. The transmit window width W_{ta}, referred to as the *advertised window*, is used for the flow control and is advertised by the receiver by sending credits to the transmitter; the transmit window width W_{tc}, called the *congestion window*, is used for the congestion control and is determined by the transmitter itself with respect to the conditions sensed in its environment. Of course, the transmit window width must fulfil both conditions, hence the one imposed by the flow control and the other imposed by the congestion control; therefore, the transmit window width must fulfil the following condition:

$$W_t \leq min\,(W_{ta}, W_{tc}).$$

(13.2)

Chapter 14
Medium Access Control

Abstract At first the mode of information interchange in classical local area networks via a common transmission medium is discussed and the need for the medium access control is pointed out. Then the peculiarities of the message transfer in local area networks are described. The topologies that are used in wired and wireless local area networks are presented, as they have a strong impact on the protocols that are to be used in such networks. Then the medium access control methods are classified. In the continuation of the chapter, only the asynchronous medium access control methods are treated. The methods which are more thoroughly described are token passing, basic random access (aloha), carrier sense, carrier sense with collision detection and carrier sense with collision avoidance. The token passing, basic random access and CSMA/CD protocols are also formally specified. There are 11 figures in the chapter, including those that formally specify the three protocols in the SDL language.

Communication networks with a limited physical size and a limited number of communicating elements are referred to as *local area networks* and denoted with the acronym *LAN*. *Metropolitan area networks* (*MAN*) and *Personal area networks* (*PAN*) also have a limited physical size and a limited number of communicating elements; however, MAN networks are larger and PAN networks are smaller than LAN networks both in physical size and in number of elements. As the differences between these three kinds of networks are more quantitative than qualitative, these differences will not be of special interest in this chapter.

The early local area networks were data networks, interconnecting small groups of computers, mainly in universities, companies or homes. These computers were, and still may be, physically connected to a common transfer medium (channel) to which all computers transmit and from which all computers receive information. Consequently, the mode of communication in such networks is a *broadcast* communication by its very nature; hence, any communicating element is able to receive anything. Because all communication processes use the same transfer resources that are offered by the common medium, the most important problem to be solved in such networks is the management of common resources in the way that any resource is

D. Hercog, *Communication Protocols*, https://doi.org/10.1007/978-3-030-50405-2_14

257

Fig. 14.1 Topologies of
local area networks

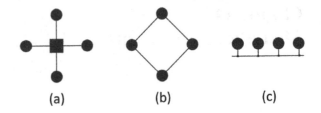

used by at most one communication process at any time. This kind of resource management is referred to as *medium access control* and denoted with the acronym *MAC*. The medium access control is the task of special *medium access control protocols* (*MAC protocols*); these protocols are the topic of this chapter. The medium access control protocols in LAN/MAN/PAN networks are so important that a separate sublayer of the data-link layer was defined in the OSI reference model (see Sect. 5.4.1).

In local area networks, the term *station* is often used for network elements.

Before we discuss the medium access methods in more detail, let us see which *topologies* are used in local area networks, hence the way in which LAN elements can be interconnected with the common medium; one must be aware that the choice of the MAC method to be used in a network strongly depends on the common medium topology of that network.

In Fig. 14.1 the three topologies which are most frequently used in local area networks are shown. In Fig. 14.1a the *star topology* is shown; in networks employing the star topology one network element is different from all the others (in the figure it is shown as a square); all information that is transferred between network elements passes through this element. The *ring topology* is shown in Fig. 14.1b; in ring networks the functionalities of all network elements are equal, however, the elements are ordered (any network element has exactly one predecessor and exactly one successor), their order being determined with their interconnection; the speed of the propagation of messages through a ring is determined with the rate of their processing and transmission in network elements, as messages are transferred from one network element to its successor along the ring. The *bus topology* is shown in Fig. 14.1c; all network elements on a bus also have equal functionalities; the speed of information transfer on a bus is, however, limited only by the laws of physics (this speed cannot be higher than the speed of light).

In Fig. 14.1 the physical topologies of local area networks are shown, hence the way in which network elements are physically interconnected. One physical topology can, however, be mapped into a different logical topology by the use of a medium access control protocol; then a MAC protocol is employed for the medium access that is more appropriate for the logical topology. For example, using a network element called *hub*, a physical star topology can be mapped into a logical bus topology; a physical bus topology can also be mapped into a logical ring topology by maintaining a logical order of network elements.

Nowadays *switches* are very often used in local area networks. Whenever a switch receives a message on one of its inputs it relays it towards the destination

Fig. 14.2 Topologies of
wireless local area networks

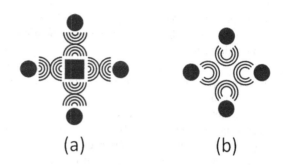

(a)   (b)

network element via one of its outputs. The operation of MAC switches is a
combination of the operations of switches and routers in classical communication
networks. What is important from the viewpoint of this chapter is that there is no
common transfer medium if every network element is connected to its own switch
port (used exclusively by this element) by a duplex connection; in such a case, the
medium access control is no more needed.

From what was told in the previous paragraph one could conclude that the
medium access control methods are obsolete and no more needed nowadays. In
reality, this is not true! In *wireless local area networks* (*WLAN*) communication is
based on the radio transmission of information; this, however, means that a common
communication medium, represented by a radio channel and its frequency spectrum
in the area where the communication is running, is used. Wireless local area
networks are extremely important nowadays, especially if one considers the radio
access networks (cells) of wide area mobile networks also to be wireless local area
networks.

Two different topologies are used in wireless local area networks; both of them
are illustrated in Fig. 14.2. In Fig. 14.2a an *infrastructured network* is shown;
essentially this is a star network, implemented with the radio transmission. The
network element of an infrastructured network that is different from the other
elements and which relays all the traffic (in the figure it is shown as a black square)
is usually referred to as *access point* or *base station*. (The term access point stems
from the fact that it provides for the access to external networks, such as the Internet.)
In Fig. 14.2b an ad-hoc *network* is illustrated; in an ad-hoc network all network
elements have the same functionality, in fact this is a wireless variant of a bus.

In what follows some characteristics of the information transfer in local area
networks will be considered.

Whether the information is transferred via a common medium or switches are
employed, any network element must possess an address in any case. In classical
local area networks these addresses are called *MAC addresses*. Often 48-bit MAC
addresses are used which were standardised by the organisation IEEE.

Classical local area networks are usually wired (network elements are
interconnected with copper cables or optical fibres). A limited size of a network
(which is usually relatively small) has two consequences: signal delays are quite
short, and bit error rates are low due to short connections. MAC protocols in classical

LANs therefore do not try to provide a reliable data transfer but rather concentrate on the common medium access control; therefore the mode of communication is connectionless.

If an electromagnetic wave propagates in a free space, its power density decreases with the square of the distance from the transmitting antenna. The powers of received signals can therefore be very low, even if the size of a network is small; due to this fact the bit error rates in wireless local area networks can be extremely high at the receiving side. There are also some other physical phenomena that impact the bit error rate, such as the interference. Therefore the MAC protocols that are used in wireless local area networks must also detect and correct errors, in addition to do the medium access control. Apart from high error rates, some other problems are also encountered in wireless networks which are going to be discussed in later sections.

14.1 Modes of Medium Access Control

Until now, many different medium access control methods have already been invented and used; they differ in the mode of medium access as well as in their properties and usefulness.

The medium access can either be centralised or distributed. If a *centralised access* is used, there is a special master network element which assigns communication resources to different communication processes. If a *distributed access* is employed, all network elements have the same functionalities and they coordinate the use of communication resources themselves.

A further division of medium access methods is into synchronous access and asynchronous access methods. A *synchronous access* makes a *circuit-oriented transmission* of information possible, hence an information transfer where the delays of all transferred messages are equal. All methods of synchronous access are centralised, as the master network element must assign any communication process a certain amount of communication resources for some time period; this process is the only one that may use the assigned resources during this time, so the delays of all messages are equal. On the other hand, an *asynchronous access* provides for a *packet-oriented transmission* of information, hence delays can change from one message to another; if this transfer mode is used, the resources of the common medium are assigned to different communication processes dynamically, with respect to their current needs; because of this, the shares of resources assigned to processes are variable. Furthermore, there are two kinds of asynchronous access, a scheduled access and a random access. While with a *scheduled access* a kind of schedule is maintained which regulates the use of common communication resources, with a *random access* the data transfer is by no means regulated, hence there is no schedule—the times when network elements begin transmitting are more or less random.

Nowadays different kinds of medium access are often combined. For example, a synchronous or a scheduled asynchronous access can be combined with a random access; in such a case resource assignments or a schedule can be based on resource reservations which are made by communication processes using a random access method.

We have already told that in wired local area networks switches are most usually used nowadays, hence there is no common medium and medium access control methods are not needed in such networks.

In local and personal networks which are most usually used in individual homes, companies and universities for the data transfer the medium access control is in most cases distributed and based on asynchronous random access; on the other hand, in the cells of wide area mobile networks operated by telecommunication operators the centralised medium access control combined with the random access based resource reservation is often used.

In this book the asynchronous medium access methods will be mostly treated, as these methods require the use of special protocols on which the network operation is based.

14.2   Synchronous Medium Access

The synchronous medium access control methods are similar to classical synchronous multiplexing methods. There are several synchronous MAC methods, differing in which communication resource is assigned to different communication processes; synchronous MAC methods are named similarly as classical multiplexing methods, only the term »multiple access« is used instead of the term »multiplex«. The kinds of synchronous multiple access methods are *frequency division multiple access* (*FDMA*), *time division multiple access* (*TDMA*), *code division multiple access* (*CDMA*) and *orthogonal frequency division multiple access* (*OFDMA*). One must, however, be aware that pure synchronous methods tend to use communication resources quite wastefully, so they are virtually no more used in modern communication networks. Nowadays these methods are mostly combined with asynchronous methods; thus, however, the communication essentially becomes asynchronous.

14.3   Scheduled Asynchronous Medium Access

Scheduled asynchronous medium access methods can be either centralised or distributed. As was told already, a kind of a *schedule* is maintained in the network which determines when which network element may transmit. If the control is centralised, the schedule is managed by the master network element; if the control is distributed, however, the schedule is coordinated by all network elements. Of course, the schedule management also requires some amount of network resources which is especially evident when the load upon the network is low. Therefore, the delays can be considerable even if the network is only lightly loaded. Hence the performance of a scheduled medium access method is not very favourable if the traffic intensity in a network is low.

14.3.1 Polling Protocols

A *polling protocol* is a simple medium access method with a centralised control. The master element determines when other network elements may transmit by polling them. Different strategies can be used to determine the order in which network elements may transmit; example strategies are *round robin scheduling* or *priority scheduling*. The master element can also determine the time period within which a network element may transmit. The protocol is extremely simple, but unfortunately it ceases to function if the master fails.

14.3.2 Token Passing Protocol

The *token passing protocol* is a scheduled asynchronous medium access protocol which is based on a predetermined order of protocol entities on a physical or a logical ring; the order of protocol entities determines the order in which they may transmit. Any protocol entity on a ring has exactly one predecessor and exactly one successor. If the order of elements is determined physically, as was shown in Fig. 14.1b, this kind of local area network is called *token ring*. However, communication elements can also be physically interconnected with a bus and their order determined logically (the logical order is set up when the network is initialised); such a network is referred to as *token bus*. Both protocols are standardised. The token ring was quite a popular technology a few decades ago and was later even succeeded by a similar protocol which was used in metropolitan area networks and named *fibre distributed data interface (FDDI)*. Nowadays the local area networks ethernet which are based on the random access technology are by far predominant; a similar random access protocol is used also in wireless local area networks.

With token passing protocols any message is propagated around the physical or logical ring, from one protocol entity to another (from a predecessor to its successor); a message is removed from the ring (is not forwarded to the next entity) by the entity that first transmitted it, so it is received by all the entities (one after another) on the ring—hence the communication is of the broadcast type. The transmission of information messages is regulated by a special control packet referred to as *token*. Just like all the other messages, the token, too, is propagated around the ring from one protocol entity to another one and thus allows network elements, in the same order, to transmit information messages. When a network element receives from its predecessor a message which is not the token it must relay it to its successor, unless it has itself transmitted this message (in this latter case it discards the message and so removes it from the ring). If the station recognises the destination address of the message as its own address, it passes the user message which is contained in it to its user. If, however, an entity receives the token from its predecessor, it may begin transmitting its message if it has already it prepared, otherwise it must relay the token to its successor. Of course, it also must forward the token to its successor after it finished its transmission. If it does this only after it has received the message it transmitted itself, this is called the *release-after-receive* method (*RAR*); in this case a single token and a single message can be present on the ring at the same time. If, however, an entity transmits the token to its successor immediately after it has finished its transmission, the *release after transmit* method (*RAT*) is used; in this case several tokens and several messages can be present on the ring at the same time. While the RAR method is appropriate for small rings with a small number of communication elements, the RAT method is usually used on large rings with a large number of elements. The RAT method can only be used in networks with a physical ring topology.

Of course, in real life the token passing protocol is not as simple as was described here. The most important function that must also be emphasised is the ring management. On a physical ring the order of communicating stations is determined by physically connecting them, each one with its predecessor and successor; in a token bus network, however, the order of stations is not physically predetermined, so the protocol must define a special procedure to be run during the network initialisation to determine the order of protocol entities. Furthermore, one of network elements must provide also for the management of the token—during the network initialisation it generates the token and then, during the network operation, it checks that the token has not been lost or doubled. A special problem that must also be taken care of is the reliability of the ring operation. Because all packets (including the token) are transferred from one network element to another, the network might cease to operate if one of its elements failed. In token ring networks this problem is solved by the availability of physical *bypasses*, while token bus networks must reconfigure themselves if a failure occurs, in order to restore the network operation. Token passing protocols also may provide some additional functionalities. Various network elements may have different priorities which determine the order of transmission, in addition to the order of network elements. The protocol also can limit transmission times and consequently limit also the message delay variations.

Nowadays the token passing protocols are much less used than they were in the past; only their basic behaviour will therefore be described here.

In Fig. 14.3 the specification of the protocol entity functionality according to the simple token passing protocol is shown. The protocol that is specified here allows exactly one token and at most one message to exist on the ring at the same time, hence the RAR method is specified. The protocol entity which is specified in this figure acts also as a token manager, so it generates the token and sends it to its successor during the initialisation; of course, only one protocol entity in the network is allowed to do this. The protocol entity is specified as an SDL process s, while the process nxt is its successor on the ring. All the other network elements on the ring have the same functionalities, only they do not generate a token during the initialisation (of course, they also have different addresses and different successors).

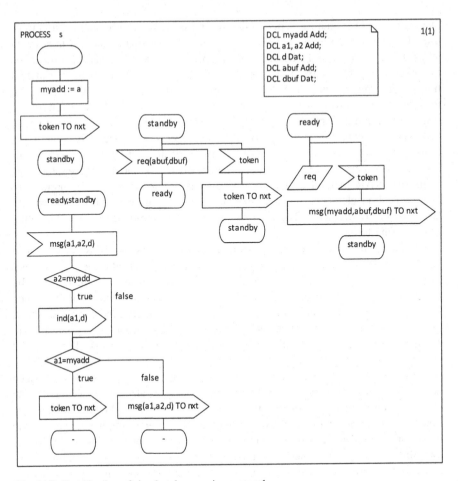

Fig. 14.3 Specification of simple token passing protocol

In the specification the data type Add is used to model addresses of communication elements on the ring and the data type Dat is used to model user messages to be transferred in the network. In the specification shown in Fig. 14.3 the definition of these data types is not shown, these data types were defined at the system specification level which is not shown here; indeed, they are even not important for us (in different networks, different formats of addresses and data may be used). Any protocol entity on the ring must possess its address; the entity specified here stores its own address in the variable myadd. The signal token models the token, the signal msg(Add,Add,Dat) (this is the way the signal is specified in the SDL system specification which is not shown here) models the information protocol data unit; the first two parameters of this signal (both of data type Add) contain the addresses of the source and the destination of this message, respectively, while the third parameter (of data type Dat) contains a user message (service data unit). A protocol entity may send the signals token and msg only to its successor. A user requests the transfer of a user message with the primitive req(Add,Dat) where the first argument contains the destination address and the second argument contains the user message. A protocol entity forwards a user message to the destination user with the primitive ind(Add,Dat) containing the source address (the first argument) in addition to the user message (the second argument).

A protocol entity can be in one of two different states. The initial state is standby; in this state the entity has no user message prepared to be transmitted, while in the state ready it has a user message and its associated destination address stored in the variables dbuf and abuf, respectively; of course, there may also be zero or more additional requests in the input queue of the process. As was already told, standby is the initial state of the process. The process changes its state to ready after it has received a request for a user message transfer from its user and has stored this request into the variables dbuf and abuf. If a new request req is received in the state ready, it must stay in the input queue and wait there to be processed in the state standby. If the process receives the token when in the state ready (when it has a new user message prepared to be transmitted), it transmits the message and returns to the state standby; if, however, it receives the token in the state standby, it forwards it to its successor. If the process receives a protocol data unit msg, the further processing depends on both addresses contained in this message. If the destination address of the message is its own address, it forwards the user message to its user. If the source address of the message is its own address, it does not forward it to its successor (thus removing the message from the ring) but rather releases the token and sends it to its successor; if the process is not itself the source of the message, it forwards the message to its successor.

Although the structure of the system is not presented here, we must emphasise that nxt is the name of the station which is the successor of the specified station.

14.4 Random Medium Access Control

Random medium access methods are always distributed methods. A communication entity which has an information message ready to transmit can in principle initiate the transmitting procedure at any time; however, it may do this more or less politely, thus more or less bothering other entities in the same network. Hence, network elements transmit their messages in random points of time, from where originates the term random access. Due to such a mode of the access to common network resources, two or even more network elements may happen to use the same network resources at the same time; in such a case a *collision* is said to have occurred. All the messages that have been affected by a collision are of course destroyed. A transmitter must detect in some way the losses that are due to collisions and retransmit the lost messages. It is not difficult to understand that the probability and therefore the frequency of collisions is the higher, the higher is the traffic intensity in the network; on the other side, the higher is the loss rate, the more are there retransmitted messages and the higher is the offered load. Evidently, there is a positive feedback in random access networks; such networks are therefore potentially unstable, their dependence of the throughput on the offered load is such as was shown in Fig. 13.1 in Sect. 13.1. In order to avoid congestions or to recover from them, should they already have occurred, congestion control procedures must also be used in random access networks. The operation of random access networks is the best if the load upon the network is low, as loss rates are also low in such conditions; because message retransmissions are rare in such cases, delays are also short if loads are low. If the load upon a network is high, collisions and hence also losses are frequent and message retransmissions are also frequent; hence, delays are long and the probability of a congestion is high.

14.4.1 Basic Random Medium Access Protocol (Aloha)

Aloha is the simplest random medium access control protocol. This also was the very first random access protocol to be invented and implemented; the name of the protocol (and this kind of networks as well) originates from the fact that it was first used to interconnect the computers of the University of Hawaii.[1] Here the original protocol and the network aloha will not be described in all details (the aloha network was naturally based on radio transmission, Hawaii being an archipelago); the basic random medium access protocol can be used in various variants and in networks based on different communication technologies and configurations. Furthermore, it served as a starting point to develop more sophisticated and more efficient random access protocols.

This protocol is not only the simplest but also the most thoughtless, not to say the most brutal, as a communication entity may begin transmitting as soon as it has prepared a packet to be transmitted, regardless of what is going on in the network at the moment. If the common medium is just being used by other entities in that moment, a collision can of course not be avoided. If the destination entity receives the packet successfully, it must immediately send an acknowledgment. An acknowledgment that has been successfully received tells the transmitter that there was no collision. If, however, the transmitter does not receive the acknowledgment in expected time (hence if it times out), it triggers a procedure for retransmission. In such a situation it of course may not retransmit the packet immediately. If a collision has occurred, all the transmitters affected by the collision sense it almost at the same time; if they all retransmitted their packets immediately, they would do it almost at the same time and a new collision would occur for sure. A transmitter that has timed out therefore chooses a random time of postponement after which it retransmits the packet. The probability that at least two transmitters choose a same time of postponement and so cause a new collision is the lower, the larger is the interval of possible postponements to be chosen from; however, one must be aware that, while

---

[1]The word aloha in Hawaiian language means a salutation, but also has several other meanings.

lowering the probability of a next collision by broadening the range of possible postponements, the average delays are also increased. The space of possible postponement times must be discrete; any two different postponement delays must differ at least by the amount that is necessary that a new collision does not occur if any two transmitters choose two different postponement times (here packet transmit times and propagation times in the medium must be taken into account). The quantum of postponement times must therefore be greater that the longest possible delay in the network.

For the congestion control the exponential backoff method, which was explained in Sect. 13.6.2, is used. The more consecutive timeouts indicate the higher load upon the network and consequently the higher frequency of collisions; the probability of further collisions must be lowered in such cases by increasing the maximum or average value of randomly chosen postponement times for packet retransmissions; in fact this is a congestion avoidance procedure. Furthermore, the maximum number of retransmissions must also be limited: if the network is heavily loaded (already congested), it is most easily relieved if communication elements cease to transmit; this is a congestion recovery procedure. Of course, a protocol entity must warn its user if it withdraws from transmitting; it is then up to the user when to repeat its request for the data transfer.

In Fig. 14.4 the specification of the functionality of a protocol entity operating according to the basic random medium access control protocol (aloha) is shown. As everywhere in this book, this specification is given in the SDL language.

The specification in Fig. 14.4 uses some constants, data types and signals which are declared at the system level; these declarations are therefore not shown in this figure; for the sake of simplicity and generality only the operation of a protocol entity is specified here as the process s.

In the specification two data types are used which are not defined here because their definitions are of no interest in this book, as they can in practice be implemented in different ways. The data type Add is the set of all addresses of protocol entities, while the data type Dat represents the set of all user messages. The constant nmax of data type Integer determines the maximum number of consecutive timeouts that are allowed; the constant dmax of data type Duration defines the maximum possible round trip time in the network and hence the necessary retransmit timer expiration time; the constant quant determines the quantum of postponement times; the constant keb defines the multiplicative factor used to increase the postponement times from one timeout to the next one.

A protocol entity uses primitives to interact with its users. A user requests the protocol entity to send a user message to another user with the primitive req(Add, Dat) (the signal declaration is written here in SDL just as it was written in the specification of the communication system), where the first parameter of data type Add contains the address of the destination protocol entity and the second parameter of data type Dat contains the user message itself. A receiving protocol entity forwards a user message to its user with the primitive ind(Add, Dat) where the first parameter contains the source address of the message and the second parameter contains the user message itself. A transmitting protocol entity reports the success of

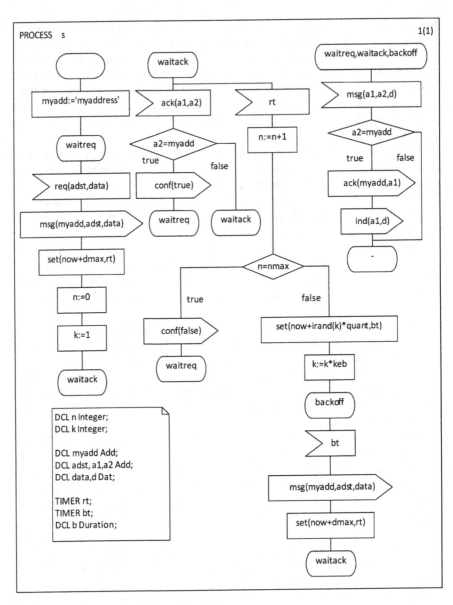

Fig. 14.4 Specification of aloha protocol entity

a message transfer to its user with the primitive conf(Boolean) where the value true of the parameter means a successful data transfer and the value false means an unsuccessful transfer. Protocol entities exchange information protocol data units msg(Add,Add,Dat) between them, where the first two parameters contain the addresses of the source and the destination, respectively, and the third argument contains a user message, and control protocol data units ack(Add,Add) where the

parameters contain the addresses of the sender and the receiver of the acknowledgment, respectively. Because all protocol entities are interconnected with a common
information transfer medium, a message (either msg or ack) which is transmitted by
any entity is received by all entities (except, of course, if the message is destroyed by
a collision); because of this an entity must verify if it is itself the destination of the
message it has received.

Of course, the protocol entity must know its own address; therefore it memorises
it in the variable myadd (in Fig. 14.4 this value is defined during the process
initialisation). Apart from several variables the process s also uses two timers; rt is
the retransmit timer, while bt is the backoff timer (regulating retransmission
postponements).

The protocol entity can be in one of three different states. In the state waitreq it
can receive a user request req; upon this it transmits the information protocol data
unit msg, sets the retransmit timer rt and initialises the counter of consecutive
timeouts n and the factor of retransmit postponement time k. Then it enters the
state waitack to wait for the acknowledgment of the successful data transfer ack. If it
receives the acknowledgment, it confirms the success to its user and returns into the
state waitreq. Of course, it must verify that it has received the awaited acknowledgment, as there may be many different messages, including many different acknowledgments, on the medium. If, however, the entity s times out (the timer rt expires), it
triggers the retransmit procedure if the timer has not yet timed out nmax times (if the
maximum allowed number of consecutive timeouts has already been reached, the
entity warns the user with the primitive conf and returns into the state waitreq). The
protocol entity begins the retransmit procedure by calculating a random integer
number between the values 0 and k, inclusive, using the function irand(k) for this,
and then postpones the retransmission by the random multiple of the postponement
quantum, using the timer bt for this purpose; it waits the expiration of bt in the state
backoff. After any successive expiration of the retransmit timer rt it increases by the
factor keb the width of the interval from where it chooses a random integer number.
Here it must be emphasised that the SDL language does not define any functions to
calculate random values, so the function irand in our specification is an informal one
in the sense of the SDL language. After the backoff timer bt has expired the entity s
retransmits the message msg. In our specification it was assumed that the user sends
another request for the data transfer only after it has received a confirmation of the
(un)success conf of the previous one; if this were not assumed, the save symbol
/req/ should be used in the states waitack and backoff to preserve an eventual new
user request in the input queue of the process.

The protocol entity can receive an information message msg in any state. It
verifies that the destination address of the received message is its own address. If
it is, it sends the acknowledgment and forwards the received user message to the user
with the primitive ind.

The retransmit timer of a protocol entity can expire because either the information
message msg or the acknowledgment ack has been destroyed by a collision. In any
case the transmitting entity retransmits the message msg. If the acknowledgment has
been lost, the receiver will receive a duplicate. Because, however, protocol messages

msg do not contain any sequence numbers or some other message identifications, there is no way for the receiver to recognise the retransmitted message as a duplicate. There is nothing wrong with this! As was already told all medium access control protocols are connectionless protocols and therefore unable to provide a reliable data transfer. If a reliable transfer of messages is needed, some protocol in a higher layer (usually in the transport layer) must provide for it.

In previous paragraphs we described the simplest random access protocol which implements both medium access control and congestion control. The alohanet network was the first local area network that enabled the packet-mode data transfer between computers. The alohanet operation was based on the radio transmission, but the common medium was used only for the data transfer from terminals towards the central computer; two frequency bands were used, one for the transfer from all terminals to the central computer (here the protocol, described in preceding paragraphs, was employed) and the other for the transfer from the central computer towards terminals (this was controlled by the central computer). Of course, the aloha protocol can also be used in wired networks with different transmission techniques and different topologies. Different variants of the protocol were also developed and deployed later.

An important step in the development of the aloha protocol was the *slotted aloha* protocol. With the slotted aloha protocol a statistical multiplexing is used. The time is partitioned into time slots of equal length; the duration of all protocol data units equals the duration of a time slot. A protocol entity may begin the transmission of a protocol data unit exactly at the beginning of a time slot; if necessary, the transmission of a protocol data unit must therefore be deferred until the beginning of the next time slot.

As was already told, the aloha protocol is efficient only if the offered load is low, while with the increasing offered load the throughput decreases (if compared with the offered load), the delays increase, and the probability of a congestion also increases; the dependence of the throughput on the offered load is such as was seen in Fig. 13.1. The slotted aloha protocol has a similar characteristic, only it is more efficient. With the aloha protocol the maximum throughput of 0.18 can theoretically be achieved in the point C of Fig. 13.1, while with the slotted aloha protocol the maximum throughput that can theoretically be achieved is 0.36. In practice, however, this throughput cannot be obtained, because a congestion would occur already at the loads that are lower than the point C and the throughput would drop.

It is interesting and instructive to ask oneself how is it possible to double the protocol efficiency by simply discretising the time. The principle is illustrated in Fig. 14.5. Before we explain the difference between the two protocols, let us assume that all protocol data units (with both protocols) have the same length, let their transmit times be T. Furthermore, the time of signal propagation between protocol entities is assumed to be much shorter than the message transmission time and therefore neglectable. Let a communication element be transmitting a message that is shown in grey on the top of both parts of the figure; hence that protocol entity transmits the message from time 0 to time T. The protocol efficiency strongly

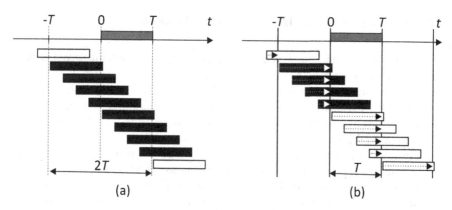

Fig. 14.5 Explanation of protocol efficiency of (**a**) aloha and (**b**) slotted aloha

depends on the time period within which no other element may begin transmitting not to cause a collision. As can be seen in Fig. 14.5a all the messages that appear within the time interval $-T < t < T$ overlap with the »grey« message at least a little (which is sufficient to cause a collision), if the plain aloha protocol is used. The messages that cause a collision are shown in black. The length of the critical time interval in case of the aloha protocol is evidently $2\,T$. In Fig. 14.5b a similar scenario is shown for the slotted aloha protocol; in this figure time slots are delimited with vertical solid lines. If the slotted aloha is used, all the messages are delayed until the beginning of the next time slot and transmitted only then; in the figure this is shown with horizontal arrows showing in the direction of increasing time. One can see in the figure that in this case the critical time interval within which no message may appear is $-T < t < 0$, the length of this interval is thence T which is a half of the critical interval with plain aloha. Well, one might object that in the case of the slotted aloha protocol more messages are stuffed into a single time slot; one must, however, be aware that a congestion occurs regardless of the fact how many messages overlap and for how long. Therefore, twice as much time is lost with collisions in the case of aloha as in the case of slotted aloha.

The aloha protocol also served as a starting point for the development of other random access protocols that are more thoughtful and more efficient; some of them will be treated in the following sections. One must, however, be aware that these protocols are »better« only in relatively small networks where propagation delays between network elements are relatively short. Consequently, in networks which are physically large (e.g. satellite networks) the aloha protocol and its variants are still used, as they are much simpler than the protocols which are going to be explained in next sections.

14.4.2   *Carrier Sense Multiple Access Protocols*

In the previous section we stated that the basic random medium access control protocol is rather inefficient, especially if the load upon the network is high; to a great extent, this can be considered due to its thoughtless and selfish nature—a protocol entity begins to transmit with no regard to what is currently going on in the network. Collisions are the consequence of this thoughtlessness, and a consequence of collisions is the loss of time as a communication resource.

A measure that can be taken to mitigate this problem is that a protocol entity listens to the medium and verifies if there is already somebody transmitting, before it begins the transmission itself. This procedure is referred to as *carrier sense*, and the medium access that uses this procedure is called *carrier sense multiple access* or *CSMA* for short. Hence, a communication element always senses the medium before transmitting and may initiate the procedure to transmit only if it »hears« no other activity on the medium. But, even if the entity senses no activity, this does not necessarily mean that the medium is indeed free. It may well happen, due to signal delays on the medium, that in the moment when one element senses the medium another one is already transmitting, only the beginning of the message that is being transmitted has not yet arrived to the entity which senses. In the worst case (if the two entities are separated by a maximum possible distance in the network) the signal needs the time $T_p = D_{max}/v$ to arrive from one entity to another, where D_{max} is the longest possible distance between any two network elements (hence the physical size of the network), and v is the speed of electromagnetic wave propagation on the medium. An entity is therefore unable to sense the activities which have begun later than T_p time units before sensing. The sensing is the less efficient, the larger (physically) is the network.

Several variants of the CSMA protocol do exist. With all of them, however, a transmitter which detects a collision (because it has timed out) initiates the procedure of message retransmission only after a random delay. As we already stated for the case of the basic random access protocol, the postponement delay must be quantised.

The *1-persistent protocol* defines the following procedure for data transmission. An entity that wants to transmit first verifies if the medium is free. If it is, the entity immediately begins to transmit; if it is not, the entity waits until the medium becomes free and then immediately begins to transmit (the name of the method originates from the fact that the probability that the entity begins transmitting immediately when possible is 1). The probability that two different network elements sense a free medium almost at the same time is low, although not zero. Much higher, however, is the probability that two or more entities simultaneously wait for the medium to become free; as soon as they sense the medium free, they both (all) begin transmitting and a collision is unavoidable. This collision is resolved with a retransmission after a collision, as was described in the previous paragraph.

The *non-persistent protocol* is another variant of CSMA protocols. An entity which wants to transmit first senses the medium to find out if it is free. If it is, it immediately begins transmitting. If it is not, the entity waits some randomly chosen time and then again senses the medium the same way as described in the previous sentence.

It is interesting to compare the performances of 1-persistent and non-persistent protocols. The efficiency of the non-persistent protocol is better because it is less aggressive in seizing the medium than the persistent protocol, which results in less frequent collisions; on the other side, a non-persistent network element in average waits longer to seize the channel than a persistent element, so the average delays are longer in case of the non-persistent protocol.

The *p-persistent protocol* is also used; with this protocol, the time is partitioned into time slots. An entity which wants to transmit a message verifies if the time slot is free. If it is free, the entity begins its transmission with the probability p, or with the probability $1 - p$ repeats this procedure in the next time slot, until it succeeds to transmit the message. If, however, it finds a time slot already seized by another entity, it repeats the procedure, which has just been described, in the next time slot.

CSMA protocols, too, are potentially unstable and have the same characteristic $S(G)$ that was shown in Fig. 13.1. One must therefore employ congestion control methods, just like with the aloha protocol. If a CSMA protocol is used, collisions are understandably less frequent than with the aloha protocol because the medium is sensed before any transmission.

14.4.3 Carrier Sense Multiple Access with Collision Detection

We told already that a collision can occur even if a communication element senses no activity on the medium before transmitting. The simplest way of detecting collisions is for a transmitter to wait either to receive an acknowledgment or to time out: the acknowledgment tells there has been no collision, and the timeout tells there has been a collision; we already explained this in the case of the aloha and CSMA protocols. (Because bit error rates are assumed to be very low in wired local area networks, the losses and consequently the timeouts that are due to bit errors are neglected.) One must, however, be aware of the following inconvenience. The collision might occur very soon after the transmitter has begun transmitting a long message; if the CSMA protocol is used, the transmitter keeps transmitting the message although this message is evidently lost, as the transmitter detects the collision only after the timeout which happens after the round trip time only; in this way a lot of precious time as the communication resource is lost. From the viewpoint of an efficient use of communication resources it is therefore much better to use another way to detect collisions: while a transmitter is still transmitting, it simultaneously senses the medium; if it senses there another activity, in addition to its own, it knows that a collision has occurred and stops transmitting immediately. The medium access method which uses this way of detecting collisions is referred to as *carrier sense multiple access with collision detection*, or *CSMA/CD* for short. Hence, CSMA/CD protocols represent an upgrade of CSMA protocols that add the detection of a collision as a simultaneous presence of two or more signals on the

Fig. 14.6 Minimum length
of messages that allows
collision detection

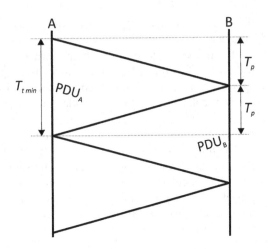

medium. The methods of detecting this presence will not be discussed here, as this procedure is carried out in the physical layer.

If a CSMA/CD protocol is used, a transmitter senses the medium before transmitting; it may use non-persistent, 1-persistent or p-persistent strategy. While transmitting it continues to sense the medium. As soon as it detects a collision it stops transmitting the message and transmits a special packet that is called *jam signal*; the purpose of the jam signal is to assure that all communication elements on the medium, which listen to it, are aware of the collision. Then the element which detected the collision starts the procedure of retransmission, while simultaneously executing the congestion control, as was already explained for the basic and the CSMA protocols.

Hence a protocol entity detects a collision as a simultaneous presence of two or more messages on the medium. This is, however, possible only if protocol data units are sufficiently long, because of the limited propagation speed of the electromagnetic wave on the medium. In Fig. 14.6 the most unfavourable scenario is shown in which both messages must still be present at the same time at the spot where the entity to detect a collision is placed. In this scenario the protocol entity B begins transmitting the message PDU_B just in the moment when the beginning of the message PDU_A, transmitted by the entity A, arrives to it; before it begins to transmit, B has not sensed any activity on the medium. The entity A can detect the collision only if the beginning of the message PDU_B arrives to it when it is still transmitting. As can be seen in Fig. 14.6, the minimum transmit time of a protocol data unit must therefore be

$$T_{t\ min} = 2 \cdot T_p = 2 \cdot \frac{D_{max}}{v}, \tag{14.1}$$

where T_p is the propagation time of the electromagnetic wave between the two most distant network elements on the medium, D_{max} is the size of the network (the distance between the two network elements that are most apart one from the other) and v is the speed of electromagnetic wave propagation on the medium. Of course, the minimum message transmit time also determines the minimum message length which is therefore

$$L_{min} = T_{t\ min} \cdot R = 2 \cdot T_p \cdot R = 2 \cdot \frac{D_{max}}{v} \cdot R, \tag{14.2}$$

where R is the nominal transmission rate on the medium. Because a protocol entity may not prescribe the necessary user message length, it must add so-called *padding* bits to a message which would otherwise be too short.

One can also reverse Eq. (14.2) and write

$$D_{max} = \frac{L_{min} \cdot v}{2 \cdot R}. \tag{14.3}$$

This equation determines the maximum allowed size of a network (the distance of the two network elements that are most apart) in the case when the protocol prescribes the minimum length of a protocol data unit.

Before we proceed to the formal specification of the protocol, we must emphasise that a transmitting protocol entity can detect a collision only within the time period $T_{t\ min} = 2 \cdot T_p$ after the beginning of transmission. If it does not sense a collision within this time frame, it means for sure that there has been no collision and the transmission is successful (in the absence of bit errors, of course). If the message is long enough so that the condition (14.2) is fulfilled, the transmitting protocol entity knows after it has finished the message transmission that this transmission has been successful. CSMA/CD protocols therefore do not use acknowledgments. Because the physical length of local area networks is relatively small, the probability of bit errors is low and can be neglected; if, however, errors nevertheless occur, the receiving protocol entity discards the corrupted messages. It is then up to the protocols in higher layers to provide for a reliable message transfer, if needed.

In Fig. 14.7 the specification of the functionality of a simple 1-persistent CSMA/CD protocol entity is shown. Essentially, this specification is similar to the specification of the basic random medium access protocol, as was shown in Fig. 14.4; here, however, the carrier sensing and the collision detection are added. Here, too, only the protocol entity is specified as the process s; the system specification which is not shown here is similar to the aloha system specification, with the addition of some signals, exchanged between the protocol entity and the medium (i.e. between the MAC and the physical layer). The specification of data types and the definition of constants are also similar to those given for the aloha system.

The carrier sensing and the collision detection are actually carried out in the physical layer; here, only the data-link layer (in fact, its medium access control sublayer) is specified. The procedures of carrier sensing and collision detection are therefore modelled somehow unnaturally with the signals sense, free, busy, done and coll (in our model these signals have the role of the primitives between the MAC sublayer and the physical layer). In this model the protocol entity senses the medium

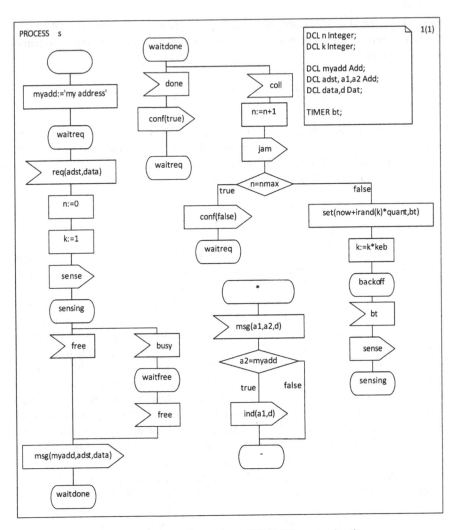

Fig. 14.7 Functionality specification of 1-persistent CSMA/CD protocol entity

by »asking« it with the signal sense if it is busy or free. The medium (the physical layer) replies either with the signal free or with the signal busy. Furthermore, the medium sends the signal free to all MAC entities whenever it becomes free which allows protocol entities to wait for the medium to become free. As soon as the transmitting of a message is finished the medium tells this to the protocol entity with the signal done, so the entity can know that the transmission has been successful. If, however, a collision has occurred during a transmission, the medium sends the signal coll to the MAC entity (this is the model of collision detection). In actual implementations of CSMA/CD network interface cards the functionalities of the medium access sublayer and the physical layer are mostly integrated, so such explicit signals or primitives are not needed. Such an integration of both layers has sense because the MAC standards specify both the medium access control and the physical layer functionality for each LAN technology separately. Consequently, different kinds of local area networks operate differently not only in the MAC sublayer but also in the physical layer.

Just like in the specification of the aloha protocol (see Fig. 14.4), here, too, an information protocol data unit is modelled with the signal msg(Add,Add,Dat), where the first two arguments contain the source and the destination address, respectively, and the third argument contains a user message. An acknowledgment is not needed. The use of the signal jam was described in the preceding text.

The primitives to be used between the process s and a higher layer entity are modelled in the same way as in the case of the aloha protocol. req(Add,Dat) is a user request for the transfer of a user message (the second argument) towards the destination (indicated with the first argument). ind(Add,Dat) is used by the protocol entity to forward a user message (second argument) to the user, whereas the first argument indicates the address of the source entity. The primitive conf(Boolean) is used by the protocol entity to report the (un)success of data transfer to the user (the value true indicates the success, while the value false indicates the failure).

Because a CSMA/CD transmitter does not expect an acknowledgment after it has transmitted a message, it does not need a retransmit timer. The fact that it has received either a signal done or coll from the medium tells it whether the transmission was successful or not. However, it needs the timer bt for the postponement of a message retransmission.

The protocol entity s can be in one of five states. In the state waitreq, which is the initial state, it waits for a user request to transfer user data; this request can only be received in this state. In the state sensing it senses the medium and waits for the answer from the medium (either free or busy). If the medium is currently busy, it waits for it to become free in the state waitfree. In the state waitdone it is transmitting; in this state, either the end of transmission or a collision can occur; the reception of the signal done tells that the transmission has successfully been terminated while the reception of the signal coll informs it that a collision has occurred. In the state backoff it waits for the expiration of the backoff timer to retransmit the message. The exponential backoff algorithm is the same as was already used for the aloha protocol.

In this specification a user is also assumed to send a next request for data transfer only after it has already received the confirmation of the success of the previous request. In the case it is informed about the failure it can decide after some time to request the transfer of the same user message again; this is, however, a decision of the higher layer in the protocol stack.

The specification of the basic random access control protocol was already explained in detail in Sect. 14.4.1; only those characteristics of the CSMA/CD specification which essentially differ from the aloha specification have therefore been emphasised here; the author therefore feels that a reader should have no trouble to understand the CSMA/CD specification given in this section. In what the SDL language is concerned, we must, however, still point out that the character »*« in a state symbol means any state which is defined in this process.

In practice, the CSMA/CD protocol is very important. The operation of the ethernet networks which nowadays represent by far the most important wired LAN technology is based on the CSMA/CD protocol. Although switches are used in most ethernet networks nowadays, which means that a common transfer medium is no more used and so the CSMA/CD access method somehow lost its importance because of this, this protocol is still the base of the operation (at least in what the frame format is concerned, if not in what the collision resolution, as there are no more collisions in switch-based networks) of many modern networks, either of those that use switches or of those which are even not local area networks because they are used for the data transport over longer distances. We also must especially emphasise that the most important protocol which is used in wireless local area networks nowadays evolved from the CSMA and CSMA/CD protocols; more about this will be told in the next section.

In the previous section we found out that the carrier sensing before transmission efficiently reduces the frequency of collisions and therefore also losses, but only in relatively small networks. In this section, however, we saw that in large networks which require long protocol data units large amounts of padding bits may need to be added to messages; this of course increases the overhead and thus decreases the efficiency of the usage of communication resources. Therefore, the random medium access control with carrier sensing and collision detection CSMA/CD is also appropriate only for small networks with short delays. Hence, in large networks with long delays, such as satellite networks, much simpler medium access protocols that are based on the aloha protocol are usually used.

14.5 Wireless Medium Access

The technology of wired local area networks has been developing for more than 30 years, and so have been developing medium access control protocols. The most successful in this area were (and still are) the ethernet networks in which, however, both medium access control according to the CSMA/CD protocol and packet switching are used. Due to many advantages offered by a wireless interconnection of communication devices and a wireless access to wide area networks, such as the Internet, technologies and protocols for the wireless medium (common radio spectrum) access are also being rapidly developed. While the methods for the access to wired media are, due to the ubiquitous use of switches, fading nowadays, the access to common wireless media stays the only possibility for the communication in wireless local area networks. The wireless medium access control is even more important due to the fact that a cell serving for the access to a wide area mobile network also can be considered a wireless common medium.

The wireless information transfer is based on the propagation of electromagnetic waves in space; hence, there are many peculiarities which are not present in wired networks. Wireless medium access control protocols must be adapted to these peculiarities, so they differ in many aspects from traditional protocols for a wired medium access. In this section especially those wireless medium access protocol mechanisms will be emphasised which differ from classical mechanisms, due to the above-mentioned peculiarities. Wireless medium access protocols will also be compared with the CSMA/CD protocol which is by far the best established wired medium access protocol.

The power density of an electromagnetic wave propagating through free space decreases with the square of the distance between the transmitting antenna and the point of observation. Therefore, the power that a receiver gets from the receiving antenna also decreases with the square of the distance between the transmitting and the receiving antenna, or with the square of the distance between the transmitting and the receiving communication element. Most usually, wireless networks are also mobile networks; the physical distances between communication elements can therefore be changing all the time, hence the received powers can also be changing and can reach values that can even be several orders of magnitude lower than the transmitted powers. Consequently, the signal-to-noise ratios can also be extremely unfavourable. Several inconvenient physical phenomena associated with electromagnetic waves propagation can also be present and can even more deteriorate the receiving conditions; a typical such phenomenon is the interference. A consequence of all these characteristics are much higher bit error rates than are common in wired communications.

In the previous section we stated the fact that in wired local area networks the absence of a collision means that a message has most probably been successfully transferred, as the losses due to bit errors are very rare. The CSMA/CD protocol therefore does not employ acknowledgments. Unfortunately, this is not true in wireless networks where bit error rates can be higher than in wired networks by several orders of magnitude, due to the problems mentioned in the previous paragraph.

In wireless networks the collision detection is practically not feasible, because the ratio of the transmitted power over the received power at a communication station's site can be so high that a communication entity cannot distinguish between a signal received from some other station in the presence of the signal transmitted by itself, and a noise.

From what has just been told it is clear that wireless medium access control protocols must employ acknowledgments of correctly received messages, in order to be able to detect both collisions and corrupted and lost messages.

Unfortunately, there are problems in wireless networks not only with high bit error rates and collision detection, but also with a reliable carrier sensing before transmission. The medium is sensed by the network element which wants to transmit; a problem, however, arises if a collision (a presence of two or more different signals at the same time) occurs at the receiver's site. This distinction is not important in wired local area networks, as the signal attenuation is not strong there and the power of a received signal is therefore not much weaker than the power of a transmitted signal. In wireless networks, however, the difference between the respective powers of a transmitted and a received signal can be very big; a signal that is transmitted by one communication entity may even happen not to be heard by all

Fig. 14.8 Hidden station
problem

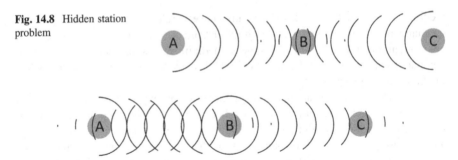

Fig. 14.9 Exposed station problem

other entities.[2] This phenomenon is the cause of the hidden station problem and the exposed station problem.

The *hidden station problem* is illustrated in Fig. 14.8. The communication element A is sending a message to the communicating element B. Meanwhile, the element C also wants to send a message to B. When C senses the medium before transmitting, it does not hear the transmission of A because A and C are too far apart, so C begins to transmit, thinking that the medium is free; unfortunately, the station B hears both A and C. A collision occurs at the spot where B is situated and both messages are destroyed, in spite of the fact that C politely sensed the medium and heard nothing before transmitting.

The *exposed station problem* is illustrated in Fig. 14.9. This problem is a kind of inverse of the hidden station problem—a communication element does not begin to transmit, although it might do so. The message which is being transmitted by the station A in Fig. 14.9 is received by B, but is not heard by C, because C is located too far from A. The station B wants to send a message to C but does not do so, because it hears the transmission of A; indeed, B might send a message to C without causing a collision, as C hears only B and not A.

In wireless medium access control the collision detection problem can be solved with acknowledgments and a retransmission timer. However, in previous sections it was shown that a transmitter receives the acknowledgment or times out only after the round trip time has elapsed, and this can be quite long if messages are long. Even if a collision occurs, the transmitter keeps transmitting the whole message, not being aware that this message will be lost, so a lot of time as communication resource is lost and the transfer efficiency deteriorates. Because collisions cannot be detected during a message transmission in wireless networks, wireless medium access control protocols try to avoid collisions if possible.

Medium access methods where other procedures of collision avoidance, in addition to carrier sensing before transmission, are also used, but collision detection during transmission is not used, are referred to as *carrier sense multiple access with*

---

[2]Although acoustic signals are of course not used in radio-based information transfer, we will however sometimes use the verb »hear« to indicate the reception of a signal.

collision avoidance, or *CSMA/CA* for short. With these methods the following procedures are employed to try to avoid collisions and thus decrease the collision probability: physical carrier sensing before transmission, logical carrier sensing before transmission, random postponement of transmission, medium access with different priorities, and reservations of medium resources with requests for transmission.

Because bit error rates are often very high in wireless media, wireless medium access protocols can use segmentation and reassembly of messages (see Chap. 9), as the probability for message corruption is the lower, the shorter is a message, as was seen in Eq. (4.1). If a long message is segmented into several short segments before transmission, a single segment may be corrupted during the transfer and must be retransmitted; however, if a whole unsegmented long message is corrupted, it must be retransmitted in its entirety, thus consuming more transfer resources. Message segmentation and reassembly is used only with long messages, because the segmentation and reassembly procedure by itself introduces an additional overhead which is not justified if messages are short.

As was already told, acknowledgments and timers are used to detect collisions and recover from them when data are transferred in wireless networks. The transmission of an information protocol data unit and an acknowledgment associated with it are considered an atomic operation (i.e. they cannot be separated), so protocols act as stop-and-wait protocols; because the signal propagation times in local area networks are short (much shorter than message transmit times), the protocol efficiency is not seriously affected by this fact and is not much lower than one (of course, in a lossless case), as can be seen in Eq. (12.8).

A protocol entity which wants to transmit a message senses the medium first; this kind of sensing is referred to as *physical sensing*. The medium must stay idle for a certain amount of time; this time period is called *interframe space* and denoted with the acronym *IFS*. If the medium is still free after IFS, the entity may begin to transmit. If, however, the medium is busy, the entity waits until the medium becomes free, then it waits for the time period IFS, and then still defers the transmission for a random amount of time. If, after all this waiting, the medium is still free, the entity may begin transmitting.

The interframe space IFS may assume different values. Protocol entities which use a shorter IFS have a higher priority for transmission than those which use a longer IFS: if an entity must wait for a longer time, it can after this time of waiting sense the transmission of another entity which waited for a shorter period and was so able to transmit sooner. Basically, three different spaces IFS are used. The interframe space *SIFS* is the shortest one and is used to transmit an acknowledgment after a successful reception of an information message; sending acknowledgments has therefore a higher priority than sending information messages. This makes the transmission of an information message and its associated acknowledgment an atomic procedure, the transmission of some other message is therefore not easily pushed in between. The interframe space SIFS is also used between the segments of a longer message that has been segmented before transmission. The interframe space of medium length *PIFS* is used for the transmission of messages which should have as short delays as possible, e.g. for the transmission of voice samples, as the transfer of this kind of information must have a higher priority than the transfer of ordinary data. The longest interframe space *DIFS* is used for the transfer of ordinary data.

In wireless local area networks the *logical sensing* of the medium is used in addition to the physical sensing. All messages that are transferred in a network also contain the information about the time needed to transfer the message which includes the duration of both the message and its associated acknowledgment; segments of a longer message contain the time needed to transfer the whole message, including all the acknowledgments. Using this information all communicating elements can calculate when the medium will be free.

An additional method for avoiding collisions, or at least for reducing the probability of collisions, is the use of short control messages RTS and CTS. Just before an entity wants to transmit a long message it may make a reservation of the medium for the message transfer. This is done by sending the message *RTS* (*request to send*) to the recipient; this message contains the information about the time needed to transfer the intended long message and its associated acknowledgment. If the recipient receives the message RTS, it replies to it with the message *CTS* (*clear to send*), containing the same information. All network elements that have heard the message RTS and/or CTS now know how much time the medium is going to be busy and do not try to transmit during that time. Of course, the message RTS or CTS also can be destroyed by a collision. However, as these messages are short no much time is lost, so not a big damage is caused by such a collision. The use of messages RTS and CTS also can solve the problem of a hidden or an exposed station. If an entity hears the CTS (even if it has not heard the RTS), it is close enough to the recipient of an information message that is going to be transmitted next, to cause a collision if it itself begins a transmission. However, if an entity hears the RTS message but does not hear the associated CTS, it is an exposed station which is far enough from the recipient of the information message to follow, so it cannot cause a collision. A CTS is a reply to an RTS, so it must have a high priority and is therefore transmitted after a short interframe space SIFS. The exchange of the related messages RTS, CTS, information protocol data unit and acknowledgment is considered an atomic operation, so the short interframe space SIFS is used between these messages, as can be seen in Fig. 14.10, where the acronyms IPDU and ACK denote the information protocol data unit and the acknowledgment, respectively. The labels DIFS and SIFS indicate the time while an entity senses the medium.

The messages RTS and CTS are used only when there are long information messages to be transferred; the use of RTS and CTS with short information protocol data units is not justified, as a collision of a short information message does not cause a big damage, while RTS and CTS themselves consume some communication resources. If the messages RTS and CTS are not used, the transfer of an information protocol data unit looks like the one which is shown in Fig. 14.11. In this case, too, the short interframe space SIFS is used between the messages IPDU and ACK.

Of course, the CSMA/CA protocol is also a random access protocol. Like with CSMA/CD, with CSMA/CA the congestion control with exponential backoff algorithm for the retransmission postponement is also used.

Fig. 14.10 Transfer of information protocol data unit with preceding control messages RTS and CTS

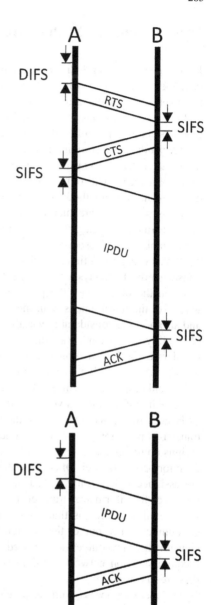

Fig. 14.11 Transfer of information protocol data unit without preceding control messages RTS and CTS

The CSMA/CA protocol is a successful replacement for CSMA/CD protocol in the environments where the collision detection during transmission is not possible. It is also used in wireless local area networks which are extensively used nowadays.

14.6   Combining Medium Access Control Methods

In this chapter several different methods for medium access control have been described. Quite often combinations of different methods are also used. One must, however, be aware that a combination of a synchronous and an asynchronous access yields an asynchronous mode of information transfer. Such combinations are popular especially in radio-based access networks of wide area mobile networks.

An example of a combination of different methods of medium access that is quite often used is the *access with reservations* method. This method will not be discussed in detail here, we will only shortly indicate what it is about. Usually scheduled or even synchronous methods are combined with random access methods so that the advantages of both are merged. The problem of scheduled methods is that the schedule must be managed for all communication elements, for those which heavily use communication resources, as well as for those that currently do not need them at all. In this way a lot of time, which could be used for information transfer, is lost, and consequently the delays are too long even when the offered load in the network is low. On the other hand, the performance of random access methods is the best whenever the traffic intensity in the network is low, while delays severely increase and also become considerably variable if the network is heavily loaded; the problem of stability also occurs. Random access methods are therefore not suitable for the transfer of the information which is sensitive to delay variation (it is about the transfer of information in real time); the examples of such information are live video and, especially, voice. One must, however, be aware that the transfer of multimedia information is very popular today.

Hence, the scheduled access methods are most appropriate for information transfer in real time. The scheduled access is based on a kind of a schedule which assigns communication resources to different communication processes to be used for information transfer. Of course, resources can be assigned most efficiently if they are assigned dynamically, based on the current needs of various communication processes for information transfer. Because, however, these needs are not known in advance, the solution is that communication processes make reservations of the communication resources they currently need. The reservations themselves do not consume much resources; as the needs for resource consumption occur randomly in a communication network, random access methods are most appropriate to make these reservations.

The medium access with reservations is most usually used in centralised networks. In such networks resources are assigned to various communication processes by a master network element, which is ordinarily referred to as a base station or an access point. Other network elements use a random access protocol to let the master element be aware of their current needs for communication resources.

The CSMA/CA protocol which was discussed in the previous section is an example of a distributed medium access with reservations if it uses the RTS and CTS control messages. Of course, there are also many other examples of such access.

Part IV
Short Presentation of Some Standard Protocols

Up to now, the main topic of this text were the methods of information transfer on which the operation of communication protocols is based; this actually is the principal goal of this book. In the chapters to follow some specific protocols will be shortly presented. This will allow readers to acquaint a better idea about why and how the already described methods are used; they will also be able to assure themselves that everything we have already discussed is not a mere theory without a practical value.

Actually, there is no distinct dividing line between the terms »communication method« and »communication protocol«. So we have frequently mentioned protocols even if communication methods were in mind. In the following chapters, however, specific protocols will be discussed. A specific communication protocol is a standardised way for solving one or more communication problems; the interchange of messages is specified by a specific protocol in all necessary details, including the abstract and transfer syntax, as well as the message exchange rules. A specific communication protocol is based on one or more communication methods, which were discussed in the previous chapters. Hence, a specific protocol can, for example, provide for connection management, reliable message transfer, flow/congestion control and even something else. A communication protocol as a standardised communication procedure specifies with all necessary details the way protocol entities exchange protocol data units between them; it, however, does not specify how protocol entities should operate internally to achieve their external behaviours. The implementations of different protocol entities can therefore differ if only they all act externally (on their input and output channels) in strict accordance with the protocol specifications and can in spite of this successfully interoperate.

A huge number of different specific communication protocols have been developed, standardised, implemented and used in recent decades. Here only a few of them will be shortly described. We had three different goals in mind when choosing them. Firstly, we will try to show the use of as many communication methods as possible with the presented protocols. Because different functionalities are implemented by different layers of a protocol stack, protocols will be presented here which are normally used in different layers that are higher than the physical layer. The communication technologies that appear nowadays and also in the near future as the most important are the technologies which are used in the so-called IP networks (including the already classic Internet). Such protocols will therefore also be emphasised in this part of the book.

Specific protocols will only be shortly presented rather than described in all details in this book; the communication methods they use will be emphasised. For any protocol discussed we will tell in which protocol stack layer it is used and which functionalities it provides. All these descriptions will be as short as possible. A reader should see this part of the book primarily as the illustration of the topics discussed so far. The readers willing to know more will find many more details about the specific protocols in the extensive literature on the topics—in books, magazine papers and, of course, in standards. The reader who understands the topics of this book and consequently the communication methods discussed here should not have much problems to study and understand the specific protocols described and specified in the literature.

In Chap. 15 we will show how some protocols evolved from the older ones in the past; the protocol HDLC and its successors are used in the data-link layer, and so is also the protocol PPP which will be discussed in Chap. 16. In Chap. 17 three protocols which are most usually used in local area networks, two of them in the medium access control sublayer and one in the logical link control sublayer, will be presented. In Chap. 18 the protocol IP will be discussed, and then in Chaps. 19 and 20 two protocols used for address translation. In Chaps. 21, 22 and 23 three transport layer protocols will be discussed, two classical ones and one newer one; in Chap. 24 another transport layer protocol will be presented which is employed to transfer information in real time. In Chap. 25 it will be shown how the secure information transfer can be assured in the transport layer of IP networks. Chapter 26 is devoted to some application layer protocols; one of them is a newer one and is used in Internet telephony, while the others are classical.

Chapter 15
HDLC Family of Protocols

Abstract HDLC is a very old data-link layer protocol, which has been adapted to new possibilities and new needs along with the development of communication technologies. So many new protocols evolved from HDLC or from one of its descendants. The main goal of this chapter is just to illustrate how a family of protocols can be developed by evolution, rather than by revolution, and to illustrate many basic concepts of data transfer in the data-link layer, discussed in the previous part of the book. At first the transfer syntax of an HDLC frame is described, as this is the basis for the transfer syntaxes of all the descendants of HDLC. Then the protocols LAPB, LAPD, LAPDm and LAPF are very shortly described. Only the error correction by the LAPDm protocol is described with some more detail, as it well illustrates an early variant of the hybrid automatic repeat request protocol. There are three figures in this chapter.

The protocol *HDLC (High-level Data Link Control)* was developed and standardised in the seventies of the twentieth century by the organisation ISO. Hence, this is not a protocol developed to be used in IP networks, although HDLC and some of its derivatives can be used and actually were used in some access networks for the Internet. In this book, HDLC is of interest primarily as a protocol that had a significant impact on the further development of communication protocols; actually, many other protocols evolved from it. In this chapter it will therefore be shown how protocols can be developed in the form of an evolution, rather than a revolution, process. One must, however, be aware that there are some deficiencies of the protocol development that is based on older protocols—the new protocol can thus incorporate some »garbage« inherited from its ancestor, though it is no more needed in the new protocol; the properties of a protocol that has been developed in this way are not as optimal as they could be, were it developed from scratch with a strictly defined goal. On the other side, the development of a protocol is quicker and less expensive if its development is based on an existing, already tested and used protocol, possibly with some reputation.

D. Hercog, *Communication Protocols*, https://doi.org/10.1007/978-3-030-50405-2_15

The HDLC protocol is a data-link layer protocol that provides for many various services of the transfer of network-layer packets either in connection-oriented or connectionless mode, so providing or not providing a reliable data transfer. In the connection-oriented mode of operation it can work either as a go-back-N or a selective-repeat protocol. It can operate either in *balanced mode* (with both communicating entities having equal functionalities) or in *unbalanced mode* (in this case one protocol entity is the master and the other one is the slave). HDLC can be used for a point-to-point communication or for a one-to-multipoint communication. It can therefore be seen that the HDLC protocol provides for many various services, something for everybody. In accordance with so many and various services, HDLC is quite a complex protocol.

At the time it was developed, HDLC was an advanced protocol and therefore served as some kind of a model for further protocol development. Many later data-link layer protocols therefore evolved from HDLC, so one can speak about the HDLC protocol family. Later protocols were developed primarily by simplifying HDLC and adapting it to various environments and various services needed in these environments. The most salient common features of all protocols in the family are the structure of protocol data units (frames) and the procedures these protocols execute.

In this chapter the protocols LAPB, LAPD, LAPDm and LAPF will be very shortly described, as today all of them are already considered obsolete; in the next chapter the protocol PPP will be described. Once more, it should be emphasised that the basic goal of this chapter is to demonstrate how protocols developed in time, along with the development of new technologies; often a descendant of HDLC evolved not directly from HDLC, but from a previous HDLC descendant.

15.1   HDLC Frame

In this section the structure (the transfer syntax) of an HDLC frame will be described on which the structures of the frames of all other protocols in the HDLC family are based. This description should be seen as an example of the transfer syntax of a bit-oriented protocol, as was already promised at the end of Chap. 8. The transfer syntaxes of other protocols in the family will not be described in detail, only the differences between the frame structures will be indicated.

In Fig. 15.1 the basic structure of an HDLC frame is shown.

HDLC is a binary-oriented protocol, so the fields of its frame contain bit sequences which in general cannot be interpreted as characters.

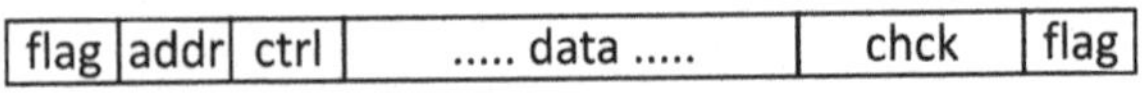

| flag | addr | ctrl | ..... data ..... | chck | flag |

Fig. 15.1 Basic structure of HDLC frame

As was already told in Sect. 8.2, an HDLC frame is delimited with the binary sequence 01111110 both at the beginning and at the end; this binary sequence is referred to as *flag*. Two successive frames can be delimited with one or more flags. To assure a transparent transfer of user data the bit stuffing procedure is used (see Chap. 8) by adding a zero to any sequence of five consecutive ones in a message at the transmitting side (of course, this is not done within a flag!); these added bits are removed by the receiver, thus preserving the original message.

The field addr is a sequence of 8 or more bits and contains an address. One must, however be aware that the presence of an address in a protocol data unit has sense only in the case of a one-to-multipoint communication, while in the case of a one-to-one communication it has no sense and is only formally present.

The field data has a variable length and contains user data, hence a protocol data unit of the higher (network) layer.

The field chck contains redundant bits which allow a receiver to detect errors in received messages. A receiver simply discards erroneous messages.

The field ctrl is the most important control field of a frame, as it contains the essential control parameters of a protocol data unit; because more such parameters are needed, the field ctrl is further divided into subfields. The length of the field ctrl can be either 8 bits or 16 bits. The protocol HDLC uses three basic types of protocol data units, namely U, I and S. *Unnumbered frames* (*U*) are used for the connection management, so these frames do not use sequence numbers (thence the name unnumbered). *Information frames* (*I*) are information protocol data units and contain user messages, in addition to the control information which includes sequence numbers and piggybacked acknowledgments. *Supervisory frames* (*S*) contain only the information that is needed to control the user data transfer within the frame of a connection, while they do not contain any user information. From what has just been told, it is clear that the field data in Fig. 15.1 is present only in information (I) frames. Frames U, I and S have different structures of the field ctrl; these structures are shown in Fig. 15.2 for the case of an 8-bit control field. The frame types (U, I, S) of the field structures are indicated on the left edge of the figure.

The leftmost one or two bit(s) indicate(s) the type of a frame. If the leftmost bit is 0, the frame is an information frame (I). If the two leftmost bits contain the sequence 11, this is an unnumbered frame (U), while a supervisory frame (S) contains 10 in the two leftmost bits.

The protocol HDLC and its descendants use several subtypes of unnumbered frames; that is why the frame subtype of a U frame is encoded with five bits contained in the subfield m in Fig. 15.2(U). A supervisory frame has four or three subtypes, so the subtype is encoded in the two-bit subfield s, as can be seen in Fig. 15.2(S).

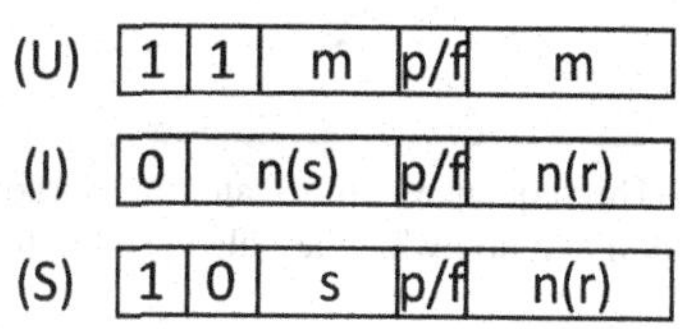

Fig. 15.2 8-bit control field of HDLC frame

(U)	1	1	m	p/f	m
(I)	0	n(s)	p/f	n(r)
(S)	1	0	s	p/f	n(r)

The control field of an information frame contains a sequence number of the frame $n(s)$ and a piggybacked acknowledgment $n(r)$.[1] The control field of a supervisory frame contains only an acknowledgment number. As can be seen in Fig. 15.2, both the sequence number and the acknowledgment number are three bits long which means that modulo 8 counting ($2^3 = 8$) is used if an 8-bit control field is used.

The bit p/f (poll/final) is used by protocol entities for the management of their interaction; this bit is important in the case of an unbalanced communication.

15.2   LAPB Protocol

In the eighties of the twentieth century the International Telecommunication Union (ITU) developed a new technology for data networks with packet mode of transfer. In the recommendation X.25 it specified and standardised the data exchange between a packet network and a packet terminal; this kind of networks was therefore dubbed X.25 networks. A three-layer protocol stack was defined for the network-terminal interface, including the lower three layers of the OSI model, namely from the physical up to the network layer. A new protocol, named *LAPB* (*Link Access Procedure—Balanced*), was developed to be used in the data-link layer; actually, this protocol was developed as a simplification of the HDLC protocol.

LAPB is a point-to-point protocol and can operate only in balanced mode (this is already indicated with its name). It is strictly connection oriented and runs as a go-back-N protocol; it therefore provides for the service of a reliable transfer of network-layer packets.

The structure of LAPB frames is such as was shown for the HDLC protocol in Figs. 15.1 and 15.2, but addr and ctrl can only be 8-bit fields. Sequence numbers and acknowledgment numbers are therefore three bits long, LAPB frames are counted modulo 8, and the maximum transmit window width is 7 (see Sects. 12.6.2 and 12.9).

The communication process according to the LAPB protocol will be described very briefly, as the connection management, the reliable data transfer and the flow control were already described more generally in Chaps. 11, 12 and 13, respectively.

LAPB protocol entities are interconnected with a physical channel where floating corpses cannot appear. A connection is therefore set up and torn down with unnumbered frames using a two-way handshake procedure. In case of a collision of requests sent by the two entities, the general rule is that a connection can only be established if both entities agree on it.

---

[1]The sequence number and the acknowledgment number are here indicated with the same marks n(s) and n(r) which are also used in the standard.

In the user data transfer phase (when a connection is established) protocol entities use the information frames to transfer user data and the supervisory frames RR, REJ and RNR to control data transfer. The go-back-N method is used to provide for a reliable data transfer. User information can be transferred in both directions simultaneously within information frames. Acknowledgments are piggybacked onto information frames, or can be sent as explicit acknowledgments RR (Receive Ready) if there are no user data to be sent. If an entity receives an unexpected (an out-of-sequence) information frame, it sends a negative acknowledgment REJ (REJect) to its peer. If a data loss cannot be resolved with a negative acknowledgment (e.g. because a reject has been lost), the transmitter retransmits frames after it has timed out. If a receiver cannot process the received frames as fast as they arrive, it can send the frame RNR (Receive Not Ready) to request its peer to temporarily stop transmitting until the transmitter receives an RR frame.

15.3   LAPD Protocol

When a new generation of telephone networks *ISDN* (*Integrated Services Digital Network*) was developed and standardised by the organisation ITU (International Telecommunication Union) a data-link layer protocol was needed to transfer signalling on the D channel of the interface between the terminals and the network. The protocol *LAPD* (*Link Access Procedure on D channel*) was defined for this purpose; LAPD was developed by adapting the LAPB protocol. LAPD is quite similar to LAPB, so only the differences between the two protocols will be emphasised here.

Basically, the LAPD frame is similar to the HDLC frame; they differ essentially in the address field and the control field which together are 32 bits long. There are two reasons for this. Up to eight terminals can simultaneously be connected to the ISDN network via a single user interface; in any of these terminals, several communication processes can run concurrently. Double multiplexing is therefore needed; any LAPD frame contains two addresses: one of them indicates the terminal and the other one indicates the network layer process within the terminal; the former address is 7 bits and the latter one is six bits long. The other important difference is that LAPD information frames are counted modulo 128, so sequence numbers and acknowledgment numbers must be seven bits long. Because LAPD, too, is a go-back-N protocol, the maximum transmit window width is 127.

The connection management and the user data transfer are both very similar to those used by the LAPB protocol.

15.4   LAPDm Protocol

The protocols HDLC and LAPD were later still adapted to transfer the signalling in the data-link layer of the wireless interface between a mobile station *GSM (Global System for Mobile communication)* and a base station of a GSM network. The new protocol was named *LAPDm (LAPD mobile)*.

Because of numerous and essential peculiarities of the communication on radio interfaces the differences between the protocols LAPD and LAPDm are substantial.

The transfer of information in the physical layer of the radio interface of a GSM system is based on the time division multiplex (TDM); any active terminal (mobile station) is assigned a separate time slot within the TDM frame. Within a time slot exactly one LAPDm frame is transferred, so the TDM structure of the physical layer itself provides for the synchronisation of LAPDm frames because the beginning and the end of a frame coincide with the beginning and the end of a time slot, respectively; a LAPDm frame therefore does not need delimiting flags. As a time slot of the TDM structure has the constant length, a LAPDm frame also has the constant length. Longer packets of the network layer must therefore be segmented. A one-bit field in the LAPDm header M indicates whether the frame contains the last (or possibly the only) segment. If the length of a network-layer packet is not divisible with the length of the user message area within a LAPDm frame, the last segment must be padded with padding bits; within the header of a frame there must therefore be the information about the true (usable) length of the user contents (without padding).

Bit error rates can be extremely high in wireless channels, especially in wireless channels of mobile networks. In such environments it is better to correct errors in the lowest layers of the protocol stack. The information that is transferred across the radio interface of a GSM system is therefore channel encoded in the physical layer; hence LAPDm frames do not contain redundant bits for error detection/correction.

In the physical layer of a GSM radio interface any active mobile station is assigned a special physical channel. The multiplexing of terminals in a GSM cell is therefore executed in the physical layer, while the protocol LAPDm provides only for the multiplexing of different processes in a terminal; in a LAPDm frame there is therefore a single address.

Although the LAPDm protocol requires the frame counting modulo 8, it however operates as a stop-and-wait protocol. The designers of this protocol considered the flow control unnecessary, so LAPDm does not use an RNR frame, like LAPB and LAPD do.

The transfer resources of a radio channel are extremely precious and expensive, hence LAPDm protocol tries to use them as effectively as possible. One of the measures used with this purpose is to merge sometimes the functionalities of unnumbered and information frames. For example, the frames which set up a LAPDm connection bear also the user information.

Both physical and data-link layers collaborate in error detection and correction process. This procedure is quite interesting as it illustrates the use of simple hybrid ARQ methods (see Sect. 12.2.3). Let us therefore have a more detailed look at this process!

As a general rule, in any network the transfer of signalling messages must be most reliable. In a GSM system, too, the transfer of signalling messages uses the most advanced and reliable method, while the transfer of voice and user messages is more simple and less reliable (but adds less overhead). Here, only the error correction used for the transfer of signalling messages will be explained because only in this case the hybrid ARQ method (which we want to illustrate here) is used. In the HARQ method used in GSM both the physical layer and the LAPDm protocol collaborate. The procedure of coding, decoding and error correction is illustrated in Fig. 15.3. The »message« in this figure is a network-layer packet (or a segment of it), the »LAPDm frame« is an original (not coded) frame, while »A« and »X« are coded messages at the transmitting side, and »Y« and »B« are coded messages at the receiving side. In the physical layer double coding is used. The inner coding (which is carried out first) is block coding, its purpose being error detection and error correction with the ARQ protocol in the data link layer; »A« in the figure indicates the result of the inner coding. The outer coding (which is carried out after the inner coding) is a convolutional coding which allows the error correction at the receiving side; »X« indicates the result of the outer coding. The receiver in the physical layer receives the message »Y« which is equal to the message »X« if no errors have occurred. If the message »Y« appears to be corrupted, the receiver tries to correct it with the convolutional decoding procedure which yields the message »B«. If the error correction was successful, the message »B« is equal to the message »A; if the error correction was unsuccessful, the error detection process in the receiver detects an error and discards the message. Hence the LAPDm receiver either receives the LAPDm frame (if there was no error or if the error correction was successful) or it receives nothing. In the latter case the message loss is detected and handled in accordance with the LAPDm protocol, the LAPDm transmitter retransmits the message.

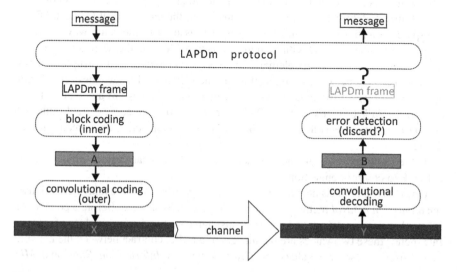

Fig. 15.3 Error correction of signalling messages in GSM

15.5   LAPF Protocol

The technology of X.25 networks was very complex, some functionalities (e.g. the error correction) were executed in more than one protocol stack layer; because of this it was not possible to achieve high speeds of transfer with this technology. As a response to these problems the *frame relay* mode of data transfer was developed. The frame relay technology evolved from the X.25 technology; it was, however, much simplified, so that the data-link and network layers of the OSI reference model (and the X.25 protocol stack) were replaced with a single layer in which the *LAPF* (*Link Access Procedure for Frame mode bearer services*) protocol was running.

The protocol LAPF itself evolved from the protocol LAPB. LAPF is a very simple protocol which implements the simple service of the user information transfer from a source terminal through a frame relay network to a destination terminal. The data transfer is connection oriented. The LAPF protocol, however, does not provide for the connection management service; for the connection management the LAPD protocol is used in a separate signalling channel (this kind of signalling is referred to as *out-of-band signalling*). Although LAPF is a connection-oriented protocol, it does not provide for a reliable data transfer; designers of this technology assumed error rates in modern wired channels to be very low (indeed, this is most usually true), so a LAPF receiver only detects errors and discards corrupted messages, rather than correcting them. However, when the frame relay technology was being developed, it was already clear that the congestion control was needed in new networks. Actually, the LAPF protocol does not execute the congestion control, it only makes it possible to be done.

LAPF uses only one frame type, namely the information frame which is used for the user information transfer. Basically, this frame has the same structure as the HDLC frame (which was shown in Fig. 15.1), only the control field ctrl is missing; a control field is not needed, as this is not an automatic repeat request protocol.

A frame is delimited with two flags, just as in the case of HDLC and LAPB; a transparent transfer of user information is also assured in the same way. A LAPF frame, too, contains a field with redundant bits which enable the destination terminal and all intermediate network elements to detect errors and discard corrupted frames.

A frame relay network is connection oriented; frame relay switches relay LAPF frames with respect to connections with which these frames are associated. In the field of a LAPF header which is officially referred to as an address field (although this term is not correct) there is the indicator of the connection associated with the frame.

In the »address field« of a header there are also three binary values which help the network to control congestions.

One of these fields is named *DE* (*Discard Eligibility*). A frame relay network uses the *admission control* method to enforce congestion avoidance. Before a connection is established the network and the user must agree on two limiting values of transfer rate (hence these two values are parameters of a traffic contract between the network and a user); these two values are *CIR* (*Committed Information Rate*) and *MIR*

(*Maximum Information Rate*). If a source transmits with an average rate that is lower than CIR, then the network transfers its frames with the lowest possible probability of frame loss. If a source transmits with an average rate that is higher than CIR, but lower than MIR, the network sets the DE bits of its frames to 1, which allows any switch in the network to discard such frames, should it experience an imminent congestion; hence the frames with their DE bits set to 1 are discarded by the network with a higher probability than the frames with their DE bits reset to 0. If, however, a source transmits with an average rate that is higher than MIR, its frames are unconditionally discarded.

In a LAPF header there are also two bits referred to as *BECN* (*Backward Explicit Congestion Notification*) and *FECN* (*Forward Explicit Congestion Notification*) which allow the network to inform connection endpoints that there is an imminent congestion in at least one network element through which the connection is established. The bit BECN is set to 1 by a frame relay switch if it experiences an approaching congestion in the direction that is opposite to the direction travelled by the frame with BECN bit set; hence, the network can thus inform about a congestion directly the source which can then immediately act by lowering its transmission speed. On the other hand, the bit FECN is set to 1 if a switch experiences a congestion in the same direction in which the frame with the FECN bit set is transferred; the destination network node can then inform the source by communicating in a higher layer of the protocol stack, because the LAPF protocol itself does not provide for this possibility.

Chapter 16
PPP Protocol

Abstract Here we present another example of a data-link layer protocol, which also evolved from the protocol HDLC; however, PPP is still much used today and is usually employed to support the access of users to the Internet. Actually, PPP is a collection of protocols that provide for different, but complementary functionalities; one of the reasons why it was included into this overview of protocols was just to illustrate such a collaboration of closely related protocols. The services and the modes of operation of the protocol PPP are explained, including its multiprotocol transfer capability. A detailed description of a PPP frame structure is given. The procedure of setting up and maintaining a PPP connection between two endpoints is also explained. There are two figures in this chapter, depicting the PPP protocol stack and the structure of a PPP frame.

The protocol *PPP* (*Point to point protocol*) was conceived as a data-link layer protocol to provide for the service of the transfer of IP datagrams over different physical *point-to-point links* between a terminal and a router or between two routers. At first, the *SLIP* protocol (*Serial line internet protocol*) was developed for this purpose; it was, however, soon replaced by the PPP protocol due to its several deficiencies. In many its traits, PPP is similar to HDLC and its descendants.

According to the OSI reference model, PPP is a data-link layer protocol; in the TCP/IP protocol stack, it is one of the network access layer protocols. Although we said in Sect. 5.4.2 that network access protocols of the TCP/IP protocol stack are not standardised in Internet standards, PPP is nevertheless defined in the standard RFC 1661 as one of protocols that can be employed to interconnect Internet network elements. PPP is capable of transferring packets of various network layer protocols, not only IP datagrams; however, it is by far most frequently used to transfer IP datagrams.

The PPP protocol does not expect some special services of the underlying physical layer, it only requires it to provide for a bidirectional bit transfer. It can therefore be used and actually is used over various physical media, such as modem connections, ADSL links, optical cables, in SDH networks... Special variations of

D. Hercog, *Communication Protocols*, https://doi.org/10.1007/978-3-030-50405-2_16

PPP were also developed to be used above ethernet networks (*PPPoE—PPP over Ethernet*) and ATM networks (*PPPoA—PPP over ATM*).

The basic service provided by the PPP protocol is the transfer of network layer packets between two endpoints. Most usually a nonreliable data transfer is provided (in HDLC terminology, this is called *unnumbered* mode, as sequence numbers are not used in this transfer mode). This transfer mode is usually used because most physical media are quite reliable nowadays (this, of course, is not true in case of radio links, especially mobile links). In any case, IP does not provide for a reliable transfer; if it is needed, it is provided by the TCP protocol in the transport layer. It is, however, also possible to use the *numbered* mode of PPP transfer; in this case, PPP messages are numbered and a reliable data transfer is provided. In any mode a PPP receiver detects corrupted messages and discards them. The PPP protocol and its components also provide for negotiations on the properties of data transfer in data-link and network layers, as well as the authentication of the entities wanting to communicate.

Actually, PPP is a collection of protocols. The basic protocol PPP provides for the data framing and the transfer of frames though a physical channel. The protocol *LCP* (*Link control protocol*) manages the data-link layer logical connection by configuring transfer parameters. The protocols *NCP* (*Network control protocol*) manage the network layer protocols by configuring their parameters. Because PPP can transfer messages of various network layer protocols, there are various NCP protocols for various network layer protocols. The most usually used NCP protocols are *IPCP* (*IP control protocol*) for the configuration of IPv4 protocol parameters and *IPV6CP* (*IPv6 control protocol*) for the configuration of IPv6 protocol parameters. The most important parameters of both these protocols are the IP addresses of both protocol entities and the address of a domain name server.

PPP frames can transfer LCP messages, NCP messages, authentication messages (various authentication protocols can be used) and network layer messages over the physical layer. In Fig. 16.1 the protocol stack is shown for the case that the IP protocol is used in the network layer; auth in this figure indicates the authentication protocol.

The PPP frame structure is based on the frame structure of HDLC and its descendants. It is shown in Fig. 16.2.

Fig. 16.1 PPP protocol stack

			higher layers
LCP	auth	IPCP	IP
basic protocol PPP			
physical layer			

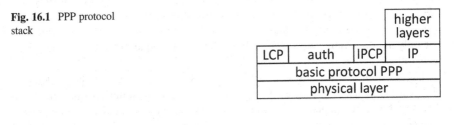

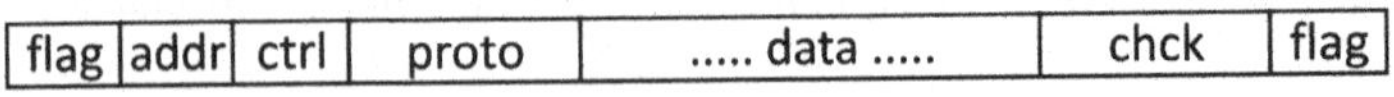

Fig. 16.2 Transfer syntax of PPP frame

Just as in the case of the HDLC protocol, PPP frames also are delimited with flags (the binary sequence 01111110) in the field flag. The cases when this binary sequence appears within a user message are resolved with the bit stuffing procedure (if the physical layer transfer is synchronous) or with the octet stuffing procedure (if the physical layer transfer is asynchronous).

In most cases, the fields addr and ctrl have constant values, except in the cases when a reliable data transfer is required.

The field proto facilitates the *multiprotocol transmission* which means that PPP frames can transport packets of various higher-layer protocols, possibly even simultaneously, over the same PPP connection. The field proto contains different standard codes for different network layer protocols which allows for the packet demultiplexing at the receiving side (the distribution of received packets to correct higher-layer protocol entities). The protocol LCP and various NCP protocols also are assigned appropriate standard codes.

The field data contains packets LCP, NCP, packets of the authentication protocol and the network layer packets for user data transfer (mostly IP).

The field chck contains redundant bits that allow the receiver to detect errors in received frames.

Before transferring network layer packets using the PPP protocol, the basic PPP connection must first be established, which allows for the framing and the message transfer over the physical link. When the PPP connection is set up LCP negotiations can be used to configure the necessary parameters of this connection. This configuration is carried out with two exchanges of LCP packets ConfigureRequest – ConfigureAck; in this way the connection parameters are determined for each direction of data transfer separately; in each of these two exchanges one protocol entity proposes parameters and options which can then be accepted or rejected by the other protocol entity. An example parameter that is often negotiated by LCP entities is the maximum length of frames. The configuration also determines whether protocol entities are going to be authenticated after the configuration has been done, and, if yes, which authentication protocol is going to be used. Then the NCP entities execute the negotiation which determines the network layer protocol parameters. As was already told, the most frequently used network layer protocols are IPCP and IPV6CP; the procedure for the network layer protocol parameters configuration is similar to the link parameters configuration using the LCP protocol. Only after the network layer protocol parameters have been configured, entities can begin exchanging user data. When the communication between users has been terminated the IPCP protocol is used also to terminate the IP session (at this occasion the IP addresses that have been temporarily assigned are released). Finally, the PPP connection is torn down using the LCP protocol.

The protocol LCP may also be used at any time during the existence of the connection, in order to test the connection or to detect that the connection has ceased to exist, e.g. with the exchange of messages EchoRequest – EchoReply.

During the connection configuration according to the LCP protocol the entities may agree upon establishing the numbered mode of communication to provide for a reliable data transfer. In such cases the LAPB protocol or the LAPD protocol is used for data transfer. As was told in Chap. 15, the most important difference between the protocols LAPB and LAPD is the modulus used to count frames, and consequently the maximum transmit window width. The transmit window width is determined during the PPP connection configuration, using the protocol LCP; if it is less than 8, the modulus 8 is used, and if it is between 8 and 127, the modulus 128 is used. If the numbered mode of data transfer cannot be agreed upon, the unnumbered mode is used.

Nowadays the protocol PPP is still massively used. Most Internet access providers employ PPP or one of its derivatives, PPPoE or PPPoA, to implement the user access to the Internet.

Chapter 17
Local Area Network Protocols

Abstract Two heavily used medium access control protocols are described here to illustrate the use of principles explained previously, namely the CSMA/CD protocol (used in ethernet networks) and the CSMA/CA protocol (employed in IEEE 802.11 networks). The evolvement of the versions of the ethernet protocol is described. The structure of an ethernet frame and the functions of its fields are explained. Then the wireless local area networks are overviewed and the functionality of the IEEE 802.11 networks is introduced. The difference between the two modes of transfer, one of them appropriate for the transfer of data and the other one appropriate for the information transfer in real time, is pointed out. The structure of an IEEE 802.11 frame and its fields is shortly described. The optional use of the congestion avoidance mechanisms is overviewed. Although LLC is not a medium access control protocol, it is presented here as an example of an interface between two adjacent protocol stack layers. The three modes of its operation are mentioned and its capability to multiplex the network-layer processes is emphasised.

In Chap. 14 medium access control methods were discussed which are used in local, metropolitan and personal area networks. In this chapter two specific random access protocols will be described in more detail which are by far most usually used in wired and wireless local area networks, respectively. Furthermore, the LLC protocol will also be shortly described; although LLC is not a medium access protocol, it supplements the services that are provided by medium access control protocols for the network layer, as was already told in Sect. 5.4.1.

17.1 IEEE 802.3 (CSMA/CD)

Among the local area network technologies which are based on wired interconnections of network elements, the highest success has been achieved by the *ethernet* networks, which were standardised by the organisation IEEE in the standard 802.3. This success was such that all other local and metropolitan area networking

D. Hercog, *Communication Protocols*, https://doi.org/10.1007/978-3-030-50405-2_17

technologies (e.g. token ring, FDDI...) have been more or less abandoned in recent decades, while the ethernet/802.3 technology keeps to be employed even after almost 40 years of use. One of the reasons for such a prosperity of ethernet networks is its flexibility—this technology has been developed and upgraded many times, while retaining the back compatibility with previous versions; this allowed for several (newer and older) versions to be used concurrently. Whenever a new version was developed and put into use, older equipment did not need to be discarded and could continue to be used. The ethernet equipment that is used nowadays is of course quite different than the equipment which was used decades ago. On one side, the latest technologies allow information to be transferred up to 1000 times as fast as at the beginning of the ethernet development. On the other side, communication elements are no more interconnected with a wired common medium, but rather with switches, so the common medium access control methods are no more needed. Furthermore, nowadays ethernet is no more only the local area network technology, as ethernet links can also be used to interconnect the elements of metropolitan and even wide area networks.

The original ethernet network technology was designed jointly by three companies, namely Digital, Intel and Xerox; although the new technology was dubbed ethernet from the very beginning, it also became to be known as DIX, according to the initials of the three originating companies. The technology was soon standardised by the organisation IEEE as the standard *IEEE 802.3*. IEEE introduced a minor modification to the DIX version; in spite of this, the name ethernet was kept and is still used today. As all medium access control standards, 802.3 also specifies the network operation in both physical layer and MAC sublayer. As new versions of the standard were developed and introduced, the physical layer specifications were modified each time and the transmission rate was increased from one version to the next one by the factor 10. While the original standard specified the transmission rate to be 10 Mb/s, the later versions employed 100 Mb/s (*fast ethernet*), 1 Gb/s (*gigabit ethernet*), and 10 Gb/s (*10-gigabit ethernet*). In what the MAC protocol is concerned, only one essential modification was made during this long development process. At first, computers were interconnected by a bus in the form of a coaxial cable and the CSMA/CD protocol was used as the medium access control method. Later, computers were interconnected first by *hubs*; however, a network where computers are interconnected by hubs is logically still equivalent to a bus, so the CSMA/CD method was still used in such networks. Nowadays network elements[1] are interconnected almost exclusively with switches. If a link interconnecting a terminal and a switch is a *half-duplex* link (allowing the transfer in both directions, but not simultaneously), the CSMA/CD method is still necessary; if, however, an interconnection is *duplex* (allowing the transfer in both directions simultaneously), the CSMA/CD method is not needed, as collisions cannot occur in such cases. It was

---

[1]Although network elements in old networks were only computers, nowadays we can also use tablets, smart phones and other smart devices – however, these devices always contain computers as their essential components; in this book, the term computer will therefore be generally avoided.

already mentioned that later versions mostly preserved the *backward compatibility* with previous versions. So with all the versions up to and including the 1 Gb/s version the CSMA/CD method can be used, which means that network elements can be interconnected with both hubs and switches; the 10 Gb/s version does not use the CSMA/CD method any more. Furthermore, all network elements must be capable of communication using the slower versions of the protocol; this was made possible with an addition to the protocol which allows network elements to negotiate the transmission rate to be used by all communicating elements.

In Fig. 17.1 the structure of a MAC frame, according to the standard IEEE 802.3, is shown. The difference between the structures of the original DIX frame and the one shown in Fig. 17.1 is not big. This difference will be explained when the frame fields that differ will be described.

The 802.3/ethernet protocol is an asynchronous protocol; in the physical layer the transfer is asynchronous as well. This means that a network element transmits nothing when it does not transmit a MAC frame. The field preamble of an 802.3 frame is seven octets long and contains a sequence of ones and zeroes which allows a receiver to synchronise its receiving bit rate with the bit rate of an incoming frame, as the clocks of different network elements never have exactly the same frequency. The one octet long field sfo differs from the preceding octets only in the last two bits which indicates to a receiver that an 802.3 frame follows. According to the original specification DIX, there is no sfo field, but the preamble is eight octets long.

The fields da and sa contain the MAC addresses of the destination and the source of a frame, respectively. In accordance with IEEE standards, MAC addresses are 6 octets (i.e. 48 bits) long. The destination address can be an individual, a group or a broadcast address (as was explained in Sect. 10.1).

The biggest difference between the formats of an IEEE and a DIX frame is in the field len. In a DIX frame there was the field type instead which determined the type of a packet contained in the user data field, i.e. it indicated the protocol associated with user data; this allowed for the multiplexing of network layer processes running according to different network layer protocols. According to the 802.3 standard, the field len contains the length of user data. In the protocol stack that is used in local area networks the LLC sublayer is positioned above the MAC sublayer (see Sect. 5.4.1), so the multiplexing of network layer processes is provided for by the LLC protocol. Nowadays this field can serve both purposes. As the largest length of the user field is 1500 octets and the standardised codes of all network layer protocols are larger than 1536, it is easy for a receiver to know which kind of information is contained in the field: if the value is greater than 1536, the contents are interpreted as the packet type, otherwise as the user contents length.

The field data contains user data, hence a protocol data unit of the higher layer. As was already told in Sect. 14.4.3 and in Eq. (14.2), a frame must be longer than a minimum length which depends on the size of a network, i.e. on the maximum

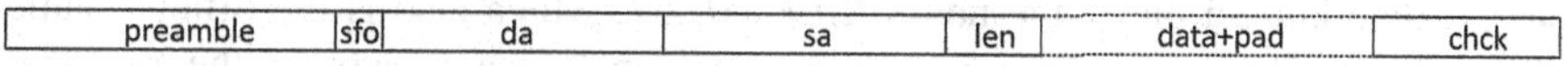

preamble	sfo	da	sa	len	data+pad	chck

Fig. 17.1 Structure of IEEE 802.3/ethernet frame

possible distance between network elements. In the original variant of the protocol this maximum network size was 2500 m and the minimum length of a frame was 64 octets (which means that the minimum length of the data field was 46 octets). In later versions of the protocol the minimum length of a frame was preserved, while the maximum size of a network was modified due to the higher transmission rates and accordingly shorter transmission times. Of course, a transmitting protocol entity can get from its user a message that is shorter than 46 octets; in such a case padding bits must be added to the user message to fulfil Eq. (14.2). As with many other protocols, the standard also specifies the maximum length of user data, 1500 octets in this case. Because the frame length of 1500 o is very short if the transmission rate is 1 Gb/s or above, the possibility of so-called *jumbo frames* was introduced into gigabit ethernet networks; however, this possibility is not standardised.

The field chck contains redundant bits that allow a receiver to detect errors in received frames; corrupted frames are simply discarded.

Let us now still have a short look at the medium access control method that is used in ethernet networks which are not built with ethernet switches. These networks employ the 1-persistent CSMA/CD protocol that was already described and specified in Sect. 14.4.3. The *binary exponential backoff method*, discussed in Sect. 13.6.2 and specified in Fig. 14.7 of Sect. 14.4.3, is used for the congestion control; this exponential backoff is called binary because the factor of the backoff increase (keb in Fig. 14.7) equals 2. The binary exponential backoff is used after the first ten consecutive collision detections; after the next six collisions the backoff is not increased, and after the sixteenth consecutive collision the transmitter ceases to transmit the frame and informs its user that the transmission has not been successful. The advantage of the 1-persistent CSMA/CD method is that its performance is satisfactory in both heavily and lightly loaded networks. If the offered load is low, the delays are not too long, as the transmissions are not deferred after the medium has become free; if the network is heavily loaded, too frequent collisions are efficiently avoided, so the network operation stays stable.

17.2 IEEE 802.11 (CSMA/CA)

As was already told in Sect. 14.5, wireless local area networks, based on the radio transmission, are more and more used nowadays; in this kind of networks, the common medium access cannot be avoided, although it is almost no more used in modern wired networks.

Several radio medium access control methods and technologies were already developed and standardised in the past. In these endeavours especially active was the organisation IEEE which developed several MAC protocols also for wireless networks. By far the most successful among them was and still is the protocol which has become known under the name *IEEE 802.11*, which also is the official name of the standard. For the use in wireless personal area networks, the protocol *IEEE 802.15*, also dubbed *Bluetooth*, has also become much used and very popular.

Other wireless medium access technologies have more or less been abandoned. The protocol 802.11 is used in the networks which are popularly known as *wireless local area networks*, *WLANs*, or *WiFi* networks (meaning *wireless fidelity*); this is the protocol we are going to describe in some more detail in this section. In addition to wireless local area networks, protocols to be used in radio access networks of wide area mobile networks have also, and still are being developed.

The role of the protocol 802.11 in wireless local area networks is similar as the role of the protocol 802.3 in wired local area networks. On one side, both protocols provide similar services; both of them are also based on the random medium access control method. On the other side, they are substantially different, which is due to many peculiarities of radio-based networks, already discussed in Sect. 14.5.

During the course of the development of the standard 802.11, several different physical layer technologies have been developed; they differ in the frequency band that is used (2.4 GHz, 5 GHz, or both of them), the line coding method, and, which is above all interesting for users, in the physical layer transfer rate. In spite of this, all these variants of the protocol use the same MAC sublayer protocol (with only a few later additions to provide for a better quality of service and higher transfer rates). In the following, the MAC (medium access control) sublayer protocol will be described in more detail; we will refer to this protocol simply as 802.11, or even just as the protocol. This protocol must therefore be capable of using the services of different physical layer protocols, which are employed with the respect to a network environment and capabilities of other communication elements in the same network. Furthermore, the protocol must be able to operate in both structured and ad-hoc networks. As has already been told, the basic mode of operation is somehow similar to the operation of ethernet networks. This mode of operation is referred to as *DCF* mode (*distributed coordination function*); as can be seen from the name, this is a distributed kind of medium access control. The DCF mode is appropriate for data transfer. The DCF mode can be upgraded with the *PCF* mode (*point coordination function*) which is based on the centralised medium access control and is appropriate for the information transfer in real time. The PCF mode uses the services of the DCF mode for its operation. While the DCF functionality must be implemented in all 803.11-compliant devices, the implementation of the PCF functionality is not mandatory. The PCF mode is not possible in ad-hoc networks, as it needs an access point for its operation.

Three basic functionalities are executed by the 802.11 protocol. As a medium access control protocol, it of course provides for the access of a communication entity to the radio channel. Because the error rates in wireless channels can be very high, the protocol must also provide for the retransmission of lost and corrupted frames. A very important service of this protocol is also to ensure a secure information transfer; the security is severely threatened in wireless networks, as anybody having an appropriate equipment at their disposal can access a radio channel. One of the functionalities the 802.11 protocol must also be able to provide is also the segmentation and reassembly of the messages that are longer than a specified threshold. Furthermore, a protocol used in wireless networks must use as little energy as possible in mobile terminals, the operation of which is mostly based on the energy provided by a battery.

The protocol 802.11 uses three main types of frames, each of them including several subtypes. The management frames provide for the mobility management of communication elements in a wireless local area network, hence for the association and disassociation of mobile terminals with the mobile network, the roaming of terminals between the cells of the network, and also for the security management (authentication of communication elements and enciphering of transferred information). The data frames are information protocol data units which also can carry various control data (such as piggybacked acknowledgments) in addition to user data. The control frames contain only control information; the most frequently used subtypes of control frames are the explicit acknowledgment ACK, and the frames RTS and CTS (see Sect. 14.5).

In Fig. 17.2 the structure of an 802.11 frame is shown which is basically the same for all frame types; the structures of various frame types can slightly differ from the one shown in Fig. 17.2. The frames of the more recent variants of the protocol contain two additional fields for the control of quality of service and higher transmission rates.

In fact, the field preamble provides for a functionality of the physical layer, as it enables the bit synchronisation of a receiver. The field plcp (physical layer convergence procedure) contains the information about the transfer in the physical layer which allows for more different transfer modes to be used in the physical layer. The field mac pdu contains a MAC frame. The structure of the MAC frame is shown in the lower part of the figure and will be described in more detail in the next paragraphs.

The field ctrl consists of several subfields and contains several kinds of control information. Among other things, the version of the protocol, the type and subtype of the frame are present in this field, as well as the information needed for the segmentation and reassembly, security management and the control of power consumption in a mobile terminal.

The field d/id can play two different roles, which depend of the type of a frame. Usually it contains the information on the duration (transmission time) of the frame—this enables the logical sensing of the medium: the entity which wishes to transmit can calculate when the medium will become free if it knows the duration of the frame which is currently being transmitted; this information includes the transfer times of both the information frame and its acknowledgment, and even the time needed to transfer all the segments and their associated acknowledgments of a message which has been fragmented. Sometimes the field d/id contains the indicator of the connection with which the frame is associated.

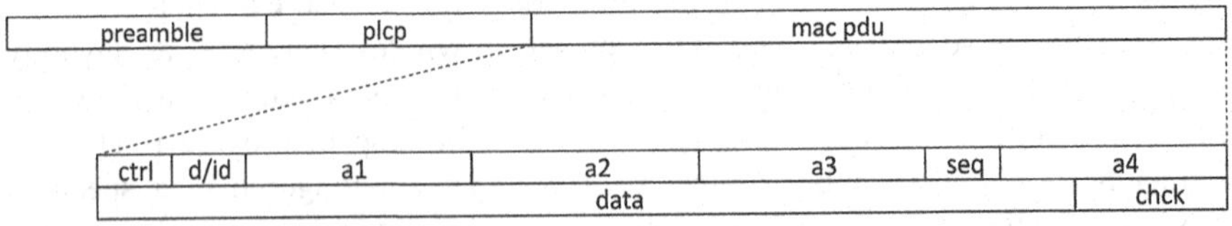

Fig. 17.2 Frame structure of protocol IEEE 802.11

The fields a1, a2, a3 and a4 contain addresses; how many and which addresses a frame actually contains depends on its type. These addresses can indicate the source and the destination terminal within the wireless network, the source and the destination terminal in another network (e.g. in the Internet with which the wireless network is connected via the access point), and/or the indication of the wireless local area network.

The field seq contains the sequence numbers of the frame and of the segment within the frame. If the original frame has been fragmented, then all the segments of this frame contain the same sequence number of the frame, but different sequence numbers of the segments.

The field data contains the protocol data unit of the higher sublayer (LLC) or a segment of that message.

The field chck contains redundant bits which allow a receiver to detect errors and discard corrupted frames.

The operation of the protocol in the DCF mode is based on the CSMA/CA (carrier sense multiple access with collision avoidance) method which was explained in Sect. 14.5. The procedure of the physical sensing of the medium depends on the technology used in the physical layer and is different in different versions of the standard 802.11. The logical sensing uses the value of the field d/id (see Fig. 17.2) of the frame that is currently being transmitted in the network.

In the DCF mode of transfer two different values of the interframe space IFS (see Sect. 14.5) are used. The shorter interframe space SIFS is employed whenever a protocol entity must quickly react to some event by transmitting a message; the frames that are sent after SIFS therefore have the higher priority. Such messages are acknowledgments, responses CTS to requests RTS, or the messages that are segments of a longer message. The longer interframe space DIFS is used before sending messages with the lower priority (such as plain user data).

Whenever a transmitter transmits an information frame it activates its retransmit timer and waits for the acknowledgment. If it receives the acknowledgment, it stops the timer, otherwise it retransmits the frame. The transfer of an information message and its associated acknowledgment is considered an atomic operation which should not be interrupted by the transmission of some other message. Therefore, a receiver that has successfully received an information frame transmits the acknowledgment with the high priority, i.e. after the short interframe space SIFS. The use of the retransmit timer and the acknowledgments has two purposes in the protocol 802.11. The acknowledgments (better to say, the absence of them) allow collisions to be detected. Besides this, retransmitting after timer expirations allows those frames which have been lost due to transmission errors to be recovered. A protocol entity may transmit a new information frame only after it has received the acknowledgment of the previous one; hence the stop-and-wait protocol is used to provide for the reliable data transfer. Because the message transfer delays are short in a local area network, the efficiency of the stop-and-wait protocol is not low. The sequence number that is contained in the field seq allows duplicates to be detected.

The 802.11 protocol is a non-persistent carrier sense multiple access protocol with collision avoidance (see Sect. 14.2). An entity that wants to transmit ordinary (low-priority) data first senses the medium. If the medium is free, it defers the transmission; if after the delay DIFS the medium is still free, the entity begins transmitting. If, however, the medium is not free, the entity waits for it to become free, then again waits for the time DIFS and then still defers the transmission for a random amount of time; if after all of this the medium is still free, it may begin to transmit. In this way the entity allows the entities with higher priority (shorter IFS) to transmit first; furthermore, after the medium becomes free, it lowers the probability of a collision with those entities which have also been waiting for the medium to become free. If the entity times out (does not receive the acknowledgment within the expected time period), it triggers the retransmission procedure and at the same time increases the average time of the retransmission backoff, according to the binary exponential backoff method (similarly as was already explained for the protocol 802.3 in Sect. 17.1).

An information frame can also be transmitted using the four-way exchange of messages RTS – CTS – information frame – acknowledgment, thus lowering the probability of a collision, even in the presence of a hidden station. Although the exchange of the control frames RTS and CTS consumes only a small amount of network resources, it is worth to be used only if the information frame is long enough.

As was already told, a transmitting entity can fragment an original long message into shorter segments, while the receiver reassembles them into the original message. Because the data transfer is reliable, the segmentation and reassembly procedure is simple. The field seq contains the sequence number of a segment and the field ctrl contains a flag indicating whether the frame contains the last segment of the original message or not. The protocol specifies the short interframe space SIFS to be used between the transfers of the segments of a same original message, so the information frames of other transmitters cannot interfere with the transfer of a whole segmented message.

The functionality PCF is based on the polling protocol. The master protocol entity which is placed in the access point of a network polls the slaves in network terminals that have sent their requests for transmission, to transmit; the round robin strategy is used. Before sending a poll, the master uses the interframe space PIFS which is shorter than DIFS and longer than SIFS; in this way the transfer of information in real time (which is sensitive to delays) has a higher priority than the transfer of data (where DIFS is used), but a lower priority than the transfer of frames which use SIFS (e.g. acknowledgments). A slave responds to a poll after the short interframe space SIFS. The systems which use the PCF functionality partition the time into so-called *super frames*; one part of a super frame is used for the PCF functionality and the other part is devoted to the DCF functionality.

The protocol IEEE 802.11 also makes possible the saving of the energy of the battery in a terminal. From time to time a terminal can enter the *sleep state* in which it temporarily switches its transmitter and receiver off, making the access point aware of this. While a terminal is »sleeping« the access point stores the frames to be sent to it. After a while the access point can inform the terminal about the frames stored for it, or the terminal itself requests them from the access point.

The 802.11 protocol also specifies the authentication of mobile terminals and the enciphering of transferred data. However, we will not tell any more about this topic here.

17.3   LLC Protocol

As was already told in Sect. 5.4.1, the data-link layer of the OSI reference model is partitioned into two sublayers for the use in local area networks; both sublayers are, however, employed not only in those protocol stacks that are based on the OSI model, but may be used also in others, including the network access layer of the TCP/IP protocol stack. Above the MAC (medium access control) sublayer the *LLC protocol* (logical link control) is used. The LLC protocol is a kind of an interface between various medium access control protocols and a protocol of the network layer. The LLC protocol was standardised by the organisation IEEE in standard IEEE 802.2.

There are two groups of services of the LLC protocol. The first functionality, and nowadays by far the most important (in many cases the only one), is the multiplexing/demultiplexing of network layer processes. Because various protocols can be used in the network layer, the LLC header contains the addresses of network layer protocol entities at both the transmitting and the receiving side, thus indicating the protocol which is used in the network layer; such addresses were referred to as service access points in Chap. 4. In the header of an LLC frame there are therefore the addresses DSAP (destination service access point) and SSAP (source service access point) which indicate the service access points at the receiving and the transmitting side, respectively. The other functionality of the LLC protocol which, however, is not much used nowadays is the classical functionality of data-link layer protocols that is quite similar to the functionality of the HDLC protocol (see Chap. 15); in this sense the LLC protocol can provide for three different services: the unreliable data transfer in connectionless mode (transfer of datagrams), the reliable data transfer in connection-oriented mode (as a sliding window protocol), and the partially reliable message transfer in connectionless mode (transfer of datagrams with acknowledgments).

In Fig. 17.3 the structure of an ethernet (IEEE 802.3) frame is shown that includes an LLC frame which in turn includes a network layer message, possibly with padding bits (the network layer massage and the padding bits are indicated as data+pad in the figure). The text 802.3 header in the figure indicates the ethernet header as was already described in Fig. 17.1 in Sect. 17.1. The fields ds and ss contain the addresses DSAP and SSAP, respectively, while the field ctr contains the control information which is relevant only if the connection-oriented mode or the

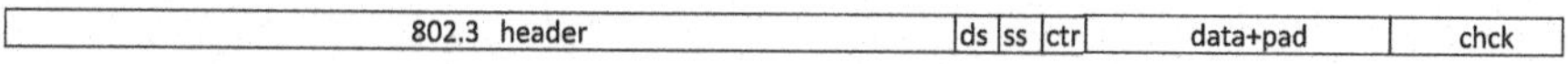

802.3 header	ds	ss	ctr	data+pad	chck

Fig. 17.3 LLC header in 802.3 frame

connectionless mode with acknowledgments is used (see the previous paragraph). Hence one can see that the LLC protocol only inserts three fields between the 802.3 header and the network layer message.

The operation of the LLC protocol can be summarised as follows. In the connectionless mode with unreliable data transfer (which is used in most cases) LLC only inserts two addresses (SSAP and DSAP), thus enabling the multiplexing and the demultiplexing. In the modes with partially reliable and especially with reliable data transfer it works like the HDLC protocol.

Chapter 18
IP Protocol

Abstract The description of the protocol IP was included into the book not only because it is one of the most important protocols nowadays but also because its transfer syntax illustrates very well the use of various fields of datagrams for the deployment of different functionalities. At first the historical development of IP is presented and its current weakest point, the lack of IP addresses, is pointed out. The functionality and the use of the IP protocol are then explained. The structure of an IPv4 datagram is described and the operation of the protocol, as related to the fields of a datagram, is described. The addressing, the prevention of the bad impact of floating corpses, the multiplexing which permits the multiprotocol transmission in the higher layer, and the segmentation and reassembly are the IPv4 functions that are explained. The protocol ICMP, which is always used concurrently with IP, is also described. Then the protocol IPv6 is shortly presented; principally, the differences between IPv4 and IPv6 are emphasised. The only figure of this chapter shows the transfer syntax of IPv4 datagrams.

The protocol *IP* (*Internet protocol*) is the basic and the most important protocol that is used in the Internet network. It is also used in private networks and local area networks; many of them are, however, most usually connected to the Internet and therefore act as access networks of the global Internet. The name of the protocol stems from this very role. Because of such a high importance of this protocol, all the networks that use the IP protocol for their operation are often referred to simply as *IP networks*. Nowadays the IP protocol can be considered an old one, as it was developed in the late seventies of the twentieth century and first standardised in 1981. The first three versions were more or less experimental ones, so only the version 4, known as *IPv4* (*Internet Protocol version 4*) was put into regular use and is still massively used nowadays. When it was developed the IPv4 protocol was overengineered, which is especially true for the size of the address space; unfortunately, due to the extremely fast expansion of the Internet and to the way IPv4 addresses are assigned to communication elements, the size of the address space turned out as its weakest point during the years. Professor Tanenbaum once wrote his famous statement that the IP protocol became the victim of its own success. In the

D. Hercog, *Communication Protocols*, https://doi.org/10.1007/978-3-030-50405-2_18

course of years the problem of the address space size was being solved in various ways, e.g. by assigning temporary addresses to hosts, as well as by using private addresses in private networks and translating them into public addresses in the so-called *network address translators* (*NAT*); this, however, brought about some other problems. It therefore turned out that the only solution which would not be only temporary would be to design a new protocol with a larger address space. Although the name of *IPv6* (*Internet Protocol version 6*) is basically the same as the name of IPv4, IPv6 is quite a different protocol which is even not compatible with IPv4. IPv6 was already developed in the early nineties of the twentieth century, but has nevertheless not yet been entirely substituted for its predecessor, although officially there are no more free IPv4 addresses.

From the very beginning the IP networks were conceived as networks with connectionless information transfer, so both versions of the IP protocol are also connectionless protocols. The connectionless operation was chosen for several reasons. One of them is the relative simplicity of connectionless protocols. The other one, which was especially important in time when IP was developed, is its high immunity to network failures, even in the case of a nuclear attack and a partial destruction of a network (one must keep in mind that IP was primarily developed to be used in the military network when the cold war was at its peak!). Although connection-oriented technologies were much more in favour in that time, the connectionless IP protocol later prevailed over classical connection-oriented technologies, due to its flexibility and simplicity. Nowadays the IP technologies predominate in the telecommunication world, not only for the transfer of data but also for the transfer of information in real time (such as voice and video).

In this book the IPv4 protocol will be described with some more detail. In what IPv6 is concerned, only the main differences between the two versions of IP will be emphasised.

18.1   Use of IP Protocol

The IP protocol runs in the internet layer of the protocol stack TCP/IP (see Table 5.2 in Sect. 5.4.2), which corresponds to the network layer of the OSI reference model. The service provided by the IP protocol is the transfer of transport layer messages from one host computer through the network to another host computer. As the network that can be seen in the internet layer consists of *terminal computers* (*terminals, hosts*) and *routers*, IP is implemented only in hosts and routers. Either a host and a router or two routers can directly communicate using the IP protocol; however, IP protocol data units are always transferred from one host through the network to another host, being processed in all intermediate routers. The pairs router–router and host–router can be interconnected with subnetworks that are based on different technologies; in this way the Internet interconnects various networks and is therefore often dubbed a network of networks. The role of the IP protocol in this environment is to glue subnetworks together and thus to merge them into a unique network. In this way the higher layer messages can be transferred from a source terminal through various subnetworks to a destination terminal. (Somehow, a host also can be considered a simple kind of a router, as it also uses a simple routing table; however, a host only can serve as a source or a destination network element, rather than a relay element.)

18.2 Functionality of IP Protocol

Hence, routers or routers and terminals can be interconnected by subnetworks based on various technologies. The IP protocol does therefore not impose any special requirements on the channels that interconnect pairs host–router and router–router. It is only important that the network infrastructure, implemented by the data-link and the network or only by the data-link layer of a subnetwork, is capable of transferring IP messages between two adjacent elements of the IP network (a host and a router or two routers).

IP is a connectionless protocol, so it does not provide for a service of reliable data transfer. IP datagrams may be lost, duplicated or reordered. Even floating corpses can occur in an IP network. This kind of communication service is referred to as *best effort service*.

Because IP networks are connectionless networks, IP datagrams are transferred through them by means of routing which is carried out by routers. Although the IP protocol itself does not execute routing, it allows it to be executed by routers. The essential mechanism that enables the routing is the addressing. IP determines the format and the meaning of addresses, so they are called IP addresses. IP also has some other mechanisms, although somehow less important, that are related to routing, such as the (partial) prevention of floating corpses and the possibility of source routing.

As the IP protocol only provides for an unreliable data transfer service, it did not offer any congestion control mechanisms at the beginning. However, the quality of service has been becoming more and more important and network congestions are not at all unusual in IP networks, so some congestion control mechanisms have also been introduced into the protocol during the course of its development.

An IP datagram may traverse various subnetworks on its way from the source to the destination host. On this way it can encounter different restrictions regarding the maximum size of protocol data units. The IP protocol must therefore provide also for the segmentation and reassembly of user messages. The protocol IPv4 allows the segmentation to be done in any router traversed by a datagram, as a datagram may always happen to be transferred from a subnetwork allowing longer packets to a subnetwork that allows shorter packets; a datagram may therefore be segmented several times on its way from the source to the destination terminal. However, segments may only be reassembled in the destination host. On the other side, the protocol IPv6 allows datagrams to be segmented only in the source host; the segmentation in intermediate routers is avoided by the source host by executing the *path discovery algorithm* before beginning to send datagrams to a certain destination. In this way the maximum segment size that can be transferred through the network without further segmentation is discovered. The operation of this algorithm is however based on the assumption that the path, which datagrams follow through the network, does not change frequently.

In the next section the IPv4 protocol mechanisms will be described with some more detail.

18.3   IPv4 Datagrams and Transfer Mechanisms

IPv4 is quite a simple protocol, far from the complexity of some connection-oriented protocols which provide for a reliable data transfer. Only a single type of IP datagrams is used; its role is to transfer a higher layer message from a source terminal to a destination terminal. The transfer mechanisms of IPv4 rely heavily on the values of some fields of an IP datagram, so both the transfer syntax of IP datagrams and the mechanisms used by the protocol will be discussed in this section.

In Fig. 18.1 the transfer syntax of an IPv4 datagram is shown. Each row in the figure represents a 32-bit word which can however be partitioned into several fields. The field widths in the figure are proportional to the actual field lengths, so the lengths of fields can easily be seen from the figure (in the figure the field lengths are not indicated for this reason, and especially because they are not considered essential in the context of this text).

Basically, an IPv4 datagram consists of three main parts: the constant length header of five 32-bit words (20 octets), the options part with variable length (indicated as opt in Fig. 18.1) and the user message of variable length (labelled data in the Figure).

The field ver in the header indicates the version of the IP protocol. This field is positioned at exactly the same place also in an IPv6 header but contains a different value; this allows a receiver to further analyse a received datagram, be it an IPv4 or an IPv6 datagram, as the formats of IPv4 and IPv6 are otherwise different.

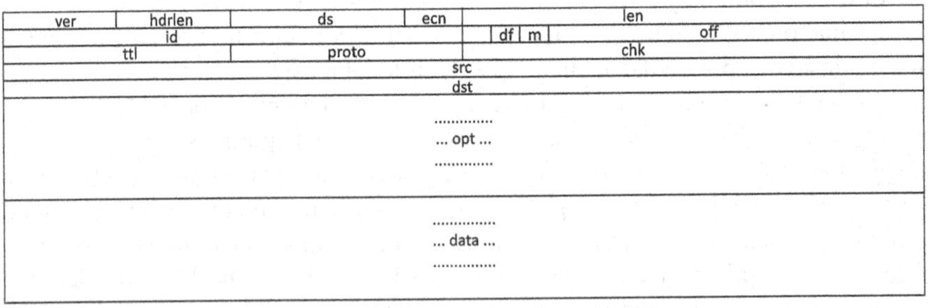

Fig. 18.1 Format of IPv4 datagram

Because the 20-octet long header of a datagram can be followed by a variable length options field, the field hdrlen contains the length of the whole header, including the 20-octet header and the options field, as the number of 32-bit words; this allows a receiver to quickly find out where the user data begin. The length of the field hdrlen (4 bits) therefore limits the total length of the options field. The field len contains the length of the whole datagram (including the header, the options and the user data) in octets, which restricts the total length of a datagram. Some lower-layer protocols (such as the ethernet protocol) may add padding bits to messages to fulfil the minimum length of a frame. An IP protocol entity can therefore receive from the lower layer an IP datagram with added padding; the field len allows it to find out also in such cases what is the datagram and what is the padding.

IPv4 uses 32-bit addresses. The fields src and dst contain the *source address* and the *destination address* of the datagram. The contents of these fields, especially the contents of the field dst, are therefore essential for datagram routing.

If, for any reason, the header of an IP datagram becomes corrupted, the datagram can consequently be incorrectly routed. The correctness of a header is therefore of utmost importance. Hence, the header of a datagram is protected with the redundant bits in the field chk which allows a receiving IP entity (either in a router or in a destination terminal) to detect errors in the header. A corrupted datagram is discarded.

The field data can contain protocol data units of various higher-layer protocols. The contents of the field proto therefore indicate in a standardised way the protocol according to which the message in the field data is composed; this allows the IP protocol entity in the receiving terminal to pass a user message to the correct higher layer entity.

It was told already that floating corpses can also occur in an IP network which is based on the connectionless operation. Most usually they are the consequence of routing errors or of instabilities of routing algorithms; due to such reasons datagrams can be catched in routing loops, circulating in them for a shorter or a longer time period. The original idea of the designers of the IP protocol was to limit the time datagrams may stay in a network, so the 8-bit field in the third 32-bit word of the IP header was named the *time to live* (ttl) field. However, due to the lack of timing synchronisation of different routers in a network it is quite difficult to precisely measure the transfer time of datagrams; therefore, the field ttl actually contains the *hop count* of a datagram, i.e. the number of hops that the datagram is still allowed to traverse; a hop is a transfer of a datagram between two adjacent routers. The source terminal assigns a certain value of ttl to a datagram; this value is decreased by 1 by each router traversed by the datagram. If this value drops to 0 in some router, this router discards the datagram and informs the source terminal about this. (This functionality of the field ttl is used by the application *tracert* (*trace route*) which finds out and records the paths traversed by IP datagrams through an IP network.)

The four fields of the second 32-bit word allow for the segmentation and reassembly of IP datagrams, as was already discussed in Sect. 9.3. Routers may only segment IPv4 datagrams if necessary, while segments may be reassembled only in the destination protocol entity IP. As we already found out in Sect. 9.3, this is not the most efficient mode from the viewpoint of data transfer; however, it requires the minimum amount of processing and memory resources in intermediate routers, and, what is even more important, a router could not reassemble a datagram if not all of its segments were transferred through it (in principle, various datagrams may follow different paths through a connectionless network)! Because a datagram can be segmented several times on its way from the source to the destination, sequence numbering of segments would not be practical. The field off (*fragment offset*) therefore contains the offset of a segment (in octets) from the beginning of the original (nonsegmented) datagram; this simplifies the reassembly of segments a lot, even if they are not received in the correct order. The one-bit field m (more) contains the value 0 in the last (or the only) segment of a datagram, and the value 1 in all other segments. Because the data transfer in IP networks is connectionless, not only the segments of a single datagram can be received in an incorrect order, but even the segments of different original datagrams can be mixed at the receiving side; any original datagram is therefore assigned a unique identifier which is copied into the field id (identity) of all its segments. A user of the IPv4 protocol may require the network not to segment a datagram; such a datagram is assigned the value 1 in the field df (don't fragment). If a router should fragment such a datagram due to the requirements of the next subnetwork, it may not do so, it discards the datagram and informs the source IP entity about it. The bit df can be used in the path discovery procedure to find the maximum length of datagrams that can be transferred nonfragmented between two hosts.

Originally, the IPv4 header contained an 8-bit field tos (type of service) to indicate to routers the desired routing method and thus the desired quality of service for the datagram (a datagram could be routed with the aim of low losses or short delays). In practice, however, routers in most cases did not consider these requirements. Therefore, this field was later partitioned into two fields. The field ds (differentiated services) allows a datagram to be assigned a quality of service class in the *differentiated services* domain of an IP network. The field ecn (explicit congestion notification) is intended to enable the congestion control with explicit feedback. A router with an almost full waiting queue can set the ecn bit in the datagrams which are transferred through it in the problematic direction, so informing the IP entity in the destination host about the potential congestion; the IP entity in the destination terminal informs about the problem the entity in the higher layer of the protocol stack, which can then request its peer entity at the transmitting side to reduce the transmission rate. This kind of congestion control has two shortcomings. Firstly, such a control works only if the TCP protocol is used in the higher layer; if the UDP protocol is used instead, this method of control is not possible, as the communication between UDP entities does not provide for the possibility of flow control. Secondly, this method of congestion control requires the collaboration of two protocols in two different protocol stack layers and is unfortunately not yet standardised.

The header of an IPv4 datagram can contain zero or more options which are encoded in the TLV format (see Sect. 8.3). An example of an IPv4 option is the path description used in the *source routing*; unfortunately, this option is practically not usable, as the restriction of the maximum total length of an IPv4 datagram header (written in the hdrlen field) does not permit the description of a realistic path through an IP network.

18.4   ICMP Protocol

In the internet layer of the TCP/IP protocol stack, the *ICMP* protocol (*Internet control message protocol*) always runs beside the IP protocol. Although ICMP messages are transferred as user messages inside IP datagrams, ICMP is not a transport layer protocol; in most cases, ICMP messages are sent by routers rather than by hosts (although hosts also can send them). Routers use ICMP messages to inform the sources of IP datagrams about the network operation and about the problems encountered in the transfer of IP datagrams. ICMP messages are also used by some applications that analyse and control the network operation. The transfer of ICMP messages is not reliable, as they can be lost together with IP datagrams which transfer them; the recipients of ICMP messages send no acknowledgments to the senders of these messages.

Now we are going to mention only a few types of ICMP messages.

The messages Echo Request and Echo Reply are used by the application *ping* to test the connectivity of a network. This application, running in a host, sends the message Echo Request to an arbitrary IP address (associated with either a host or a router); the addressee replies with the message Echo Reply. If both entities are not able to communicate the sender of the Echo Request of course does not receive the answer Echo Reply. Nowadays there is often a problem with this application because many entities simply discard Echo Request messages for security reasons.

The ICMP message Parameter Problem informs a source that the datagram it has sent contains an error (either concerning the length or the contents of the datagram).

If a router is not able to send an IP datagram to a destination host it sends the ICMP message Destination Unreachable to the sender of the datagram.

It was told already that a router discards the datagram whose time to live (ttl) field has been decreased to the zero value; when this happens the router informs the source of the discarded datagram about this with the ICMP message Time Exceeded. The Time Exceeded message, together with IP datagrams with different initial ttl values, is also used by the application traceroute to find out the path through the IP network which is followed by IP datagrams between a source and a destination.

A router whose input queue is growing full, thus threatening with a congestion, can send the ICMP message Source Quench to the sources of datagrams which are transferred though the router in the endangered direction; in this way the router can request from datagram sources to reduce their transmission rates. The Source Quench message is therefore used for the congestion control with explicit feedback.

18.5   IPv6 Protocol

It was already told that IPv4 is a rather old protocol. However, its huge success shows that it was designed with care and a good measure of foresight. However, it was not at all possible to foresee 40 years ago, when the IP network was still in its age of infancy, that the number of addresses $2^{32} = 4,294,967,296$ could ever be insufficient! However, nowadays there are officially[1] no more IPv4 addresses available! In spite of this, the IPv4 protocol is expected to be still in use for quite some time in the future and will so have to coexist and also cooperate with the protocol IPv6, partly because of the conservatism and the comfort of network users and operators, but also because it is not possible to replace all the installed hardware and software overnight, for both technical and economic reasons.

The primary goal of the development of the protocol IPv6 was the increase of the address space. However, several other modifications were introduced into the protocol at the same time, especially with a quicker and a more efficient processing of datagrams in routers in mind. As a matter of fact, the size of the Internet has been increasing at a higher pace than the processing speed of routers, so the datagram processing nowadays already imposes a very high burden on routers.

Like IPv4, IPv6 is also a connectionless protocol. In fact, this is a new protocol which is quite different from IPv4 and is even not compatible with it. The fact that the field in a datagram in which the version of the protocol is indicated is equally long and positioned at the same location within the datagrams of both protocols, allows both protocols to coexist in networks for a longer period of time, as the receiver of a datagram can read the protocol version at first and only then decode and analyse the rest of the datagram. Of course, there are some problems with the transit of datagrams through mixed IPv4/IPv6 networks; however, the solving of these problems is not the topic of this book.

The IPv6 protocol will not be treated in detail in this section. Rather, only some most important differences between the two protocols will be described.

The most important difference between IPv4 and IPv6 is of course the length of IP addresses and consequently the size of the address space. While the IPv4 addresses are 32 bits long, the length of IPv6 addresses is 128 bits! While IPv4 addresses can be configured with the aid of the protocol *DHCP* (*Dynamic host configuration protocol*) and the DHCP servers, the IPv6 addresses do not need DHCP because they can be configured automatically. IPv6 addresses can be individual addresses (indicating a definite single entity IP), group addresses (indicating a definite group of IP entities) and anycast addresses (that indicate any entity within a definite group of addresses).

---

[1]The organisation IANA (Internet Assigned Numbers Authority) assigns groups of addresses to regional and local organisations which then assign them to institutions, companies and individuals; so regional and local organisations still can have free IP addresses, though there are no more free addresses at the global level.

The channel coding of IPv4 headers requires a lot of processing, as some fields of the header (the ttl field in the first place) are modified in each router; each router must therefore verify and recalculate the redundant bits. Furthermore, if the correctness of a message is important for an application, the correctness of the message must nevertheless be verified at a higher layer. For these reasons, an IPv6 header is not protected with redundant bits.

Although the IPv6 header is twice as long as the IPv4 header (because of the addresses that are four times as long), it nevertheless contains less fields, is more simple and requires less processing. Instead of options an IPv6 datagram can contain so-called *extension headers* which are coded in a format that is easier to be processed. An IPv6 entity can process only those extension headers that it considers important, while skipping the others; this is especially important because some extension headers are to be processed exclusively by the destination entity, and the others by all the entities, including routers. Among other things, extension headers contain the information that is related to routing, security and datagram segmentation and reassembly.

We mentioned already that the transfer of IPv4 segments is not very efficient. The protocol IPv6 therefore allows the segmentation to be done only in a source terminal. However, before sending data, the largest length of datagrams that can be transferred through a network must therefore be found out, using the path discovery procedure. IPv6 segments are reassembled in the destination entity. The information needed for a correct reassembly is contained in one of extension headers.

When the IP protocol and the Arpanet network (the predecessor of the Internet) were conceived many decades ago, the network was small and the access to it was strictly limited; hence, nobody then considered the communication security problem seriously (to tell the truth, most of the hackers of this day were then even not yet born!). Until today, the network has expanded enormously and virtually everybody can gain access to it; the network and the communication security has therefore become an extremely important topic. When the IPv6 was designed the security mechanisms were built into the protocol itself. These mechanisms are referred to as *IPSec* and can also be used with the protocol IPv4. The information that is necessary to assure a secure communication is included into IPv6 extension headers.

In modern IP networks streams bearing the information in real time are more and more frequently transferred, concurrently with plain data. The transfer of different kinds of information requires different qualities of service. IPv6 therefore allows various datagrams to be associated with different information *flows*. The routers in the network treat all the datagrams that are associated with the same flow the same way (e.g. providing them with as short delays as possible, or with as low losses as possible), thus providing all the datagrams of the flow with approximately the same quality of service.

A new protocol *ICMPv6* was also developed to be used along with IPv6. It plays a similar role as ICMP with IPv4.

Chapter 19
ARP Protocol

Abstract The ARP protocol is presented in this chapter as an example of the address translation protocol, which is especially suitable to be used in local area networks and is nowadays heavily used in the access networks of the Internet to translate IP addresses into MAC addresses, although it can also be used in different environments. The role of this protocol in such an environment is described. Its operation is related to the address translation method, which is appropriate to be used in local area networks and which was specified in a previous chapter. The role of the RARP protocol is also described.

In Sect. 10.3 the translation between addresses, denoting a same protocol entity, was treated. In this chapter the protocol *ARP* (*Address resolution protocol*) will be described. This protocol is used in the TCP/IP protocol stack to translate network layer addresses into corresponding data-link layer addresses. The protocol is capable to translate the addresses of any network layer protocol into the addresses of any data-link layer protocol, as the types of both protocols (network and data-link layer) and the lengths of both addresses are recorded in the ARP header. The protocol is especially suitable to be used in the networks in which the broadcast type of communication is used (whatever is transmitted by any entity is received by all the other entities).

The ARP protocol is most usually used in local area networks to translate IP addresses into MAC addresses. This mode of ARP operation will also be described here.

In the Internet or any other IP network the network elements are assigned IP addresses and MAC addresses independently. IP addresses are assigned by network administrators at various levels of network management, while MAC addresses are hardware defined and therefore unmodifiable and are assigned by hardware producers. For the communication, however, both kinds of addresses are needed. The routing of datagrams through an IP network is based on IP addresses; however, in an access network or in a network interconnecting routers which is based on the LAN technology, the transfer of MAC frames is based on MAC addresses. The translation of IP addresses into MAC addresses is therefore necessary.

D. Hercog, *Communication Protocols*, https://doi.org/10.1007/978-3-030-50405-2_19

The ARP protocol data units are messages of the type request–answer. They are transferred through a local area network as user messages within MAC frames.

In principle, the protocol ARP operates as was described in Sect. 10.3.1. An entity wanting to know the MAC address of another entity with the known IP address transmits a request in the broadcast mode (so that all the other entities in the local area network hear the request); this request contains both addresses (IP and MAC) of the inquirer and the IP address for which the corresponding MAC address is required. This request is answered only by the entity which owns the given IP address. To reduce the number of necessary inquiries in the network and also the time needed to acquire a translation, any communicating entity in a LAN maintains an ARP table which contains a number of recently used translations; the lifetime of ARP table entries is limited.

The protocol *RARP* (*Reverse address resolution protocol*) also exists. It is used by diskless computers; after it boots up, a diskless computer knows only its MAC address (which is built into the hardware of its network interface card), but it does not know its IP address. It therefore requests its own IP address with a RARP request from a server which knows it.

The ARP protocol is not used with IPv6. The *ND* (*Neighbor discovery protocol*) is rather employed which is a part of the protocol IPv6 itself.

Chapter 20
DNS: Distributed Directory and Protocol

Abstract At first the domain names are described that are used in the Internet in addition to the IP addresses, and the need for the translation of the domain names into the IP addresses is exposed. Then the domain name system is presented that meets this need. The structure and the operation of the domain name system are described and related to the two variants of the address translation protocols to be used in wide area networks, which were treated and formally specified in a previous chapter.

In Chap. 19 a protocol for the address translation in a local area network environment was presented; in this chapter another address translation protocol will be described which is used to translate names into addresses in the Internet environment. This is the protocol *DNS* (*Domain Name System* protocol).

In IP networks *domain names* are used besides IP addresses. Domain names are hierarchically organised. Any domain name must correspond to an IP address. While IP addresses are 32-bit addresses (in case of IPv4, of course), domain names are character sequences and consequently easier to be read by humans. Human users therefore most usually use domain names; however, the routing of messages through the network is based on IP addresses. The translation of domain names into IP addresses is therefore necessary for the operation of IP networks. The storage of domain names and their translation into IP addresses is the task of the domain name system which consists of many *domain name servers*. The domain name system is a distributed database which is (partly) hierarchically organised; the hierarchy of domain name servers follows the hierarchy of domain names themselves. The operation of the domain name system is based on the operation of domain name servers (from now on in this chapter we will say just servers) and on the *DNS protocol*.

An application which needs a translation of a domain name into the corresponding IP address requests it from the entity that is referred to as *resolver*. If the resolver does not find the translation in its *cache* (temporary storage of translations) it requests it from a server. The configuration of a resolver must therefore include the IP address of at least one domain name server. If the server

D. Hercog, *Communication Protocols*, https://doi.org/10.1007/978-3-030-50405-2_20

knows the translation it returns it, otherwise it requests it from another server which again either knows the translation or queries it from yet some other server, until the resolver receives the answer. A server must always answer the resolver with the requested translation (unless for some reason the translation cannot be obtained). On the other hand, one server may respond another server either with the requested translation or with the address of yet another server which might know the translation. Hence, both versions of the address translation protocol which were explained in Sect. 10.3.2 are in use. The variant 1 from Sect. 10.3.2 (when a server replies with the address of another server if it does itself not know the translation) is referred to as *iterative query* in the DNS system, while the variant 2 (when a server always replies with the requested translation) is called *recursive query*. Hence, a resolver always uses the recursive query, while a server may employ either a recursive or an iterative query with another server.

The DNS protocol is a protocol of request–reply type. DNS messages are most usually transferred as user messages inside UDP datagrams, as a resolver can repeat its request if it does not receive any response in the expected time. If more than one translations are requested at the same time between two servers, DNS messages can also be transferred as the payload of TCP segments.

Chapter 21
TCP Protocol

Abstract TCP is presented here as an example of a complex protocol which provides for several different functionalities that were discussed in previous chapters, and also because it is one of the protocols which are most heavily used in the Internet nowadays. The functionality and the services provided by TCP are explained. The structure of a TCP segment and its fields is described. Then the operation of the protocol is presented with reference to the communication methods described in previous chapters of the book. The connection setup and release are described and commented. The reliable data transfer, which operates as a selective-repeat protocol, is also presented. A special care is devoted to the treatment of the flow/congestion control that is based on the variable window width control with both explicit and implicit feedback. There are two figures in this chapter, one of them displaying the structure of a TCP segment and the other one showing the connection establishment procedure.

The protocol *TCP* (*Transmission control protocol*) is one of the most important protocols on which the operation of the Internet is based from the very beginning. Hence, this protocol is already about four decades old which is really a venerable age for a communication protocol. It is, however, true that this protocol has been developed, improved and updated all this time. At the beginning it was defined together with the protocol IP; both protocols, however, soon separated one from the other. Maybe this is also one of the reasons why the protocol stack used in the Internet was named TCP/IP (the name it has preserved until today); the other reason, perhaps even more important, is that the protocols TCP and IP have been, and still are, the most frequently used protocols in the Internet. In any case, TCP and IP are nowadays two distinct protocols, so speaking about the »protocol TCP/IP« is wrong (although this can unfortunately be heard or read quite often)!

At first, TCP was relatively a simple protocol. However, the Internet network has been expanding all the time, so new problems have also been emerging with time; these problems have been solved by adding ever new functionalities and mechanisms. Nowadays, TCP is therefore a very complex protocol.

D. Hercog, *Communication Protocols*, https://doi.org/10.1007/978-3-030-50405-2_21

In this chapter the functionality and the most important mechanisms of the TCP protocol, such as we know and use today, will be shortly described.

Although its successor (this should be the protocol SCTP) was already developed and also put into use, TCP is still massively used nowadays.

21.1   Use of TCP Protocol

It was already explained that TCP is one of the most important protocols that are used in the Internet network. It runs in the transport layer of the TCP/IP protocol stack (see Sect. 5.4.2 and Table 5.2), which means that the TCP protocol is used for a direct exchange of messages between network terminals (hosts). In other words, TCP is not implemented in intermediate network elements, such as routers[1]. The protocol TCP is an automatic repeat request (ARQ) protocol and therefore provides for a reliable data transfer between hosts; therefore, it induces strongly variable delays into information transfer. For this reason, TCP is used when a reliable data transfer is needed, but cannot be used to transfer information in real time (such as voice or live video) which is sensitive to the delay variation.

21.2   Functionality of TCP Protocol

While the channel that interconnects TCP protocol entities is unreliable, the TCP protocol itself implements a reliable data transfer service for the higher layer. The quality of service provided for by TCP is therefore much better than the quality of service TCP itself receives from the lower layer. This is the very reason for the high complexity of the TCP protocol.

The channel that interconnects TCP entities is the network layer of the Internet network with the IP protocol working in it. IP is a connectionless protocol and does therefore not provide for a reliable information transfer. IP datagrams can be lost or reordered, even floating corpses may occur. TCP must therefore provide for the solution of all these problems. IP datagrams may traverse on their way from a source to a destination terminal various subnetworks which allow different maximum

---

[1]Nowadays many routers and other network elements allow network administrators to remotely manage, control and administrate them from remote computers, using special management software for this purpose. The management software that is installed in a network element uses the whole protocol stack, including the transport and the application layer, to be able to communicate with managing devices; hence the TCP protocol must also be used by them. However, the components that are used for the management act as network terminals. A router in the proper sense of this word, whose function is to route IP datagrams, does not use TCP. In spite of this, the whole device, consisting of a router, a switch, a web server, a management software and possibly still other components, is usually dubbed simply a »router«.

lengths of messages to be transferred. Although IP is able to segment and reassemble packets that are too long, excessive segmentation introduces additional overhead and thus decreases the transfer efficiency. At the time a TCP connection is set up, both TCP entities can therefore agree on a *maximum segment size MSS* (maximum length of TCP messages) and thus reduce the need for segmentation in the internet layer.

TCP is a connection-oriented protocol. An established TCP connection implements a channel that interconnects application layer protocol entities in host computers. A TCP connection allows a bidirectional (duplex) transfer of information. A TCP connection is globally (within the whole Internet network) uniquely defined with two pairs of addresses: the pair IP address–port number at one side of the connection and the pair IP address–port number at the other side of the connection. While the IP address indicates a network element or, more precisely, a network interface of a network element, the *port number* indicates the application layer protocol entity and thus the application in a host. The pair IP address–port number is referred to as *socket*; hence, a TCP connection is unambiguously defined with two sockets.

The basic unit of user information which is transferred with the TCP protocol is an octet. The service that is provided by the TCP protocol to the higher (application) layer is the reliable transfer of a sequence of octets. Although a TCP entity receives a message as a sequence of octets from the application layer entity above it, this message is not necessarily transferred through the network in its original form; TCP can segment the message into several fragments, or it merges more messages to be transferred in a single TCP protocol data unit, it even can transfer each octet separately (this is of course not very efficient!); what is important is that the receiving user receives the same octet sequence (without losses, duplicates and in the correct order) as was sent by the transmitting user. A transmitting application layer entity must therefore format its messages so that the receiving application layer entity can reconstruct the original messages from the received sequence of octets.

The TCP protocol performs the flow control between both TCP entities, thus preventing congestions to occur in network terminals. Because congestions may occur also in intermediate network elements (routers), TCP also executes the congestion control; this control includes both congestion avoidance and congestion recovery. The flow/congestion control is very important to provide for a reliable data transfer—if a congestion occurred, the network might cease to operate and consequently data transfer would be impossible.

21.3   TCP Segments

A TCP protocol data unit can contain a sequence of user octets which is not necessarily an integral user message; a TCP message is therefore called a *segment*. Formally, there is only a single type of a TCP segment; in fact, there are several types of segments and several types can be merged into a single segment. In this book a simplified description of TCP segments will be given.

21.3.1   Abstract Syntax

The abstract syntax of a TCP segment will be given for each segment type separately. In the next list the abstract syntaxes will be shown for individual segment types, where the type of a segment will be indicated with uppercase letters, and the most important parameters of a segment will be listed within parentheses.

- SYN(s,a,w)
- FIN(s,a,w)
- RST(s,a,w)
- ACK(a,w)
- I(s,a,w,data)

The segments SYN and ACK are used to establish a connection. The segments FIN and ACK are used to release a connection. The segment RST is used either to reject a new or to reset an existing connection. The segment I is an information protocol data unit that contains user octets. The segment ACK is also used to explicitly acknowledge the correct reception of user data.

The parameter s contains the sequence number of an information segment. A reader may be surprised to know that the segment SYN, too, contains a sequence number (as well the segment FIN). The parameter s in a SYN segment determines the *initial sequence number*, with which the counting of user octets begins after a connection setup; the initial sequence number is determined for each direction of user data transfer separately, independently one from the other. A SYN segment is assumed to contain one user octet (although it really contains none); the first information segment therefore contains a sequence number that is by one higher than the sequence number contained in the SYN segment sent by the same protocol entity. The sequence number of an information segment is higher than the sequence number of the pervious segment sent by the same entity by the number of user octets contained in the previous segment. One must be aware that user octets are counted rather than segments with TCP, and the sequence number s contained in an information segment is actually the sequence number of the first user octet of that particular segment.

The parameter a contains the acknowledgment number; the sender of an acknowledgment lets its peer know which user octet it is expecting to receive next. Cumulative acknowledgments are used. Acknowledgments may either be piggybacked onto information segments, or explicit acknowledgments are used. Explicit acknowledgments are used in the connection management procedure as well. A segment may contain an acknowledgment or not.

The parameter w contains the *advertised window* width; this is the credit which is sent by a receiver to its peer transmitter in the sense of the flow control with the variable window width method (see Sect. 13.6.4). The last user octet that a receiver is prepared to receive is the octet with the sequence number $a - 1 + w$.

The parameter data in an information segment is the sequence of octets that forms a user message.

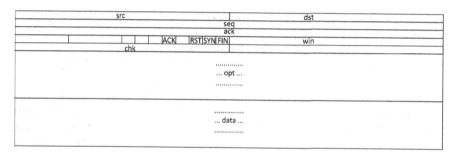

Fig. 21.1   Transfer syntax of TCP segments

We will find out the ranges of values that the parameters of a segment may contain when the corresponding fields of a TCP segment will be discussed.

21.3.2   Transfer Syntax

In Fig. 21.1 the (incomplete) transfer syntax of a TCP segment is shown; only those fields are labelled in this figure which will also be described in the following text. The format is partitioned into 32-bit words (represented as lines in the figure); some of the words are further partitioned into shorter fields.

In the fourth 32-bit word there are eight 1-bit fields that are usually referred to as *flags*; here, however, the meaning of only four flags will be described. The contents of these flags determine the type of a segment; furthermore, it depends on these flags whether some other fields have their meanings or are meaningless. If the value of the flag SYN is 1, the type of the segment is SYN; the flag FIN indicates that the segment is a FIN segment; in an RST segment the value of the RST flag is 1. If the value of the flag ACK is 1, the field ack (the third 32-bit word) contains an acknowledgment number; if the ACK bit is 0, the field ack has no meaning and the segment contains no acknowledgment. A segment of the type SYN or I can either contain an acknowledgment or not, depending on the value of the flag ACK in the segment.

The field data contains user octets and has a variable length; hence, this field is included only in information segments. The maximum allowed length of the field data is determined with the parameter MSS which may be agreed upon by the two protocol entities at the time of the connection establishment (if it is not, the default value is in force). If the value of the flag ACK of an information segment is 1, this segment contains a piggybacked acknowledgment, otherwise there is no acknowledgment. If there is no data field and the value of the flag ACK is 1, this segment is an explicit acknowledgment.

The field seq contains the sequence number of the segment, or, more precisely, the sequence number of the first user octet in the field data. As can be seen in

Fig. 21.1, both the sequence number and the acknowledgment (contained in the field ack) are 32-bit numbers; hence, user octets are counted modulo 2^{32}.

The field win contains the width of the advertised window (the credit); this is the number of user octets that the receiver (the sender of the credit and the acknowledgment) is still able to receive after the last acknowledged octet.

It was already told that a TCP connection is unambiguously determined with two pairs IP address–port number. The fields src and dst in the first 32-bit word of the header of a TCP segment contain the port numbers of the applications at the source and the destination end of the connection, respectively, while both IP addresses are contained in the header of the IP datagram which contains the TCP segment as its payload. The port numbers are 16-bit values.

The field chk in the fifth word contains the *checksum*; these are redundant bits which allow a receiver to detect corrupted segments and to discard them. A TCP transmitter calculates the checksum from the TCP segment and a part of the header of the IP datagram that transfers this segment; the combination of the headers TCP and IP is referred to as *pseudoheader*. In this way a TCP receiver can detect even those errors which are due to incorrect transfer of a segment through the IP network. The computation of a checksum from the combination of the headers of two protocols which are running in two different layers of the protocol stack would be considered unprecedented in the orthodox world of OSI; in the TCP/IP world, however, it is not at all unusual.

The usual length of a segment header is 20 octets, as can be seen in Fig. 21.1. If necessary, the header can be added optional fields which are labelled opt in the figure. An example option which is often used at the time a new connection is being established is to determine the maximum segment size MSS.

21.4   Information Transfer Mechanisms

21.4.1   *Connection Setup*

TCP uses a three-way handshake to establish a connection, because floating corpses may appear in the lower layer of the protocol stack TCP/IP (see Sect. 11.3). The parameters which are used in the three control messages that set up a connection are the initial sequence numbers to be used for data transfer in both directions of the connection (see Sect. 21.3.1). The initial sequence numbers to be used in both directions are independent one from the other; they are determined as quasi-random numbers which reduces the probability that the procedure of establishing a new connection could be disturbed by a floating corpse from a previous connection. As we already know from the treatment of sliding window protocols, the acknowledgment numbers that are transferred in one direction refer to the sequence numbers of information octets which are transferred in the opposite direction. At the time a new connection is set up, both protocol entities can also agree upon additional parameters of the connection, such as the maximum segment size MSS.

Fig. 21.2 TCP connection
establishment

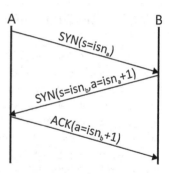

In principle, the connection setup procedure is such as was shown in Fig. 11.13. The role of the message offer in that figure is played by the segment SYN which contains the initial sequence number to be used for the data transfer from the service initiator to the service recipient side (in fact, the sequence number of the first user octet is by one higher than the sequence number contained in the SYN segment). The recipient replies to this offer with another SYN segment which corresponds to the message accept in Fig. 11.13; this SYN determines the initial sequence number for the data transfer from the recipient to the initiator side of the connection (here, too, the sequence number of the first user octet is by one higher than the number contained in this second SYN). Additionally, the second SYN also contains the acknowledgment of the first SYN. Hence, the first SYN does not contain any acknowledgment, while the second SYN contains the acknowledgment number that is by one higher than the sequence number in the first SYN. The role of the message check in Fig. 11.13 is played by the TCP segment ACK (an explicit acknowledgment) which contains the acknowledgment number. The procedure of a successful connection setup is shown in Fig. 21.2. The parameters s and a in this figure represent the sequence number and the acknowledgment number, respectively, as was already discussed in this section. The values isn_a and isn_b are the initial sequence numbers for the data transfer from A to B and from B to A, respectively; they are independent one from the other.

All these segments also contain the advertised window width; this, however, is not shown in Fig. 21.2, because this is not essential for the connection setup procedure itself; this is rather the determination of the value that should be considered during the user data transfer phase.

21.4.2 Connection Release

A TCP connection is released for each direction of data transfer separately, because one protocol entity might still have user data to be sent when the other entity wants to release the connection. A two-way handshake is used to release each direction of the connection; this two-way handshake consists of TCP segments FIN and ACK. Altogether, four segments are therefore used to tear a TCP connection down.

21.4.3   User Information Transfer

The protocol TCP provides for a reliable transfer of octet sequences. It operates as a selective-repeat protocol with cumulative acknowledgments and a single retransmit timer. This is important because a TCP receiver may happen to receive an unexpected segment not only because a segment has been lost but also because one segment has managed to reach the receiver through the connectionless-oriented network sooner than another segment which was transmitted before it; in such an environment the go-back N protocol would not be efficient, but the selective-repeat protocol uses the transmission channel efficiently, because the receiver is capable to store the message that has come »too early« in its receive buffer. As was already told, the basic unit of user information is octet, so user octets are counted rather than information segments.

   Formally, the TCP protocol does not use negative acknowledgments; in fact, however, it uses them. When a TCP receiver receives an expected information segment it sends an acknowledgment, telling its peer transmitter which user octet it is expecting to receive next; in other words, an acknowledgment contains the sequence number that indicates the left edge of the receive window. If there are no losses and TCP segments are not reordered in the connectionless channel, the receiver receives information segments in the correct order and the transmitter receives ever new acknowledgments. Because the acknowledgments are cumulative, the receiver even does not need to immediately acknowledge every correctly received information segment, but can send an acknowledgment only after it has received a few (usually two) expected segments. If, however, a segment has been lost or segments have been reordered and the receiver consequently receives an unexpected segment, the receiver retransmits the same acknowledgment which it has already sent, immediately after it has received an unexpected segment. If the transmitter receives the same acknowledgment up to three times in line (one original and at most two repeated acknowledgments), it considers the segments to have been reordered; however, after it has received a fourth equal acknowledgment in line (one original and three repeated ones), it assumes that a segment was lost because the reorder of so many segments in a connectionless network is not very probable; hence it retransmits the information segment that is positioned at the beginning of its transmit window in such a case. The reception of four equal acknowledgments in line is therefore treated as a negative acknowledgment by the TCP transmitter.

   Unfortunately, this kind of interpretation of repeated acknowledgments is not sufficient to guarantee the logical correctness of the protocol. A TCP transmitter must therefore also employ a retransmit timer. As the retransmit timer expiration time is usually much longer than the round trip time, most losses are resolved by retransmissions after the reception of three repeated acknowledgments. In case of an increased loss rate, however, several consecutive segments may not reach the receiver which therefore does not send repeated acknowledgments, and consequently the transmitter times out.

The determination of the round trip time and consequently also the timer expiration time is not an easy task. As was already told, the channel that interconnects two TCP protocol entities is the IP network which is a network with packet-oriented connectionless transfer of data. In such a network, the times packets spend in waiting queues, and consequently the delays of TCP segments, are random and depend strongly on the load imposed on the network. Even the statistical characteristics of delays cannot easily be predicted, as they depend on the time of day, the day of week and on the characteristic of user traffic. Hence TCP transmitters measure the round trip time; the measured value can vary substantially with time, so they must be averaged with a low pass filter, such as was shown in Sect. 12.12 and in Eq. (12.21). Besides the average value of the round trip time a TCP transmitter also assesses the standard deviation of the round trip time; the calculation of the timer expiration time is based on the assessment of the round trip time and with a substantial safety margin also the assessment of the standard deviation.

One more peculiarity of the TCP protocol must be mentioned here. The receive buffer of a TCP receiver also acts as the waiting queue between the TCP receiver and the application layer protocol entity above it. When a TCP receiver receives an expected segment it does not forward the user octets that are contained in it to its user (the application layer entity above it), but only informs it about the new state of the receive buffer; it can slide the receive window only after the user retrieves user octets from the beginning of the buffer. This mode of operation is of course not in accordance with the general principles of protocol stack design, as the two adjacent layers are not strictly separated; such incompatibilities with the OSI reference model principles are, however, characteristic and also generally accepted for the protocol stack TCP/IP and its protocols which are used in the Internet.

21.4.4 Flow/Congestion Control

When data are transferred through the Internet network using the TCP protocol, congestions can occur both in a receiving terminal and in intermediate network elements. In a receiving terminal, congestions can be due to the lower speed of the receiving TCP entity or the lower speed of the receiving application protocol entity and the receiving user, according to what was already told in the last paragraph of the previous section. On the other hand, congestions in the intermediate network are caused by full waiting queues in routers which operate in the lower (internet) layer of the protocol stack. The TCP protocol must therefore employ both the flow control and the congestion control.

The flow control is based on the variable window width control method with explicit feedback (see Sect. 13.6.4). The explicit feedback about the state of the receiving window is implemented with the parameter w in the abstract syntax of a TCP segment, which is referred to as the advertised window width. As can be seen in Fig. 21.1 and its associated text, a receiver informs its peer about the advertised window width with a 16-bit value in the field win which means that the maximum width of the advertised window is $2^{16} - 1$ octets. If transfer delays are very long and transmission rates are high (such cases are satellite communications), this value may not be sufficient to fulfil Eqs. (12.17) and (12.18) and hence to provide for the efficient use of the channel. When a new connection is established, both TCP entities can therefore use a special option to agree upon a factor with which the value of the field win is to be multiplied; this option was added to the protocol later because 16-bit values were more than sufficient in the initial version of the protocol.

The congestion control is based on the variable window width control method with implicit feedback. In the classical Internet which is based mainly on wired interconnections between network elements, losses that are due to bit errors are very rare, so the majority of segment losses are caused by full waiting queues and congestions in IP routers; TCP transmitters therefore interpret losses as the indications of a congestion. A TCP transmitting entity maintains a special parameter referred to as *congestion window* to manage the congestion control. If a light congestion occurs in the network, a segment or two may be lost but the others succeed to reach the receiving host; as was already told, the TCP receiver sends duplicate acknowledgments and the transmitter retransmits the segment from the beginning of its window after it has received the third repeated acknowledgment, but at the same time it reduces its congestion window width to the half of the width before the loss detection. The purpose of this act is to prevent a heavier congestion to occur, so this is a congestion avoidance measure. If, however, a heavy congestion develops in the network, the receiver receives no segments and therefore does not send duplicate acknowledgments, so the transmitter times out and resends the oldest unacknowledged segment; in this case the congestion recovery is needed, so the transmitter radically reduces the width of its congestion window to the length of a single segment. Furthermore, after each consecutive timer expiration the transmitter doubles the timer expiration time which is referred to as the exponential backoff (see Sect. 13.6.2). Of course, no congestion lasts for ever, so a transmitter must, after it has reduced its congestion window, increase its width gradually. In case of a congestion avoidance, the congestion window width is increased linearly with time after it was reduced by the factor of two. In case of a congestion recovery, however, the congestion window width is first increased exponentially for some time and then linearly. Any increasing of the congestion window is of course interrupted if a new loss is detected.

At the time a new connection is established, a TCP transmitter of course cannot know the state of the network, whether it is already congested or not. It therefore employs the congestion window width that equals the length of a single segment at first; it then increases it exponentially with time. This precaution measure is referred to as *slow start*.

In many modern networks which include wireless communication channels between the network and its terminals (such as mobile phones, tablets or laptops), the assumption that bit errors are very rare often does not hold; bit error rates can even be extremely high in the radio-based part of a network. In such environments the congestion control method that was developed for the classical Internet and has been explained in this section can even be counterproductive. The TCP research community is still searching the solutions of this problem.

A TCP transmitter must of course take both flow control and congestion control into account. The transmit window width must therefore be equal to

$$W_t \leq min\ (W_a, W_c), \tag{21.1}$$

where W_a and W_c are the widths of the advertised window and the congestion window, respectively, and the function *min* returns the argument with the lower value.

21.5 TCP Protocol Today

Although the protocol is already relatively old, it has been developed, upgraded and adapted to the development of the Internet all the time. We have already mentioned the options which can be added to segment headers. The efficiency was also somehow improved with the use of selective acknowledgments. A lot of time was also spent to develop and improve various congestion control algorithms. This can be understood easily. On one hand, the congestion control is becoming more and more important due to the more and more heavy load imposed upon the network; on the other hand, the congestion control which uses the implicit feedback is executed solely in transmitters, so it does not need the interaction between transmitters and receivers and therefore does not need to be standardised, which facilitates the introduction of ever new algorithms a lot.

Although the SCTP protocol, which is to become the successor of TCP, was already developed, TCP is still massively and successfully used nowadays and is still indispensable when data are transferred through the Internet. One of the important reasons for this is the conservatism of most users, and the other one is that many advantages of SCTP will only be noticeable when it will be used jointly with IPv6.

Chapter 22
UDP Protocol

Abstract This chapter is very short because UDP is quite simple. The differences between TCP and UDP are emphasised. The environments where the protocol UDP is most usually used are explained, and its appropriateness for the information transfer in real time is emphasised. Its ability to multiplex different application layer processes in the Internet is declared as its most important functionality. The transfer syntax of UDP datagrams is described and depicted in the only figure of this chapter.

The protocol *UDP (User Datagram Protocol)* is also one of the basic protocols that has been used in the Internet since its beginning. UDP, too, was conceived and used already many years ago when the Internet was still in its age of infancy. The name of the protocol originates from the fact that it is, unlike TCP, a connectionless (datagram) protocol.

UDP is a very simple protocol, so there is no much to be told about it. This chapter is therefore going to be much shorter than the previous one, in which TCP was described.

22.1 Use of UDP Protocol

Like TCP, the protocol UDP is also used in the transport layer of the protocol stack TCP/IP. UDP is a connectionless protocol; hence it is not an ARQ protocol and therefore cannot provide for a reliable transfer of data. It is very simple and therefore uses only a modest amount of memory and processing resources, so it can be used in systems which do not have much resources at their disposal. Usually it is also used to support short transactions, which consist of the transfer of only one or two messages, as the connection setup and release would impose a too high overhead in such cases. When simple data, such as telemetric data, are transferred periodically UDP can also be used, as the loss of a message is not critical is such environments, but delays are. If, however, UDP is used, because of its simplicity, for the transfer of data which may not be lost, it is up to the application layer protocol to provide for the reliability of data transfer. In present time UDP is very much used for the transfer of information in real time (such as voice or video), as it does not introduce long and variable delays into the transfer.

22.2   Functionality of UDP Protocol

UDP is a connectionless protocol that does not provide for a reliable data transfer, nor does it perform a flow/congestion control. The functionality of this protocol is therefore not at all reach, so it can be so simple.

The UDP protocol is a kind of the interface between the internet layer, in which the IP protocol is running, and application layer protocols. In this sense it plays a very important role; it allows the messages that are exchanged between various pairs of applications to be multiplexed. It provides this service in the same way as TCP does, namely with port numbers and sockets. UDP is, unlike TCP, message oriented; a message that it receives from a higher layer application entity is normally transferred through the network in a single UDP datagram.

22.3   UDP Datagrams

Just like the protocol itself, UDP datagrams also are simple. Only a single datagram type exists; its task is to transfer an application layer message.

In Fig. 22.1 the transfer syntax of a UDP datagram is shown. A UDP datagram contains the 16-bit port numbers of the source and the destination, as well as the checksum value that is calculated by the transmitter over the UDP datagram and the pseudoheader, as was already explained for the case of TCP.

22.4   UDP Protocol Today

Nowadays the UDP protocol is even more important than decades ago, because the real-time information (e.g. voice or video) is more and more frequently transferred through IP networks. For the transfer of information in real time, UDP is much more

Fig. 22.1 Transfer syntax of UDP datagrams

suitable than TCP; while TCP as an automatic repeat request protocol introduces a heavily variable delay into the information transfer, which is very unfavourable when the information must be transferred in real time, UDP does not introduce a variable delay into the information transfer. Furthermore, the information that must be transferred in real time often contains a lot of redundancy, so eventual losses are not critical.

Chapter 23
SCTP Protocol

Abstract The protocol SCTP is presented in this chapter as a possible successor of TCP. The chapter therefore focuses on the most important differences between TCP and SCTP. The services SCTP offers to the higher layer are listed and its message orientation is emphasised. The salient features of SCTP, which are not possessed by TCP, are emphasised. The most important novelty that was brought about by SCTP is multihoming; the advantages of multihoming are explained. Other peculiarities of SCTP, not possessed by TCP, are also multistreaming, improved security and simplified message structure.

At first the protocol *SCTP (Stream Control Transmission Protocol)* was developed as a very reliable replacement for the protocol TCP to be used primarily for the transfer of signalling in IP networks. As IP networks have been becoming more and more widespread and intensively used, the need for a very reliable transfer of various kinds of data, not only signalling, has also been rising. Consequently, SCTP has emerged as a possible general successor of TCP.

The protocol SCTP is similar to TCP in many aspects, as both of them use similar algorithms. In this short chapter only those properties of SCTP will therefore be emphasised which differ essentially from the properties of TCP.

Like TCP, SCTP also runs in the transport layer of the protocol stack TCP/IP. The channel that interconnects SCTP protocol entities is therefore implemented with the connectionless and hence unreliable IP network. SCTP offers the application layer the service of a reliable message transfer; an application layer message is therefore transferred entirely, in a single SCTP message. This is an important difference between SCTP and TCP, as TCP has some problems because of its octet-oriented nature.

SCTP is a connection-oriented protocol, too. The connection between two SCTP protocol entities is referred to as *association.*

One of the most important novelties that was introduced by the protocol SCTP with respect to TCP is *multihoming.* This means that messages associated with a single SCTP association may follow various paths through the network from the source application to the destination application, as any communication element can

D. Hercog, *Communication Protocols*, https://doi.org/10.1007/978-3-030-50405-2_23

use several IP network interfaces simultaneously and therefore have several active IP addresses simultaneously. An SCTP socket is determined by a single port number and one or more IP addresses. One of the paths through the network (with its associated IP addresses) may be primary, while the other paths (with different pairs of IP addresses) are alternative paths and are used by the association in case of problems in the primary path. The total traffic of the association can also be distributed over all the paths of that association; this feature is referred to as *load sharing*. The use of load sharing can improve the load distribution over the whole network.

Another important characteristic of the protocol SCTP is *multistreaming*; this means that several streams can simultaneously be maintained between two applications within the frame of a single association. The order of messages that are associated with a single stream is preserved, while the order of messages which are associated with different streams need not be preserved. This property of the protocol is useful in cases when packets of one stream are lost, retransmitted, and consequently delayed, which does not affect the transfer of packets associated with other streams.

The protocol SCTP is more secure than TCP. A four-way handshake is therefore used to establish an association which includes the exchange of cookies. A three-way handshake is used to release an association.

The format of SCTP packets is simpler than the format of TCP segments; an SCTP message header contains less mandatory fields, but there are more optional fields which are TLV encoded (see Sect. 8.3). Consequently, the processing of a message header is quicker, as a receiver processes only those fields which are actually present in the message. Furthermore, an SCTP packet can contain several *chunks* (information elements) which can be associated with different streams, but with the same association. An SCTP packet has a common header, and each chunk can additionally have its own header; due to this characteristic, the total overhead of a packet can be lower.

The algorithms used for the reliable data transfer and the flow/congestion control are similar to those that are used by the protocol TCP.

Chapter 24
Protocols RTP/RTCP

Abstract The protocol RTP is presented here as the protocol that supports the transport of real-time information over packet-oriented networks, the service that is becoming extremely important nowadays. Although the protocol RTP itself does not improve the quality of service offered to the higher layer, it allows the higher layer to do it; the mechanisms that RTP uses for this purpose are explained. The protocol RTCP, which is always running in parallel with RTP and supports the session management, although it is not a signalling protocol, is also described.

In Chapter 22 it was told that UDP is used to transfer the information in real time, because it does not introduce variable delays into information transfer, and also because a strictly reliable information transfer is not needed, at least not in such an extent as was defined in Sect. 12.1. We told also that UDP is a very simple protocol with its only important functionality being the multiplexing of application processes by means of port numbers.

The fact that a reliable transfer of information in real time is not needed, however, does not mean that one should be content with any quality of service. When some kinds of information, such as voice or video, are transferred through an unreliable network, it is advantageous to detect lost and reordered packets at the receiving side. Although UDP does not introduce variable delays into information transfer, the packet-oriented transfer in the lower internet layer does, due to variable queuing times in IP routers; in spite of this, receiving applications must receive packets in real time, hence with constant delays, so delays must be equalised at the receiving side.

Usually these problems are solved by receivers in the application layer; this solving must, however, be supported by a transport layer protocol. Because UDP does not provide for this facility, the protocol *RTP (Real-time transport protocol)* is used in the transport layer in addition to UDP when the information must be transferred in real time. Although RTP is a user of UDP (RTP messages are transported as the payload of UDP datagrams) and runs above UDP in the protocol stack TCP/IP, it is nevertheless considered a transport layer protocol.

Hence, the protocol RTP allows for the detection of lost and reordered messages (it makes both possible by counting messages and including sequence numbers into

D. Hercog, *Communication Protocols*, https://doi.org/10.1007/978-3-030-50405-2_24

RTP headers); furthermore, it adds the information about times messages are transmitted into RTP headers, thus facilitating delay equalisation. Because the real-time information is often transferred between many conference session participants, the identification of transmitters is also transferred by RTP, thus allowing the speakers to be easily recognised by the listeners.

The functionalities that have been described in the previous paragraph are implemented with several fields in the header of an RTP message. The field sequence number allows for the detection of lost and reordered packets. The field timestamp facilitates the delay equalisation at the receiving side. The field source identifier indicates which information source has sent the contained user information.

Along with RTP, the protocol *RTCP* (*RTP control protocol*) is always used. Although the protocol RTCP transfers the information that facilitates the management of a session for the transfer of information in real time, this is not a signalling protocol. When the user information is to be transferred through the IP network, the protocol SIP is usually used as a signalling protocol.

Chapter 25
Transport Layer Security

Abstract This is the only chapter in this book dealing with the security of information transfer. The need for secure information transfer in modern communication networks, IP networks in particular, is emphasised; furthermore, the need for security provision directly between end users, which can only be provided in the transport layer or above, is explained. The protocols SSL and its successor TLS are then shortly presented. The structure of TLS, consisting of several protocols, is described. The procedure that is carried out to set up a secure connection between a web client and a web server is described. A version of TLS that does not require a reliable data transfer in the lower layer is also mentioned. There is a figure in this chapter that illustrates the protocol stack needed for the secure access to a web browser.

Often the information which is transferred through a communication network is sensitive in the sense that it may not be understood or even heard by everybody; furthermore, one wants to know for sure with whom they really communicate. The secure information transfer through communication networks is therefore of utmost importance nowadays.

In Chap. 18 it was mentioned that a secure information transfer through IP networks can be provided with the mechanism IPSec. This mechanism is built into IPv6, but can also be used (and actually is often used) with IPv4. Often, however, it is also desirable to securely communicate directly between two terminal communicating entities and therefore between two applications. This kind of security can only be assured in the transport layer or above it.

Immediately above the transport layer a secure communication can be provided by the protocol *TLS* (*Transport layer security*); the ancestor of this protocol was the protocol *SSL* (*Secure sockets layer*) which was at first developed primarily to provide for a secure communication in the World Wide Web (this protocol was first developed by the developers of the web browser Netscape Navigator). Although there are no big differences between the protocols SSL and TLS, they are not compatible. Nowadays many distributed applications that want to be secure, such

D. Hercog, *Communication Protocols*, https://doi.org/10.1007/978-3-030-50405-2_25

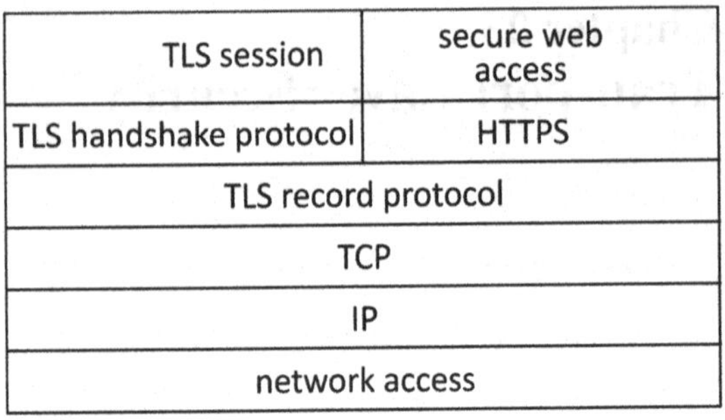

Fig. 25.1 Protocol stack for
provision of security in
transport layer

TLS session	secure web access
TLS handshake protocol	HTTPS
TLS record protocol	
TCP	
IP	
network access	

as World Wide Web or electronic mail, use the protocol TLS (or the last version of SSL) between the transport and the application layers.

The protocol TLS establishes a session between both applications; at the setup time they can mutually authenticate each other (they verify their identities) and negotiate the parameters of a secure connection. During the user information transfer phase they assure the confidentiality of the transferred information and verify the authenticity of transferred messages.

As was already mentioned, the protocol TLS runs between the transport and the application layers of the protocol stack TCP/IP. TLS assumes a reliable transfer of octets in the transport layer, so it is mostly used above the protocol TCP. In fact, TLS is not a single protocol; there are actually two protocols which run one above the other. In the lower layer acts the *TLS record protocol* which provides for the encryption and the authentication, possibly also for the compression of transferred messages; in the higher layer the *TLS handshake protocol* runs which provides for the authentication of both protocol entities and the negotiation on the parameters of a secure connection (such as the encryption algorithms and keys).[1] The protocol stack that provides for a secure access to the World Wide Web is shown in Fig. 25.1. Here one can see that a secure connection provided by the TLS record protocol has two users: one of them is the TLS handshake protocol (which needs a secure data transfer itself), and the other one is an application layer protocol, the protocol providing a secure web access (HTTPS in our case).

The protocol HTTPS differs from its non-secure version HTTP only in the fact that HTTPS does not use the services of TCP directly, but rather through a secure connection TLS. HTTP and HTTPS use different well-known port numbers; hence, an application (such as a web browser) can tell whether it wants to communicate via a secure or a non-secure connection simply by choosing the appropriate well-known port number at the server side.

When applications wish to communicate through a secure connection both TLS entities first authenticate each other (although this authentication is not mandatory, in practice the client always authenticates the server, and sometimes the server also

---

[1] Actually, two more protocols are used in the higher layer besides the TLS hanshake protocol; for the sake of simplicity, however, they are not described here.

authenticates the client); sometimes the human user must also participate in the authentication process. After the authentication has been successfully done, the client and the server must agree on the parameters of the secure connection, such as the algorithms for encryption and authentication, and determine the necessary keys. Only then can they establish the secure connection to be used for the secure transfer of user data between applications. Within the frame of a TLS session several secure connections with the same security parameters can be set up, in order to avoid the time-consuming phase of the authentication and the parameter negotiations.[2]

A variant of TLS called *DTLS (Datagram transport layer security)* also exists; the protocol DTLS uses the services of the transport layer protocol UDP. DTLS was developed primarily to be used with the application layer protocol SIP (see Sect. 26.5). Of course, DTLS can also be used with other application layer protocols which use the transport layer services of unreliable information transfer.

---

[2]For this very reason it is advisable from the security point of view that, after one has finished the work over a secure connection, the web browser is closed and consequently the TLS session is terminated.

Chapter 26
Some Application Layer Protocols in IP Networks

Abstract In this chapter some protocols are presented to illustrate different services that the application layer protocols can provide for distributed applications. Most of them are classical, though still massively used protocols for data transfer in the Internet, while one is a newer (although not very new) signalling protocol, used in applications that are becoming more and more popular nowadays. The character-oriented nature of these protocols is emphasised. The MIME standard is mentioned as the way of additionally encoding nontextual information to be transferred with a character-oriented protocol. The virtual terminal and the protocols telnet and SSH that support it are then described. The remote file system management and the FTP protocol are discussed next. The electronic mail system and its associated protocols SMTP and POP3 are also discussed, and then still the World Wide Web and the protocol HTTP. Finally, the IP telephony system and its signalling protocol SIP are treated, along with its associated protocol SDP. The two figures of this chapter describe the protocol stack used in IP telephony and an example IP telephone session.

In the application layer of the protocol stack TCP/IP many protocols are used which provide for numerous functionalities adapted to the needs of various distributed applications. These protocols are therefore application oriented as they provide exactly those services which are required by particular applications. The set of messages of an application layer protocol is therefore similar to a catalogue of services offered by the protocol to corresponding applications.

The protocol stack TCP/IP does not include the session and presentation layers (according to the OSI reference model), so application layer protocols in IP networks must also provide for the functionalities of these two layers. In spite of this, these protocols are relatively simple, at least if compared with protocols of the data-link and transport layers which must also provide for a reliable data transport. The application layer protocols which need a reliable data transfer in lower layers employ the services of the TCP or SCTP protocol. Those protocols which do not require a reliable data transfer or transmit data periodically, use the simpler protocol UDP. If,

D. Hercog, *Communication Protocols*, https://doi.org/10.1007/978-3-030-50405-2_26

however, an application layer protocol uses UDP because of its simplicity, but nevertheless needs a reliable message transfer, it must provide for it by itself.

In the application layer of the IP network there can be a great number of communication processes that are running concurrently; even in a single computer, which is identified with a single IP address in the IP network, there can be several applications, possibly of the same kind, which are running concurrently. Each of these communication processes is unambiguously identified within the whole network with a pair of *sockets*: a pair IP address–port number on one side of the session and a pair IP address–port number on the other side of the session. Because IP addresses are transferred within the headers of IP datagrams and port numbers are transported within the headers of transport layer messages, the multiplexing of applications can be said to be executed in both the internet and the transport layer of the protocol stack TCP/IP.

The application layer protocols of the protocol stack TCP/IP are of the type *client–server*. Usually clients initiate a communication process by sending communication requests to servers which in turn reply to these requests. Servers therefore use the standardised *well-known port numbers*, while clients employ the temporary *ephemeral port numbers*. The consequence of such unbalanced communication in the application layer is that the protocol entities in servers operate differently from the protocol entities in clients.

Numerous application layer protocols in the Internet are character oriented and employ the matched tag coding method to encode protocol data units (see Sect. 8.4). However, application layer protocol messages must often transport also binary contents (such as images), so such contents must additionally be encoded to look similar to character sequences; the MIME encoding (see Sect. 8.5) is most usually used for this purpose.

In this chapter only a few characteristic application layer protocols of the protocol stack TCP/IP will be described with the aim that a reader get some feeling about the nature of these protocols. There is of course a very extensive literature available for a more thorough study of these and many other protocols.

26.1 Telnet/SSH

The protocol T*elnet* allows a human user to work on a remote computer, using a virtual terminal which is not directly connected to the target computer; the *virtual terminal* is an application which communicates with the target computer through the IP network, forwarding users' commands to the target computer and responses to them from the target computer to the remote computer. Telnet does not support the transfer of graphics but only the transfer of characters, so it only supports the implementation of a character-oriented (alphanumeric) virtual terminal.

The protocol Telnet is one of the oldest protocols among those that are used in the Internet still today. Its name was coined from the words telecommunications network protocol, as it was the only tool that could be used to access remote computers at the time it was developed. Nowadays this protocol is of no use for the great majority of users because it does not allow a *graphical user interface (GUI)* to be implemented, while most users require just that. For that purpose other protocols, such as the *X Windows* protocol, can be used today. An *alphanumeric user interface* is used today almost exclusively by the administrators who want to manage those servers which do not have a graphical interface. There is still another, in fact even more severe, limitation of the protocol Telnet; this is the security problem, as Telnet does not provide for the encryption of transferred data, including the user name / password pair with which a user must usually be authenticated by the server. This presents a dangerous threat for the remote computer and its users. Telnet has therefore almost entirely been replaced by the protocol *SSH* (*secure shell*) which provides the same functionality as Telnet does, but additionally also the safe authentication of a user and the remote computer, as well as the encryption of transferred data.

The basic task of the protocol Telnet is the transfer of characters or character sequences (lines) between the user application of the virtual terminal and the corresponding application on the server. The user application sends the characters that are typed by a user on their keyboard to the server, and writes or draws the characters that are sent by the server on the user's screen; on the other hand, the server application (the program Telnet on the server) forwards the characters it has received from a user to the operating system on the server and sends to the terminal those which it has received from the operating system. Telnet therefore allows a user to directly access the operating system of a remote computer by means of alphanumeric commands. Furthermore, Telnet allows a user to set some parameters of the transfer.

Telnet messages are transferred between a client and a server through a connection that is implemented by the TCP protocol. As soon as a TCP connection is established, a Telnet session is also set up. A Telnet session is released as soon as its underlying TCP connection is torn down on the request of one of the protocol entities. During the course of a session, user characters are transferred between the client and the server. At the beginning of a session or during a session some control data can be exchanged with which the client and the server can negotiate session parameters. Control data are always preceded by a special character (the null character or the sequence of eight zeros) which prevents the receiver to interpret control data as the user information. The parameter negotiation is the functionality of the session layer according to the OSI reference model.

Various servers, as well as various clients, may employ different rules for character encoding. The protocol Telnet provides for the correct interpretation of transferred characters by specifying the standardised code for the transfer of characters; it is up to the transmitting and the receiving protocol entity to translate its local code into the transfer code and vice versa, respectively. This is the functionality of the presentation layer according to the OSI reference model.

A multitude of client applications exist that implement virtual terminals and use the protocols Telnet and SSH. Most server operating systems also contain server applications based on the protocols Telnet and SSH. For security reasons, however, the Telnet protocol entities are disabled in most server systems.

26.2   FTP

The protocol *FTP* (*File transfer protocol*) provides for the service of the management of *files* and *file systems* on remote servers, as well as the transfer of files between servers and clients (and vice versa). At the beginning of a session the server authenticates the client with its user name and password; the extent of the access of the client to the remote file system and remote files of course depends on who the user is. Some servers also allow the anonymous access to the file system to all clients, hence without authentication.

FTP, too, is one of the oldest protocols used in the Internet; it was defined and used already before the protocols IP and TCP emerged as two separate protocols. Just like Telnet, FTP also has security problems, as it does not encrypt transferred data, including the user authentication at the beginning of a session. Other versions of the protocol that are more secure are therefore usually used nowadays.

The protocol FTP requires a reliable data transfer in the underlying transport layer, so it relies on the services of the protocol TCP. As FTP provides for two different kinds of service, it employs two different channels, hence two different TCP connections with two different well-known port numbers at the server side. The control channel is used for the session management and for the control of the services that are to be provided for, while the data channel is used for the transfer of larger quantities of data, primarily for the transfer of files. Usually a separate data channel is established for the transfer of a file and is released after the transfer is concluded.

After the underlying TCP control connection has been established, the FTP session begins with the authentication of the user. Then the client can send to the server requests for various services to be executed (such as the change of the default directory on the server, the renaming of files and directories on the server, or the file transfer in either direction); any such request is contained in an FTP protocol message sent from the client to the server. The server executes the services, if possible, and informs the client about the outcome.

The establishment of a data channel (a TCP connection to be used for a file transfer) is normally initiated by the server; this means that the client must first inform the server about its ephemeral port number to be used for the connection. As an exception, the client may also initiate the setup of a data channel; this is especially advantageous if the client is positioned behind a network address translator (NAT). The transfer of a file can be carried out in one of two possible modes. The *ASCII mode* is appropriate for the transfer of text files; for this case the protocol defines the character coding method to be used for the data transfer, and both protocol entities translate the code used in the local computer to the transfer code or vice versa, if necessary, of course. We might mention an interesting difference between the coding methods used by different operating systems: this is how the end of a text line is indicated in text files (this is different even in such common operating systems as Windows and Unix). The *binary mode* of file transfer is appropriate for the transfer of binary files, such as images or executable programs; these files are plain sequences of octets, without any generally defined interpretation, and must therefore be transferred without any transformation, octet after octet. Providing the ASCII mode of file transfer is the functionality of the presentation layer, according to the OSI reference model.

As was already mentioned, an FTP session is controlled by the requests that are sent by the client, while the server sends its replies to the requests; both requests and replies are sequences of characters. The replies of the server inform the client about the outcome of service executions. They are three-digit numeric standardised codes, which, however, may be followed by non-standard textual explanations (these explanations can therefore be composed in various languages and are sometimes a little bit comic). Only numeric codes are essential for the protocol, while the textual explanations can be forwarded by the application at the client side to the human user.

Files may be transferred in three different ways. The *stream mode* is most usually used. In this mode a file is transferred as a sequence of octets; of course, TCP may merge subsequences of these octets in its segments, but this is not the business of FTP, so FTP is not aware of this. If the stream mode is used, the sender releases the data connection after it has finished the file transfer, thus telling the receiver that the file transfer is over. The *block mode* and the *compressed mode* are more complex; in these modes a file is partitioned into blocks, which are then transferred across the channel, each one with its own header; in these modes the sender must explicitly inform the receiver when the file transfer is terminated. In the compressed mode a file is compressed before being transferred.

Many applications for file transfer and file system management that are based on the protocol FTP have already been developed and used. Some of them have an alphanumeric, and the others have a graphical user interface. Nowadays almost all operating systems include the protocol FTP and at least one application that uses this protocol.

Many alternative protocols have also been developed and used. The oldest among them is the protocol *TFTP* (*Trivial file transfer protocol*) which is much simpler than FTP by itself, and besides this it also uses the services of the protocol UDP (which is also very simple, as we already know) instead of TCP in the transport layer; TFTP is therefore appropriate to be used in computers which have very small amounts of memory at their disposal or are even diskless. Because UDP in the lower layer does not provide for a reliable data transfer, TFTP must do that itself; it does this in the simplest possible way—as a stop-and-wait protocol. Many other alternatives to FTP are also in use with the very important advantage that they provide for a secure communication, including the user authentication and encrypted file transfer; examples of these protocols are *SFTP* and FTPS which use SSH and TLS to assure security.

26.3   Electronic Mail Transfer Protocols

One of the oldest and still today one of the most important (if not even the most important) distributed applications that are used in the Internet is the *electronic mail*. Nowadays the electronic mail has modified our lives (unfortunately, not necessarily in positive direction—because of electronic mail we often neglect direct and friendly relations with other people, not to mention that the quality of our language suffers due to the frequent use of non-standard words and slang!). While the electronic mail system at the beginning allowed only texts in the English language to be transferred, nowadays one can send and receive texts in various languages, as well as non-text documents, such as pictures or executable programs.

In the electronic mail system, electronic messages and documents are not transferred directly from senders to their final destinations, but indirectly through *mail servers*. Users employ applications that are called *user agents*, while mail servers represent *transfer agents*. A user agent executes various tasks. It allows electronic messages to be composed, edited and read; it makes possible sending electronic messages, intended for other users, into the electronic mail system; it allows users to manage their *mailboxes* (a mailbox is the storage for electronic messages intended for a specific user); it also supports the transfer of electronic mail from the user's mailbox on the server to the local mailbox in the local computer. The first task is executed locally and therefore does not need any protocol for its operation. To transfer electronic mail from a user agent to a mail server or between two mail servers, the protocol *SMTP* (*Simple mail transfer protocol*) has been used from the beginning; SMTP is also one of the oldest protocols that are used in the Internet. In order to manage the mailbox on a server and to transfer electronic mail from a server to a local mailbox, either an older and simpler protocol *POP3* (*Post office protocol version 3*) or a less simple but more powerful protocol *IMAP* (*Internet message access protocol*) can be used. All these protocols rely on the services of the protocol TCP in the transport layer. In this section a little bit more will be told about the protocols SMTP and POP3.

Many different user agents have been developed and are in use, which employ the protocols SMTP, POP3 and IMAP for the communication with mail servers. In recent years user agents which in fact are web applications, therefore running on mail servers and accessed by users through the Internet, using web browsers, have become more and more popular; special, possibly non-standard protocols are often used for this. This kind of user agents is usually referred to as *webmail*.

The RFC standards also specify the format of electronic messages, which does, however, not depend on the protocols for mail transfer. At the beginning, these protocols were intended to be capable of transferring text messages, hence sequences of characters. Electronic messages which contain binary (non-text) contents must therefore be additionally encoded according to the standard *MIME* (*Multipurpose internet mail extensions*).

The protocols SMTP and POP3 both operate as protocols of the type client–server. Both are character oriented which means that protocol data units are sequences of characters; even if a message contains binary contents, it formally looks like a text message after it has been additionally encoded using MIME. In case of both protocols, a client sends requests to the server, and the latter executes the requested tasks if possible; in any case, the server replies with a message which informs the client about the outcome of the operation. As was already told for the case of the protocol FTP, the replies of a server are standardised numeric codes, which can optionally be followed by non-standard textual explanations. Most protocol messages are one text line long (excluding user electronic messages defined by users, of course).

The transfer of a user message (an electronic letter) poses a special problem. This message is in most cases several text lines long. SMTP transmitter always terminates the user's message with a line containing a single period, thus indicating the end of the user message. The protocol SMTP provides for the transparent transfer of a message by adding to any period at the beginning of a line within a user's message an additional period; whenever the receiving entity detects two consecutive periods at the beginning of a line of a user message, it removes one. Thus, a period at the beginning of a line within a user message cannot be misinterpreted as an end-of-user-message marker.

The task of the protocol SMTP is to send one or more electronic messages to a mail server. At the beginning of an SMTP session the SMTP entity must therefore specify its own mail address and one or more destination mail addresses. After it has sent all messages, it must terminate the SMTP session, thus terminating also the TCP connection.

At the beginning of a POP3 session the server first authenticates the client and its user, because it depends on the user identity which mailbox on the server may be accessed. The client can then query the number of messages in the mailbox and their sizes, it can retrieve messages from the server, or even delete messages on the server.

26.4 World Wide Web and Protocol HTTP

Nowadays, the *World Wide Web* (*WWW*) is along with the electronic mail the most widespread distributed application in the Internet. The World Wide Web can be seen as a network of information resources which reside on *web servers*, and the users of this information represented by *web clients*; the most usual web clients, albeit not the only ones, are *web browsers*. Information sources must of course have their addresses which are referred to as *universal resource locators* (*URL*). The operation of the World Wide Web is based on the application layer protocol *HTTP* (*Hypertext transfer protocol*); the principal task of HTTP is the transfer of web documents from web servers to web clients, and possibly also in the opposite direction. The protocol HTTP is very general, so it is possible to transfer with it various kinds of information between servers and clients. One must, however, be aware that HTTP is a protocol for the transfer of files, while the interpretation of these files is completely up to client applications.

The World Wide Web and the protocol HTTP were »invented« towards the end of the eighties of the twentieth century in the nuclear physics research organisation CERN as a tool for the interchange of documents between researchers of this organisation. At the beginning the web was a set of interrelated *hypertext* documents; from then on, the web, that has been becoming more and more worldwide, and the protocol HTTP have been developed constantly, and so has also been developed the language for the description of hypertext documents *HTML* (*Hypertext markup language*). While the protocol first allowed only hypertext documents to be transferred from servers to clients, nowadays it allows various kinds of documents to be transferred in both directions; it also allows some web documents residing on servers to be supplemented with the information which is added by users into so-called web forms.

HTTP is a protocol of the client–server type. The messages which are sent by a client to a server are called *requests*, while the messages a server sends to a client are referred to as *responses*. Because HTTP is an unbalanced protocol, requests and responses are completely different messages. A request specifies the service a client expects from the server; such a service is referred to as *method* in the HTTP terminology. The most important methods are GET, where a client requests the transfer of a file from the server to the client, PUT, requesting the transfer of a document from the client to the server, POST, which requires the transfer of data to supplement a document on the server that has previously been acquired with the method GET (this allows a user to fill web forms), and DELETE which requests a document on the server to be deleted. Besides specifying the method to be executed, requests still contain some other parameters (such as the address of the requested document with the method GET), or even user data (such as the web document to be uploaded with the method PUT). A response contains a three-digit standard code which specifies the (un)success of the method execution and may optionally be followed by a non-standard text explanation, and that followed by various parameters which of course depend on the method that is being executed. Reponses are partitioned into various response groups. The first digit of a three-digit code determines the group into which fits a response (such as success and error), the second digit determines the subgroup, and the third digit the subsubgroup. This allows a client which does not understand the code completely to know at least approximately what is going on, with the respect to the code group. A response also can contain a user message (a response to request GET of course contains the requested document itself!).

Basically all HTTP messages are encoded as character sequences. The header of a message can consist of several text lines which are separated with the sequence of ASCII characters *CR* (*carriage return*) and *LF* (*line feed*); different lines of the header specify different parameters. The user message is separated from the header with an empty line. A user message, if present, does not necessarily contain text; non-text messages are encoded according to the standard MIME.

HTTP is a protocol of the application layer which uses a TCP connection as the channel interconnecting HTTP entities, because it requires a reliable data transfer in the lower layer. A basic transaction that is carried out by HTTP is the exchange of a request and its associated response. In the early versions of the protocol, the client and the server established a separate TCP connection for each transaction and released it after the transaction had been finished. Users, however, often want to retrieve several documents one after the other from the same server (e.g. the basic hypertext document first and then still its adjoined pictures); the original mode of operation unfortunately wasted too many time and transfer resources, as the setup and the release of a connection require the transfer of three and four TCP segments, respectively, as we already know from Sections 21.4.1 and 21.4.2. Nowadays the protocol HTTP therefore allows a TCP connection to have a longer duration and to transfer several transactions between the same pair client–server; such a connection is released only after it has been inactive for some time. The data transfer over such connections is also faster in average, as we already told in Sect. 21.4.4 that any TCP connection is slow immediately after its establishment because of the slow start mechanism.

HTTP is a connectionless protocol, as both protocol entities are unaware of their states and of the history of their communication prior to the current transaction; such protocols are often referred to as *stateless protocols*. With pure stateless protocols one transaction does not depend on the previous transactions between the same protocol entities. Sometimes it is nevertheless desirable that one transaction depends on the preceding transactions; applications can solve this problem with the use of *cookies*.

Because of the very widespread use of the World Wide Web, and especially because of its use in such activities as commerce and banking, the security of transactions carried out in the World Wide Web is becoming more and more important. Several variants of the protocol HTTP which assure the secure transfer of data have therefore been developed. The best known and the most used is the variant *HTTPS (Hypertext transfer protocol secure)* which employs the protocol TLS or SSL between the TCP and HTTP protocols.

26.5   Internet Telephony and Protocol SIP

One of the oldest communication services[1] which has expanded all over the world and has found its users both between individuals and companies, is the transfer of voice, referred to as *telephony*. The technology of voice transfer, voice connection control and telephone network design has been developed from the second half of the nineteenth century on, and did not see any essential modifications during a century or so. In the second half of the twentieth century, however, radical changes appeared and were introduced into the telephony. The digital transfer of voice and the digital control of voice connections were first introduced which both improved

---

[1]We say »one of the oldest« because the telegraphy was in fact the first communication service; telephony, however, has been and still is the most widespread communication service.

the quality and lowered the price. The rapid development of computer technologies stimulated the processing, storage and transfer of various kinds of information, such as voice, images, video and data, and also the combining of various kinds of information; the combination of two or more kinds of information is called *multimedia information*, or just *multimedia* for short. *Integrated communication networks* have begun to be developed to allow different kinds of interrelated information to be transferred concurrently. Due to the rapid rising of the importance of the protocol IP, the transfer of various kinds of information, such as voice and multimedia, over the protocol IP, began to prevail in the telecommunication world; nowadays this mode of information transfer tends to supersede other information transfer technologies. The terms *internet telephony* and *VoIP* (*voice over IP*) emerged to denote the transfer of voice and multimedia information over the protocol IP. Towards the end of the twentieth century and at the beginning of the twenty-first century an additional trend has emerged in the area of telecommunications, namely the *mobility*, which allows users and their terminals to change their physical locations and their physical addresses with them.

A VoIP network provides the same services as a classic telephone network, and much more. A connection that allows two or more user terminals to communicate and thus exchange user information between them is referred to as *session*. Within the frame of a single session one or more streams bearing the information of various kinds (such as voice and video) can be transferred. Two or more users may participate in a session. During the course of a session new users may join it and existing users may leave it. A VoIP network also supports the mobility; this means that users can change their physical locations and can establish new sessions or join existing sessions regardless of their current locations, if only they can access the network. A VoIP network also provides for the service of *instant messaging*. Presently the VoIP technology enjoys such a success that it is considered to be the technology of the future generations of communication networks, including mobile networks.

Just like in classic telephone networks, in VoIP networks two different protocol stacks are also used to transfer the session-related user information and the *signalling* that controls the session. The user information is transferred in the user plane, while the signalling is transferred in the control plane (see Sect. 5.3). Several protocols used in these protocol stacks are different. In Fig. 26.1 the protocol stacks of the user plane and the control plane in a VoIP network are shown as well as both the user part and the control part of an application. The protocol RTP and its associated protocol RTCP, already described in Chap. 24, are used for the transfer of user information in real time, while the signalling protocol *SIP* (*Session initiation protocol*) is used to control a session. SIP is the main topic of this section.

While UDP is almost always used to support the transfer of user information in real time, the protocol SIP is conceived to be able to use the transport-layer services of UDP, TCP or SCTP for the transfer of signalling; because, however, SIP is capable of providing for a reliable data transfer by itself, employing acknowledgments and timers, it most usually uses the services of UDP due to its simplicity. Such a configuration is shown in Fig. 26.1.

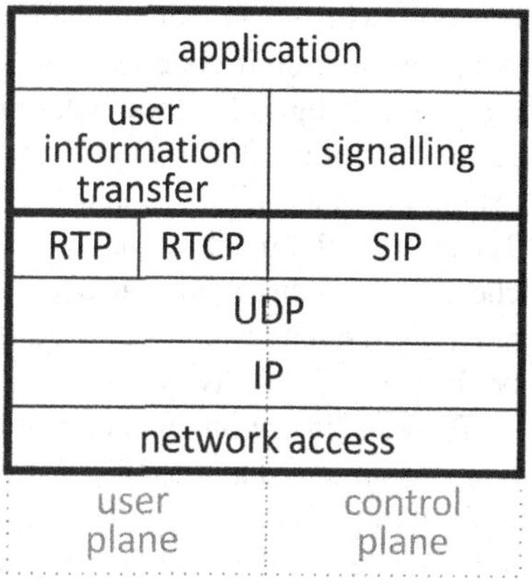

Fig. 26.1 Protocol stack in
VoIP networks

application		
user information transfer		signalling
RTP	RTCP	SIP
UDP		
IP		
network access		
user plane		control plane

In a VoIP network terminals also can have at least two addresses: the IP address is a physical address on which the routing of packets through the IP network is based, while the domain name is a logical address. In general, the domain name is referred to as *universal resource identifier* or *URI* in VoIP networks.[2] While the universal resource identifier URI of a user does not change in time, the IP address can change due to the change of the user's physical location. The address translation is thence very important in VoIP networks, as it supports the mobility. Here also domain name servers (DNS) are used for this purpose, as was already told in Chap. 20. Furthermore, so-called *location servers* are used in VoIP networks to support mobility; location servers maintain databases in which the relations between URI identifiers and their corresponding IP addresses are recorded; of course, these records dynamically change with the mobility of users. Location servers must therefore be able to communicate with some other servers of a VoIP network; however, the SIP protocol is not used for this purpose.

Many applications have already been developed to offer the service of IP telephony, that are based either on the protocol stack shown in Fig. 26.1 or on some other, often non-standard, protocols. Nowadays all »smart«[3] phones, as well as tablet and personal computers, provide for such applications.

Like most application layer protocols in IP networks, SIP is also a character-oriented protocol with its set of protocol data units being quite similar to the set of messages of the HTTP protocol, which actually served as a model to the designers of SIP. SIP, too, is an unbalanced protocol of the client–server type. This means that a client sends its requests to a server which in turn responds to them. Like in the protocols SMTP and HTTP, a SIP request also specifies a method which determines the operation to be executed, and contains various parameters and also the body of a

---

[2] In Sect. 26.4 we already mentioned the resource address URL which is a special case of URI; another special case of URI is the electronic mail address.

[3] The author dislikes the term »smart« phone, although it is already well established in both technical and daily used terminology, that is why this term is written in quotation marks here.

message which can contain some additional information. Responses specify a standard three-digit code and optionally a non-standard textual explanation. Here, too, the first digit of a code determines the group of responses, while the other two digits determine the response more specifically. Responses $1xx$ are temporary responses which let the client know that its request is being processed; responses $2xx$ tell that the method has been executed successfully; responses $3xx$ inform the client that a redirection is needed; $4xx$ are responses to invalid requests; responses $5xx$ inform about a server error; and $6xx$ are responses to the requests which cannot be fulfilled in the network.

The most important elements of a VoIP network are of course terminals which are often also referred to as *IP phones*, or simply phones, as most users are not interested in the protocol IP, or even they have never heard of it. More specifically, a phone is implemented as the software in a terminal device, usually called *user agent*. Often terminal devices include much more functionalities beyond the functionality of a phone; these can be »smart« phones, tablets, portable or desktop computers, and also various telemetric and other devices. The task of the SIP protocol is to establish, maintain, modify and release sessions between two or more terminals. Contrary to most applications in IP networks which operate strictly according to the client–server principle, the application providing a voice or multimedia communication is a symmetric one; all terminals are equal, any one can initiate a session establishment or accept the invitation to a session. Any user agent must therefore contain two SIP protocol entities—a client and a server.

In principle, two user agents can set up a communication session by themselves, without the aid of any other network elements, if only they are able to communicate directly[4] and if they know the IP addresses of each other. Besides user agents, several types of servers are also used in VoIP networks, which are collectively referred to as *SIP servers*, although they may also provide for other services. Often terminals are not connected to the public IP network directly, but through a special gateway that is called a *proxy server*. A proxy server relays SIP messages between a user agent and the external world (the rest of the VoIP network); in this sense it acts as a router for SIP messages. In this endeavour it can use the services of name servers on behalf of a user agent and forwards a requested IP address to the user agent (in Chap. 20 this was referred to as recursive query). Towards the external world a proxy server therefore acts on behalf of user agents. Because a proxy server relays SIP messages in both directions, it also must contain both a client and a server protocol entity. A proxy server that is positioned on the boundary between a private and a public network or between the internal network of an operator and the external network can also verify if a user agent may or may not establish and/or accept sessions. It was already told that the IP address of a user agent can change with time due to the user's mobility. Therefore, a user agent must register in the network before

---

[4]There can be administrative obstacles preventing a direct communication between terminals, such as the barriers between the networks of different operators which require their clients to pay for communication services.

it wants to communicate or whenever its IP address changes. It does so by sending a registration message to the server called *registrar*; upon receiving a registration message, the registrar writes the translation URI–IP address into the above-mentioned database in the location server. The proxy server can query the IP address of a certain user either from the location server or from the *redirect server*. The redirect server returns the address where the translation URI–IP address can be acquired from; this is an example of the iterative query (see Chap. 20). The registrar and the redirect server are true servers, as they run only the server-side protocol entities. Although a registrar communicates with a location server, too, it does not use the SIP protocol for this purpose.

The RFC standards specify several methods (requests) for the SIP protocol; here, however, only a few most important ones will be shortly described. The request REGISTER is used by a user agent (on behalf of a user) to register with the network; this request is sent to a registrar which in turn provides for the appropriate translation between the user's URI and IP address to be recorded in the database of the location server; consequently, the user can be called by other users even if they know only its URI, but not its IP address. The establishment of a new session is initiated with the request INVITE which can be sent from a calling user agent directly to the agent of the called user (if the caller knows the IP address of the called agent), or indirectly through one or more proxy servers. A user agent which is already involved in an active session can use the request INVITE to invite still another user to join the conference session. After a client has sent an INVITE request, it waits for the response of the server of the called agent informing it that the call is accepted. Before the user information begins to be transferred in both directions, both participants must be aware that the session is established, so the initiator informs the recipient that it has received the response with the message ACK, as the response could also have been lost or corrupted in the network. One must be aware that this precaution would not be necessary if the reliable protocol TCP were always used in the transport layer. A session release can be initiated by any participant by sending the request BYE to the other agent. If, however, there are more than two participants in a conference session, a user agent can send the request BYE to other participants to tell them that it is leaving the session. The requests MESSAGE transfer instant messages.

The user information that is exchanged between users within the frame of a session can be the information of different kinds (voice, video...), and the information of a certain kind can even be encoded according to different standards. It is therefore essential that all the users verify before the beginning of the user information transfer that they will all be able to correctly interpret the transferred information. A request INVITE therefore contains in its user information part the description of the information that is going to be transferred during the session. This description is encoded in the format *SDP* (*Session description protocol*). Hence the protocol SDP is nothing else but a language that is used for the description of the information which is to be transferred. An SDP message is always transferred as a part of an SIP message.

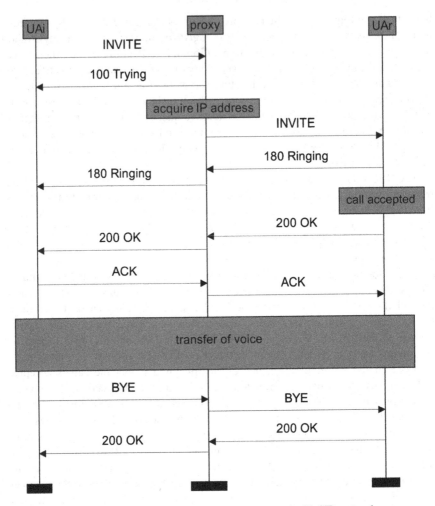

Fig. 26.2 Example of session between two user agents managed with SIP protocol

In Fig. 26.2 a simple example of a session for the transfer of voice (an IP telephone connection) between two terminals with the user agents UAi and UAr is shown; besides the two user agents one proxy server (labelled as proxy in the figure) is used; the location server that is also queried is not shown in this figure. The user agent UAi of the initiating user sends the request INVITE, containing the URI identifier of the called user and the session description besides other parameters, to the proxy server. The proxy queries the location server for the IP address corresponding to the URI of the called user and forwards the request INVITE to the agent of the called user. Because the procedure of a connection setup can take some time, the proxy informs the initiator that the session establishment is underway with the response 100 Trying. As soon as the called agent receives the request INVITE, its telephone begins to ring and the initiator UAi is made aware of this with the response 180 Ringing which is again relayed by the proxy. When the called user accepts the call the user agent UAr informs about this the initiating agent UAi with the response 200 OK (which is again forwarded by the proxy). Finally, the calling agent acknowledges the reception of the response; upon the exchange of the messages OK–ACK the session is established and the transfer of voice can begin. It is important to mention that the proxy server only helps to manage the session; the voice is transferred directly between the two terminals (only through intermediate IP routers, but not through the proxy), as both terminals now know the IP addresses of each other. The session release is initiated by one of the user agents which sends the request BYE, while the other one answers with the response 200 OK. (As was already told, standard response codes are used, 100, 180 and 200 in our case, while textual explanations are not standardised and can read differently.)

The secure session management can also be provided for using the version *SIPS* (*Secure Session Initiation Protocol over TLS*) instead of SIP; in this case all SIP messages are encrypted.

SDL Keywords and Symbols

In Chap. 3 the SDL language was presented; in the continuation of the book this language was used several times for the specification of some communication methods and some communication protocols. The aim of this Appendix, which contains a concise overview of the most important constructs of the SDL language, is to allow a reader an easier reading and understanding of these specifications.

Let us first explain the font styles that are going to be used in the column »Keyword/symbol« to present the language elements which are given in the textual form. Those words that are always written in exactly the same way as are given here (they are called keywords) will be shown here using the regular font. The language elements which are shown here in the *italic* font are substituted in actual specifications by appropriate specific language elements; e.g. if the specification of a *value* is required at some place in a specification, a specific value of the appropriate type shall be used at that place in the concrete specification.

To describe the syntax of optional or optionally repeating language elements the following notation will be used: the language element or the sequence of language elements which is written within square brackets ([...]) is optional (it may or may not be present); the language element or the sequence of language elements which is written within braces ({...}) may have zero or more repetitions (may be present zero or more times in succession).

Keyword/symbol	Parameters	Description	Section
		Text frame: textual definitions, declarations	3.2
SYNONYM *n t = v*;	*n*: Name of constant *t*: Data type *v*: Value	Definition of constant	3.2

(continued)

© The Editor(s) (if applicable) and The Author(s), under exclusive license to Springer Nature Switzerland AG 2020

D. Hercog, *Communication Protocols*, https://doi.org/10.1007/978-3-030-50405-2

Keyword/symbol	Parameters	Description	Section
SYNTYPE $st = t$ CONSTANTS $a : b$ ENDSYNTYPE;	st: Name of subtype t: Data type a: First value b: Last value	Definition of subtype; it contains values of type t from a to b, both inclusive	3.2.2
NEWTYPE t array (ti, tv) ENDNEWTYPE;	t: Name of type ti: Type of indexes tv: Type of values	Definition of array type; both type of data values and type of indexes are specified	3.2.2
DCL $var\ t\ [: = v]$;	var: Name of variable t: Data type v: Initial value	Declaration of variable with optional value initialisation	3.2.2
TIMER t;	t: Name of timer	Declaration of timer	3.4
TIMER $t\ (dt)$;	t: Name of array of timers dt: Type of indexes	Declaration of array of timers	12.11
SIGNAL $s\ [\ (\ t\ \{,\ t\})\ ]$;	s: Name of signal t: Data type	Declaration of signal which brings zero or more values	3.2.3
SIGNALLIST $sl = s\ \{,\ s\}$;	sl: Name of signallist s: Signal	Declaration of signallist	3.2.3
BLOCK b $n(N)$ struct OR funct	b: Name of block n: Page number N: Number of pages $struct$: Structure $funct$: Functionality	Definition of block instance	3.3
b	b: Name of block	Instance of block	3.3
p	p: Name of process	Instance of process	3.3
k [ss]	k: Name of channel ss: List of signals	Unidirectional channel	3.3
k [ss] [ss]	k: Name of channel ss: List of signals	Bidirectional channel	3.3
PROCESS p $n(N)$ trans	p: Name of process n: Page number N: Number of pages $trans$: Finite state machine	Definition of process instance	3.4
◯		Start symbol (initialisation of process)	3.4
st	st: Name of state	State symbol	3.4
$>s\ [(v\{,v\})]$	s: Name of signal v: Name of variable	Input symbol (reception of signal)	3.4
$>\ t$	t: Name of timer	Input symbol (reception of timer expiration)	3.4
e	e: Name of input event	Save symbol	3.4
✕		Stop symbol	3.4

(continued)

Keyword/symbol	Parameters	Description	Section
s[(x{,x})] [TO p]	s: Name of signal x: Expression p: Identifier of process	Output symbol (transmission of signal)	3.4
v:=x	v: Name of variable x: Expression	Assignment symbol	3.4
a:=(. v .)	a: Name of array v: Values of array elements	Initialisation of array	3.2.2
set(now+d,t)	d: Timer expiration time t: Name of timer	Timer activation symbol	3.4
reset(t)	t: Name of timer	Timer deactivation symbol	3.4
x / c ..... c	x: Expression c: Conditions	Decision symbol	3.4

References

1. Stallings, W.: Data and Computer Communications, 8th edn. Pearson – Prentice Hall (2007)
2. Tanenbaum, A.S.: Computer Networks, 5th edn. Prentice Hall (2011)
3. Halsall, F.: Data Communications, Computer Networks and Open Systems, 4th edn. Addison-Wesley (1996)
4. Spragins, J.D., Hammond, J.L., Pawlikowski, K.: Telecommunications: Protocols and Design. Addison-Wesley (1991)
5. Pujolle, G.: Les Réseaux, Edition 2008. Eyrolles (2007)
6. Sharp, R.: Principles of Protocol Design. Springer (2008)
7. Holzmann, G.J.: Design and Validation of Computer Protocols. Prentice Hall (1991)
8. Gauda, M.G.: Elements of Network Protocol Design. John Wiley & Sons (1998)
9. König, H.: Protocol Engineering. Springer (2012)
10. Popovic, M.: Communication Protocol Engineering, 2nd edn. CRC Press (2018)
11. Lai, R., Jirachiefpattana, A.: Communication Protocol Specification and Verification. Kluwer (1998)
12. Sarikaya, B.: Principles of Protocol Engineering and Conformance Testing. Ellis Horwood (1993)
13. Bochmann, G.V.: Protocol specification for OSI. Computer Networks and ISDN Systems. **18**, 167–184 (1989/90)
14. Yoeli, M., Kol, R.: Verification of Systems and Circuits Using LOTOS, Petri Nets, and CCS. Wiley – Interscience (2008)
15. Babich, F., Deotto, L.: Formal Methods for Specification and Analysis of Communication Protocols, IEEE Communications Surveys & Tutorials, December 2002
16. Milner, R.: A Calculus of Communicating Systems. Springer-Verlag (1980)
17. Hoare, C.A.R.: Communicating Sequential Processes (2015). http://usingcsp.com/cspbook.pdf
18. Dubuisson, O.: ASN.1—Communication between heterogeneous systems (2000). http://www.oss.com/asn1/resources/books-whitepapers-pubs/dubuisson-asn1-book.PDF
19. Belina, F., Hogrefe, D., Sarma, A.: SDL with Applications from Protocol Specification. Prentice Hall (1991)
20. Doldi, L.: SDL illustrated: visually design executable models, self-published (2001)
21. Mitschele-Thiel, A.: Systems Engineering with SDL: Developing Performance-Critical Communication Systems. John Wiley & Sons (2001)
22. Bræk, R., Haugen, Ø.: Engineering Real Time Systems: an Object-Oriented Methodology Using Sdl. Prentice Hall (1993)
23. SDL Forum Society, http://sdl-forum.org/index.htm
24. Higginbottom, G.N.: Performance Evaluation of Communication Networks. Artech House (1998)

25. King, P.J.B.: Computer and Communications Systems Performance Modelling. Prentice Hall (1990)

26. Hercog, D.: Performance of communication protocols using multiplexed channels. Electrotech. Rev. **82**(1/2), 51–54 (2015)

27. Hercog, D.: Mapping SDL protocol specification into protocol performance simulator based on Watkins simulation library, Proceedings of the 5th EUROSIM Congress on Modeling and Simulation, 6-10 September 2004, ESIEE Paris, Marne la Vallée, France (2004)

28. Fall, K.R., Stevens, W.R.: TCP/IP Illustrated, Volume 1: the Protocols, 2nd edn. Addison-Wesley (2012)

29. Farrel, A.: The Internet and its Protocols—a Comparative Approach. Elsevier (2004)

30. Hercog, D.: Generalization of the basic sliding window protocol. Int. J. Commun. Syst. **18**(1), 57–75 (2005)

31. Hercog, D.: Timer expiration time of selective repeat protocol. Electron. Lett. **45**(20), 1027–1028 (2009)

32. Hercog, D.: Selective-repeat protocol with multiple retransmit timers and individual acknowledgments. Electrotech. Rev. **82**(1/2), 55–60 (2015)

33. ISO – International Organization for Standardization, https://www.iso.org/standards.html

34. ITU – International Telecommunications Union, https://www.itu.int/en/ITU-T/publications/Pages/default.aspx

35. ETSI – European Telecommunication Standards Institute, http://www.etsi.org/standards

36. IEEE – Institute of Electrical and Electronics Engineers, http://standards.ieee.org/

37. IETF – Internet Engineering Task Force, RFC Editor, https://www.rfc-editor.org/search/rfc_search.php

38. ITU, Specification and Description Language – Overview of SDL-2010, Recommendation ITU-T Z.100, ITU, 2016, https://www.itu.int/rec/T-REC-Z.100/en

39. Rand, D.: PPP Reliable Transmission, RFC 1663, IETF (1994). https://tools.ietf.org/pdf/rfc1663.pdf

40. Wikipedia, the free encyclopaedia, https://en.wikipedia.org/wiki/Main_Page

41. Cinderella, Cinderella SDL 1.4 (software), http://www.cinderella.dk/

Index

Printed in the United States
by Baker & Taylor Publisher Services